EPISTLES OF THE BRETHREN OF PURITY

EPISTLES OF THE BRETHREN OF PURITY

The Ikhwān al-Ṣafāʾ and their *Rasāʾil*

An Introduction

Edited by
Nader El-Bizri

Foreword by
Farhad Daftary

OXFORD
UNIVERSITY PRESS
in association with
The Institute of Ismaili Studies

OXFORD
UNIVERSITY PRESS

Great Clarendon Street, Oxford OX2 6DP

Oxford University Press is a department of the University of Oxford.
It furthers the University's objective of excellence in research, scholarship,
and education by publishing worldwide in

Oxford New York

Auckland Cape Town Dar es Salaam Hong Kong Karachi
Kuala Lumpur Madrid Melbourne Mexico City Nairobi
New Delhi Shanghai Taipei Toronto

With offices in

Argentina Austria Brazil Chile Czech Republic France Greece
Guatemala Hungary Italy Japan Poland Portugal Singapore
South Korea Switzerland Thailand Turkey Ukraine Vietnam

Oxford is a registered trade mark of Oxford University Press
in the UK and certain other countries

Published in the United States
by Oxford University Press Inc., New York

First published 2008

British Library Cataloguing in Publication Data
Data available

Library of Congress Cataloguing in Publication Data
Data available

ISBN 978–0–19–955724–0

1 3 5 7 9 10 8 6 4 2

Typeset and printed in Lebanon
on acid-free paper by
Saqi Books

The Institute of Ismaili Studies

The Institute of Ismaili Studies was established in 1977 with the object of promoting scholarship and learning on Islam, in the historical as well as contemporary contexts, and a better understanding of its relationship with other societies and faiths.

The Institute's programmes encourage a perspective which is not confined to the theological and religious heritage of Islam, but seeks to explore the relationship of religious ideas to broader dimensions of society and culture. The programmes thus encourage an interdisciplinary approach to the materials of Islamic history and thought. Particular attention is also given to issues of modernity that arise as Muslims seek to relate their heritage to the contemporary situation.

Within the Islamic tradition, the Institute's programmes seek to promote research on those areas which have, to date, received relatively little attention from scholars. These include the intellectual and literary expressions of Shiʿism in general, and Ismailism in particular.

In the context of Islamic societies, the Institute's programmes are informed by the full range and diversity of cultures in which Islam is practised today, from the Middle East, South and Central

Asia, and Africa to the industrialised societies of the West, thus taking into consideration the variety of contexts which shape the ideals, beliefs and practices of the faith.

These objectives are realised through concrete programmes and activities organised and implemented by various departments of the Institute. The Institute also collaborates periodically, on a programme-specific basis, with other institutions of learning in the United Kingdom and abroad.

The Institute's academic publications fall into several distinct and interrelated categories:

1. Occasional papers or essays addressing broad themes of the relationship between religion and society in the historical as well as modern contexts, with special reference to Islam.
2. Monographs exploring specific aspects of Islamic faith and culture, or the contributions of individual Muslim figures or writers.
3. Editions or translations of significant primary or secondary texts.
4. Translations of poetic or literary texts which illustrate the rich heritage of spiritual, devotional and symbolic expressions in Muslim history.
5. Works on Ismaili history and thought, and the relationship of the Ismailis to other traditions, communities and schools of thought in Islam.
6. Proceedings of conferences and seminars sponsored by the Institute.
7. Bibliographical works and catalogues which document manuscripts, printed texts and other source materials.

This book falls into category two listed above.

In facilitating these and other publications, the Institute's sole aim is to encourage original research and analysis of relevant issues. While every effort is made to ensure that the publications are of a high academic standard, there is naturally bound to be a diversity of views, ideas and interpretations. As such, the opinions expressed in these publications must be understood as belonging to their authors alone.

Epistles of the Brethren of Purity

The *Epistles of the Brethren of Purity* is published by Oxford University Press in association with The Institute of Ismaili Studies, London. This bilingual series consists of a multi-authored Arabic critical edition and annotated English translation of the *Rasāʾil Ikhwān al-Ṣafāʾ* (ca. tenth-century Iraq), which is initiated with the present introductory volume of studies on the Brethren of Purity and their *Epistles*, and will be supplemented with a companion volume of analytical and critical commentaries on the corpus.

Contents

*Acknowledgements*

It is my delightful duty to gratefully thank all the eminent authors who contributed to the present volume. I am also most thankful to the Editorial Board of the Epistles of the Brethren of Purity series, and to the distinguished members of its Advisory Board for their continual support. Moreover, I express my deep gratitude to the editors at Oxford University Press and to the staff at SAQI Books (Beirut and London) for their valuable efforts in producing this volume of studies and the series it introduces. Profound thanks must also go to my esteemed colleagues at The Institute of Ismaili Studies, London, for their warm encouragement and their generous sponsorship of this major academic collaborative project. I would like to record here my indebtedness to Dr Farhad Daftary and to Professor Azim Nanji for their constant support of this initiative. Special thanks go to Dr Isabel Miller and Ms Tara Woolnough for their thoughtful copy-editing care. Thanks are also due to Ms Wendy Robinson for her administrative assistance at the early stages of this project, and to Mr Kutub Kassam for his editorial recommendations.

Nader El-Bizri
London, May 2008

Foreword

The corpus of the fifty-two epistles known as the *Rasāʾil Ikhwān al-Ṣafāʾ*, and generally translated as the *Epistles of the Brethren of Purity*, has been in existence for a millennium now. Representing a unique tradition in Islamic literature and cultural heritage, much controversy has surrounded the interconnected issues of the authorship, the date of composition, and the particular Islamic affiliation of the group of mediaeval authors who anonymously composed this encyclopaedic work. Modern scholarship on the Ikhwān al-Ṣafāʾ and their vast corpus dates back to the nineteenth century. After the pioneering studies of Gustav L. Flügel (1802–1870), Friedrich H. Dieterici (1821–1903), and Paul Casanova (1861–1926), a number of other eminent scholars, including especially Yves Marquet, Alessandro Bausani (1921–1988), Samuel M. Stern (1920–1969), Abbas Hamdani, Carmela Baffioni, and Ian Richard Netton, amongst others, have made significant contributions to the ongoing discourse on this subject. With a few exceptions, notably Professor Hamdani, scholars now generally agree that the *Rasāʾil* was secretly produced in Basra, southern Iraq, by a coterie of men of letters, around the middle of the fourth/tenth century. Furthermore, based on extensive internal textual evidence, there is consensus on the Shiʿi, and, more specifically, Ismaili affiliation, or at least inclination, of the Ikhwān.

Certain reliable authorities of the fourth/tenth century, including in particular the philosopher Abū Ḥayyān al-Tawḥīdī (d. 414/1023),[1] actually named certain contemporary men of letters and secretaries of the Būyids of Iraq as belonging to the circle of literati in Basra who

1 All dates are CE, unless otherwise indicated; where two dates appear (separated by a slash), the first is the hijri (AH) date, followed by CE.

were broadly affiliated to the Ismaili movement and who also composed the *Rasāʾil Ikhwān al-Ṣafāʾ*. In reply to questions put to him around 373/983 by his patron Ibn Saʿdān (d. 374/984–985), vizier to the Būyid emir Ṣamṣām al-Dawla Abū Kālījār Marzubān (r. 372–388/983–998), Abū Ḥayyān al-Tawḥīdī names these authors in his *al-Imtāʿ wa'l-muʾānasa* (ed. A. Amīn and A. al-Zayn [Cairo, 1939–1944], vol. 2, pp. 3–6), including the *qāḍī* Abu'l-Ḥasan ʿAlī ibn Hārūn al-Zanjānī, who may have been the leader of the group. Abū Ḥayyān al-Tawḥīdī's account, brought to the attention of modern scholars by Stern, was essentially corroborated by ʿAbd al-Jabbār ibn Aḥmad al-Hamadānī (d. 415/1025), the Muʿtazilī theologian and chief *qāḍī* of Rayy who was another well-informed contemporary authority. ʿAbd al-Jabbār (*Tathbīt dalāʾil al-nubuwwa*, ed. ʿA. ʿUthmān [Beirut, 1966–1969], pp. 610–611) specifically attributes Shiʿi (Ismaili) affiliation to the *qāḍī* al-Zanjānī and other members of the group.

Ismaili tradition, as summarised by Idrīs ʿImād al-Dīn ibn al-Ḥasan (d. 872/1468), the Ṭayyibī *dāʿī* and historian, in his *ʿUyūn al-akhbār* (ed. M. al-Ṣāghirjī [London and Damascus, 2007], vol. 4, pp. 525–564), actually ascribes the *Epistles* to Aḥmad ibn ʿAbd Allāh, one of the early Ismaili Imams of the *dawr al-satr* ('period of concealment') in Ismaili history preceding the establishment of the Fāṭimid caliphate in 297/909. However, the available evidence does not substantiate this ascription, nor does it support such early dating for the *Rasāʾil*'s composition. Even Professor Hamdani, who dates the *Epistles* to an early period (260–297/873–909), coinciding with the final decades of the *dawr al-satr*, attributes their authorship to a group of pre-Fāṭimid Ismaili *dāʿī*s rather than to any of the hidden Ismaili Imams.

On the other hand, Samuel Stern and Wilferd Madelung, who have contributed significantly to our understanding of early Ismailism and who also subscribe to the majority view of dating the corpus to the middle of the fourth/tenth century, postulate that the authors actually belonged to the dissident (Qarmaṭī) wing of Ismailism. In their view, the Ikhwān al-Ṣafāʾ were motivated in their encyclopaedic undertaking by a desire to reunite different dissident Ismaili communities, including the Qarmaṭīs of Bahrain who possessed a powerful state in eastern Arabia during 286–470/899–1077, on a common doctrinal basis. If the *Rasāʾil* was produced in the fourth/tenth century, then that was the period when Basra was under the virtual domination of the Qarmaṭīs of Bahrain and as such, the Ikhwān probably had at least the tacit approval of the neighbouring Qarmaṭīs. The Ikhwān do, in fact, place their teachings under the auspices of an inaccessible Imam, a view then upheld by the dissident Ismailis. The Stern–Madelung hypothesis may also explain why the *Rasāʾil* does not seem to have had any influence on the contemporary Ismaili teachings endorsed by the Fāṭimid *daʿwa* headquarters in Ifrīqiya or Egypt, and is not therefore cited in the classical Ismaili texts of the Fāṭimid period. It was about two centuries after its composition that the *Epistles* began to acquire an important place in the literature of the Ṭayyibī Ismailis of Yemen, after their introduction by the *dāʿī* Ibrahīm ibn al-Ḥusayn al-Ḥāmidī (d. 557/1162) in his *Kitāb Kanz al-walad* (ed. M. Ghālib [Wiesbaden, 1971], pp. 111ff.). Be that as it may, there is strong contemporary and internal textual evidence supporting the Ismaili connection of the authors who, professing high esteem for ʿAlī ibn Abī Ṭālib and the Ahl al-Bayt, collectively and anonymously compiled the *Rasāʾil Ikhwān al-Ṣafāʾ*.

In accord with their ecumenical spirit, the authors of the *Epistles* thought it quite legitimate to adopt all 'the science and wisdom' of the ancient philosophers in producing their own synthesis of the knowledge of the time. They drew on a wide range of pre-Islamic sources and traditions, which they combined with Islamic teachings, especially as upheld by the Shiʿis belonging broadly to the Ismaili movement. There are, for instance, traces of early Babylonian astrology, and many elements of Judaeo–Christian, Iranian, and Indian origin. Above all, the *Rasāʾil* reflects the influences of diverse traditions of Hellenistic wisdom then available to Muslims through numerous texts translated from Greek into Syriac-Aramaic and then into Arabic. Characterised by a type of numerical symbolism in the Pythagorean fashion, the *Rasāʾil* is permeated with Hermetic, Platonic, Aristotelian, Neopythagorean, and, especially, Neoplatonic, ideas and doctrines. It was, indeed, Neoplatonism, with its distinctive doctrine of emanation and hierarchism, which was the dominant Greek philosophical influence on the Ikhwān al-Ṣafāʾ, similar to the significance of Neoplatonism in the metaphysical systems of contemporary Ismaili *dāʿīs* of the Iranian lands such as Abū Yaʿqūb Isḥāq ibn Aḥmad al-Sijistānī (d. after 361/971).

It is clear from the *Epistles* that the Ikhwān espoused a highly tolerant and enlightened attitude towards religion and the wide range of other subjects that were covered in their writings. In an original fashion, the anonymous authors attempted to harmonise religion and philosophy, or revelation and reason, for the ultimate purpose of guiding man to purify his soul and attain salvation; this was similar to the attempts of Abū Yaʿqūb al-Sijistānī and Ḥamīd al-Dīn al-Kirmānī (d. ca. 411/1020), amongst other Ismaili *dāʿīs* of the Iranian lands, who interfaced their theology with different philosophical traditions when elaborating their own distinct versions of philosophical Ismailism. Both groups, Iranian *dāʿīs* and the Ikhwān, evidently wrote for the ruling elite and the educated classes. As a result, they aimed at maximising the intellectual appeal of their message by adopting the then-most-modern and intellectually fashionable scholastic and philosophical terminologies and themes without compromising the essence of their religious message.

So far, three independent main editions of this compendium of mediaeval knowledge have appeared in print, in Bombay (1305–1306/ca. 1888), Cairo (1347/1928), and Beirut (1376/1957). However, none of the available printed texts, each based on relatively late or defective manuscripts, represents a critical edition of the seminal text of the *Rasāʾil*, despite the existence of a well-established manuscript tradition here. The publication of a critical and reliable edition of the *Epistles* is a desideratum, the realisation of which has, thus far, remained overdue. There is also general agreement that a critical edition of the *Rasāʾil* corpus would provide the necessary foundation for further progress in Ikhwān studies and also, perhaps, for settling some of the outstanding controversial aspects of the work. It was in response to such challenges that this project was initiated at the Department of Academic Research and Publications of The Institute of Ismaili Studies, London, with the objective of producing a critical Arabic edition of the *Epistles*, together with a complete, annotated English translation, on the basis of a large number of early and reliable manuscripts. Drawing on the expertise of numerous scholars in the field, this represents the most comprehensive, and academically complex, collaborative projects undertaken by The Institute.

It remains for me to express my deepest gratitude to all the scholars who variously participated in this exciting and challenging textual endeavour, and to Dr Nader El-Bizri who, over the years, has indefatigably and most ably steered the organisation and publication of our Ikhwān al-Ṣafāʾ series, hereby initiated with the present volume.

Farhad Daftary
The Institute of Ismaili Studies

Prologue

The name 'Ikhwān al-Ṣafā' wa-Khillān al-Wafā'' is the pseudonym of the affiliates of a tenth-century (AH fourth-century) esoteric coterie of Mesopotamia, which they coined in order to conceal the identity of their active brotherhood during that tumultuous epoch.[1] This secretive *nom de plume* has been customarily rendered in the secondary and tertiary English sources as 'The Brethren of Purity and the Friends of Loyalty', or in a shorter form as 'The Sincere [*or* Candid] Brethren'. The exact chronology of the organisational activities of the Ikhwān al-Ṣafā' remains a matter of scholarly debate; some situate their historic activities at the eve of the Fāṭimid conquest of Egypt (969), while others place the founding of this association at an earlier period. Nevertheless, a relative consensus has been achieved amongst scholars in terms of affirming that the adepts of this fraternity operated principally from the southern Iraqi city of Basra, while also having a significant active branch in the capital of the ʿAbbāsid caliphate, Baghdad.

It is believed that the name of this clandestine urban organisation was inspired by a figurative appellation used with reference to a group of 'amicable doves', as suggested in the classic fable *Kalīla wa-Dimna*, which had been available in Arabic at that time since its early adaptation by the poet ʿAbd Allāh ibn al-Muqaffaʿ (d. ca. 757).[2]

1 The name may also be vocalised in Arabic as Ikhwān al-Ṣafā' wa-*Khullān* al-Wafā'.

2 It is possible that the name was inspired simply by pre-Islamic Arabic idioms or conventions of speech. And yet, as the Ikhwān themselves asserted, this name designates a 'property that befits those who are worthy of carrying it in truth [*bi'l-ḥaqīqa*]', and is applicable not only 'by way of a metaphor [*bi'l-majāz*]'. See *Rasā'il Ikhwān al-Ṣafā'*, ed. Buṭrus al-Bustānī, 4 vols. (Beirut: Dār Ṣādir,

The anonymous members of the Ikhwān al-Ṣafā' did also occasionally refer to themselves as Awliyā' Allāh, namely, the 'Righteous Friends of God', with a view to asserting their constant obedience to divine decrees, and as an expression of their loyal gathering, with pure souls, within a professed, utopian, mystical *polis* (i.e., *dawlat al-khayr*; 'reign of goodness').

The precise identity of the lettered adepts of this intriguing faction of urbanites remains hitherto shrouded in mystery. Nonetheless, their textual legacy, handed down over generations as a crucial part of the history of ideas in Islam, attests to the principal role they played in the evolution of intellectual history in mediaeval Islamic civilisation; and this is judged to be the case even against the unfair criteria that are usually deployed by their harshest critics when assessing their oeuvre.[3] This historic impress on Islamic thought has been secured through the dissemination of the Ikhwān's erudite teachings as principally embodied in their renowned and widely circulated compendium, entitled *Rasā'il Ikhwān al-Ṣafā'* (*Epistles of the Brethren of Purity*).[4] The anonymity of the learned authors of this corpus, as sustained by the conflicting historical reflections on their credal affiliations, along with the conditions of authorship surrounding the compilation of their epistles in the form of a gnostic and philosophically oriented 'encyclopaedia', as well as the dating of its composition, all continue to be unresolved questions that perplex many mediaevalists and historians in the field of Islamic studies.

1957), Epistle 42, vol. 3, pp. 411–412 (references to the *Rasā'il* made hereafter are based on this edition).

3 For instance, as Michot shows in his chapter of the present volume, the Mamlūk theologian Taqī al-Dīn Ibn Taymiyya conceded that the *Rasā'il Ikhwān al-Ṣafā'* exercised an intellectual influence on the two most representative figures of Islamic thought: Ibn Sīnā (Avicenna) and al-Ghazālī.

4 The principal editions of this compendium that are available in print consist of the following: *Kitāb Ikhwān al-Ṣafā' wa-Khullān al-Wafā'*, ed. Wilāyat Ḥusayn, 4 vols. (Bombay: Maṭba'at Nukhbat al-Akhbār, 1305–1306/ca. 1888); *Rasā'il Ikhwān al-Ṣafā'*, ed. Khayr al-Dīn al-Ziriklī, with two separate introductions by Ṭāhā Ḥusayn (1889–1973) and Aḥmad Zakī Pasha (d. 1934), 4 vols. (Cairo: al-Maṭba'a al-'Arabiyya bi-Miṣr, 1928); *Rasā'il Ikhwān al-Ṣafā'*, ed. with introduction by Buṭrus Bustānī, 4 vols. (Beirut: Dār Ṣādir, 1957), see note 2 above; and an additional version of the *Rasā'il Ikhwān al-Ṣafā'*, ed. 'Ārif Tāmir, 5 vols. (Beirut: Manshūrāt 'Uwaydāt, 1995).

The chronological dividing line between the earlier and the later compositional dates of the *Rasāʾil Ikhwān al-Ṣafāʾ* is set around 909, the year of the founding of the Fāṭimid caliphate in North Africa. For instance, Abbas Hamdani has proposed ca. 873 as a *terminus a quo* and 909 as a *terminus ad quem* for the chronology of the *Epistles*, while Yves Marquet has situated the compilation of this compendium within a much longer and later historical period, stretching between 903 and 980.[5] However, the historical narratives surrounding this text entice some scholars to date its composition around 961–980, while demarcating the year 986 as a reasonable *terminus ad quem*.[6] A prudent way of accommodating these variations in the chronology would be to permit a flexible designation of the *Rasāʾil* as a classical text that dates back to the tenth century, and which, in that era, potentially circulated within the geopolitical domain of the Fertile Crescent, including the Levant.

Affiliation and Identity

The most common account regarding the presumed identity of the Brethren of Purity is usually grounded on the authority of the famed *littérateur* Abū Ḥayyān al-Tawḥīdī (ca. 930–1023), as presented in his *Kitāb al-Imtāʿ waʾl-muʾānasa* (*Book of Pleasure and Conviviality*, dated ca. 981).[7] In reply to a question put to him by Ibn Saʿdān, the vizier of the Būyid (also known as Buwayhid) governor Ṣamṣām al-Dawla ibn ʿAḍud al-Dawla, al-Tawḥīdī noted that the leaders of

5 For Hamdani's argument, see his chapter in the present volume. The necessary limitations of this introduction mean that we will not now survey all the scholarly opinions concerning the dating of the *Rasāʾil*, given that these are accounted for in some detail in the various chapters of the present volume. However, we should at least signal that this matter has been discussed previously, with varying levels of expertise, by scholars of the calibre of: ʿĀdil ʿAwā, Alessandro Bausani, David R. Blumenthal, Paul Casanova, Friedrich H. Dieterici, T. J. De Boer, Susanne Diwald, ʿUmar Farrūkh, Muṣṭafā Ghālib, Ḥusayn Hamdānī, Louis Massignon, Jamīl Ṣalībā, Samuel M. Stern, ʿĀrif Tāmir, and ʿAbd al-Laṭīf Ṭībāwī, amongst others.

6 See de Callataÿ's chapter in the present volume.

7 Abū Ḥayyān al-Tawḥīdī, *Kitāb al-Imtāʿ waʾl-muʾānasa*, ed. Aḥmad Amīn and Aḥmad Zayn, 2nd ed. (Beirut: Manshūrāt Dār Maktabat al-Ḥayāt, 1965).

this coterie were identified as the men of letters known by the following names: Abū Sulaymān Muḥammad ibn Maʿshar al-Bustī (also bearing the nickname al-Maqdisī); the *qāḍī* Abū al-Ḥasan ʿAlī ibn Hārūn al-Zanjānī; Abū Aḥmad al-Mihrajānī (also known as Aḥmad al-Nahrajūrī); and Abū al-Ḥasan al-ʿAwfī. Furthermore, it is claimed that they all were the senior companions of a secretarial officer at the Būyid regional governorate of Basra known as Zayd ibn Rifāʿa, who seems to have been one of the personal acquaintances of Tawḥīdī, and who was reportedly an affiliate of the Ikhwān al-Ṣafā and a servant of its ministry. The *qāḍī* of Rayy, ʿAbd al-Jabbār (ca. 936–1025), corroborated Tawḥīdī's report and also mentioned an obscure astronomer named Abū Muḥammad ibn al-Baghl as another suspected member of the brotherhood. Tawḥīdī's story was reaffirmed by figures like al-Bayhaqī (d. 1169), al-Khwārizmī (d. 1220), the historiographer Ibn al-Qifṭī (d. 1248), and al-Shahrazūrī (d. 1285), while being alluded to by the theologian Ibn Taymiyya (d. 1328).[8] The same was also related by Ḥājjī Khalīfa (or, Kātib Chalabī [Kâtip Çelebi]; d. ca. 1657) after its partial reproduction by Abū Sulaymān al-Manṭiqī (al-Sijistānī). Moreover, some scholars rely on this line of conveyance with a view to questioning the classical and modern claims that the Brethren of Purity were *closely* connected with the ninth- and tenth-century Ismaili communities of the Fertile Crescent.

According to Tawḥīdī's version, and those associated with it, the Basran Zayd ibn Rifāʿa served the Būyids (of Persian and Shiʿi Daylamī descent from territories on the fringes of the Caspian Sea), who operated as the regent guardians of the ʿAbbāsid caliphate after having instituted their oligarchic rule by military means. This dynasty represented a Shiʿi tradition in Islam without belonging to any particular branch, and whilst it controlled the ʿAbbāsid chancelleries bureaucratically, its leaders did not assume the title of caliph, instead reviving the Persian designation '*Shāhanshāh*', or 'King of Kings'. Moreover, the Būyids did not openly challenge the suzerainty of the ʿAbbāsids, but functioned as their militarised allies in the historic imperial conflict against the Fāṭimids. With this in mind, some scholars tend to question the presumed Ismaili lineage of the Ikhwān, or at least the association with

8 See the respective chapters of Hamdani and Michot in the present volume.

figures like Ibn Rifāʿa who were connected with the Būyid secretariat, and who would have indirectly served the ʿAbbāsids. And yet, based on Tawḥīdī's story, Ibn Rifāʿa's *loyalty* was the subject of suspicion, at least in the way it instigated inquiries regarding his character by the vizier Ibn Saʿdān, who apparently performed these on behalf of the Būyid governor Ṣamṣām al-Dawla. In some sense, this state of affairs seems to reflect the antinomies that animate political allegiances. Being a servant of the Būyids, as was supposedly the case with Ibn Rifāʿa, does not seem to be in accord with the Ikhwān's stance of dissidence in challenging the sovereignty of the ʿAbbāsid caliphate. This ostensible incongruity may ultimately cast doubt on the internal coherence of Tawḥīdī's report.

It is believed that the ʿAbbāsids were politically and ideologically opposed to organisations that professed doctrinal teachings akin to those advocated by the Ikhwān.[9] This hostile political milieu would have classed the Brethren of Purity as *persona non grata*. In addition, it would have pressured them into becoming organised as an underground faction that deliberately concealed the identity of its 'revolutionary' members, in order to secure their personal safety as well as drawing a benefit from the consequent doctrinal ambivalence which expedited the dissemination of their ideas and the conduct of their operational activities. Given the altercations surrounding the chronology of their textual heritage, and the clandestine character of their coterie, the precise *madhhab* of the Ikhwān and the exact lineage of the compilers of their famed *Rasāʾil* remain vexing and unsettled matters of scholarship.

Some have wondered whether the Ikhwān were exponents of Sunni or Shiʿi traditions in Islam. And, arguably, if their opus does contain certain motifs that are attributable to concepts associated with Sunni legacies, there is still no consensus as to which school the Ikhwān would have belonged. There is also no accord on the question of whether the *Rasāʾil* displayed Muʿtazilī or specifically Sufi affinities. Furthermore, if the Ikhwān are classed as being Shiʿi, as most schol-

9 This observation is advocated by, for example, Muṣṭafā Ghālib and ʿĀrif Tāmir.

ars argue,[10] it is ultimately unclear whether they can be definitively classified as Ithnāʿasharī (Twelvers) or Ismaili. To further complicate these speculations, even if the Ikhwān are seen to be Ismailis, it is not proven whether they had specific associations with the Fāṭimids or the Qarmaṭīs, or whether they reflected any of the pre-Fāṭimid proclivities of Ismailism.[11] There have even been certain 'orthodox' historical interpretations of their oeuvre that regarded their 'exaggerated' reliance on ancient Greek sources with antagonism, and took it as indicative of the non-Abrahamic character of their teachings; hence the charge of heresy against the Ikhwān. All of these conjectures regarding the credal lineage of the Ikhwān continue to unfold in terms of the particulars of controversies that at times take the shape of *ideological* debates and risk impacting academic inquiries with their polemics.

The praise that is noted in the *Rasāʾil* (Epistles 42 and 52) with respect to the sovereignty of the *khulafāʾ al-rāshidīn* ('the righteous caliphs') — namely, Abū Bakr (r. 632–634), ʿUmar ibn al-Khaṭṭāb (r. 634–644), and ʿUthmān (r. 644–656) — along with admiration of al-Ṣaḥāba ('the Prophet's Companions') and the esteem shown to ʿĀʾisha,[12] are each understood by some as signs of a Sunni parlance and temperament. Nonetheless, certain scholars see the apparent Sunni 'signs' as motifs that can be plausibly attributed to Shiʿi practices of *taqiyya* ('precautionary dissimulation').[13] Furthermore, in the same context where the *khulafāʾ al-rāshidīn* receive mention,[14] the Brethren praised highly the Ahl al-Bayt ('Prophet's Household'); and moreover they put forward celebratory notes regarding the festival of Ghadīr Khumm.[15] There are also textual examples that hint at Sufi leanings in the *Epistles*, for instance, what is noted in the *risāla* on the essence of mystical love (*Fī māhiyyat al-ʿishq*; Epistle 37) or what is encountered

10 This is a view shared by most of the scholars mentioned in note 5 above.

11 Regarding the various debates surrounding the affiliation of the Ikhwān al-Ṣafāʾ, refer to the respective chapters of Poonawala and Hamdani in the present volume.

12 *Rasāʾil*, Epistle 9, vol. 1, pp. 358–359.

13 For instance, this interpretation figures in some of the writings of Ghālib and Tāmir.

14 Ibid., Epistle 52, vol. 3, p. 408.

15 Ibid., Epistle 50, vol. 3, p. 268.

in certain older manuscripts of the *Rasāʾil* (like the MS Atif Efendi 1681; dated 1182), which attribute the transcript to utterances of the Sufis (*min kalām khulaṣāʾ al-ṣūfiyya*). While some scholars dismiss this Sufi character as the result of later scribal interference with the content of these epistles,[16] others accept it, though without any agreement as to whether it belongs to Sunni expressions of spirituality or to Shiʿi gnosis.

Concerning the more widely accepted indications of the Shiʿi ancestry of the compilers of the *Rasāʾil*, it is clear that the Ikhwān venerated the persona of the Imam ʿAlī ibn Abī Ṭālib (r. 656–661), and that they also heeded the progeny of the Prophet as 'guided Imams' (*al-aʾimma al-muhtadīn*).[17] Moreover, in the final epistle of the compendium, on the essence of magic (*Fī māhiyyat al-siḥr*; Epistle 52), the Ikhwān appeal to the prophetic oral tradition, by metaphorically stating that the Prophet Muhammad was the city of knowledge (*madīnat al-ʿilm*) and that the Imam ʿAlī was its gateway (*bāb*), and consequently asserting that those who desire accessibility to this *madīna* have to necessarily pass through its portal.[18] In addition, the Ikhwān were repulsed and pained by the massacre of Karbalāʾ in Iraq, though of course this deeply rooted sentiment is shared ecumenically, with a sense of collective trauma and grief, amongst the vast majority of Muslims. Nonetheless, the Ikhwān specifically commemorated ʿĀshūrāʾ, and they showed a profound sense of respect and mourning *in memoriam* of those slain on the tenth of the month of Muḥarram 680 — the day the Imam al-Ḥusayn ibn ʿAlī ibn Abī Ṭālib was martyred — while considering this catastrophic episode in the history of Islam as having been marked by irredeemable disgraces.[19] Despite these significant indicators of a Shiʿi temperament in their faith, some hold that the Ikhwān criticised the Twelver Shiʿi doctrine of the awaited Imam, which is associated with the eschatological figure of al-Mahdī (*al-imām al-fāḍil al-muntaẓar al-hādī*), thus indicating that, in the Ikhwān's view, this spiritual guide

16 Critical reflections on this question are presented in Poonawala's and Hamdani's chapters in the present volume.

17 Ibid., Epistle 17, vol. 2, p. 408; ibid., Epistle 33, vol. 3, p. 211.

18 Ibid., Epistle 52, vol. 4, p. 460.

19 Ibid., Epistle 50, vol. 4, p. 269.

did not remain concealed.[20] In this context, the Ikhwān seem to have been expressive of some form of credal resonance with Ismaili beliefs, which claimed that the historical figure of the Mahdī appeared publicly in North Africa (Ifrīqiya) upon the founding of the Fāṭimid dynasty.

The Ismaili Connection

While many tend to assert, with a relative sense of confidence, that the *Rasā'il* can be classed as part of the literature of Shi'i traditions in Islam, the Ismaili character of this voluminous text confronts the inquirer with several dilemmas. Based on the historical accounts of Ismaili missionaries (*du'āt*), such as Ja'far ibn Manṣūr al-Yaman (d. ca. 957) and 'Imād al-Dīn Idrīs ibn al-Ḥasan (d. 1468), some scholars emphasise the pre-Fāṭimid Ismaili provenance of the *Rasā'il*. Others even link this opus to the ancient provincial town of Salamiya in northeast Syria, by attributing the *Epistles* to the early Ismaili Imams Aḥmad ibn 'Abd Allāh ibn Muḥammad ibn Ismā'īl ibn Ja'far (or, al-Taqī [al-Mastūr])[21] or his father, 'Abd Allāh ibn Muḥammad ibn Ismā'īl ibn Ja'far (or, Wafī Aḥmad), while also indicating that the *Rasā'il* was secretly disseminated in mosques during the reign of the 'Abbāsid caliph al-Ma'mūn (r. 813–833). It is furthermore reported that the compilation of the *Rasā'il* dates back to the times of the Imam Ja'far al-Ṣādiq (d. 765), and that this endeavour continued during the period of concealment and dissimulation (*dawr al-satr wa'l-taqiyya*) through the efforts of successive hidden Imams (*al-a'imma al-mastūrīn*). According to the various reports, along with their folkloric appendages, the *Rasā'il* would

20 Ibid., Epistle 42, vol. 3, p. 523.

21 'Imād al-Dīn Idrīs ibn al-Ḥasan, *'Uyūn al-akhbār wa-funūn al-āthār*, vol. 4, ed. Ma'mūn al-Ṣāghirjī (Damascus and London: Institut Français du Proche Orient and The Institute of Ismaili Studies, 2007), pp. 525–564. This Ṭayyibī Musta'lī *dā'ī* attributed the composition of the *Rasā'il* to the Imam Aḥmad ibn 'Abd Allāh ibn Muḥammad ibn Ismā'īl ibn Ja'far, who supposedly drafted them, and arguably ordered their dissemination, all in response to the prolific intellectual activities that were associated with the patronage of the court of the 'Abbāsid caliph al-Ma'mūn. However, this thesis is refuted by Aḥmad Zakī Pasha in his introduction to the *Rasā'il* in the Cairo edition of 1928 (see pp. 36–43), while scholars such as Ghālib and Tāmir argued in favour of it.

have been compiled over a very long period stretching from the eighth century to the tenth century.

In an attempt to confirm the Ismaili character of the *Rasāʾil*, some interpreters emphasise the Ikhwān's belief that religion (*ʿilm al-dīn*) carried two aspects: one, exoteric *qua* manifest (*ẓāhir jalī*), and the other, esoteric *qua* concealed (*bāṭin khafī*). These interpreters would also add that this distinction between the respective ascription to ritualistic-cum-prescribed forms of worship and the philosophically oriented modes of devotion requires an esoteric hermeneutics (*taʾwīl*) of the sacred scriptures, such as is found in the *Rasāʾil*, in addition to exegesis (*tafsīr*).[22] And yet, among those who affirm the Ismaili affiliation of the Ikhwān, there remain differences as to the particulars of this association. While some assert the Fāṭimid character of the brotherhood, others maintain that it was possibly connected with the Qarmaṭīs (Carmathians) of Bahrain, or that the *Rasāʾil* was expressive of pre-Fāṭimid Ismailism. Nevertheless, the Ikhwān's legacy is rarely mentioned in Fāṭimid literature; rather, it found its widest reception among the Ṭayyibī Ismailis in Yemen, after the corpus was re-introduced to the Yemeni community by the *dāʿī* Ibrāhīm ibn al-Ḥusayn al-Ḥāmidī (d. 1162).

The elements that reinforce the claim of the Ismaili character of the *Rasāʾil* did not result historically in a sympathetic reception of the Ikhwān's teachings.[23] Some classical historians in Islam, and even modern scholars, accentuated the Ismaili aspects of the *Rasāʾil* with a view to criticising the Ikhwān. Such critics proclaim that the Brethren deployed religious leitmotifs instrumentally, and with calculated utilitarian intent, in order to further ideological and political ends that aimed at subverting ʿAbbāsid governance, rather than as genuine expressions of Islamic piety. Following this critical reading, it has also been argued that the Brethren's legacy belongs to the hermetic tradition of a 'resigned reason' (*ʿaql mustaqīl*), which uses *ʿirfān* ('gnosis') to undermine rationality and the epistemic basis of *burhān* ('reasoned/

22 *Rasāʾil*, Epistle 42, vol. 3, pp. 511–514.

23 For an example, refer to Ibn Taymiyya's views in Michot's chapter of the present volume.

demonstrative proof').[24] They furthermore adopt Tawḥīdī's view that the Brethren dealt with all the arts and the sciences with insufficient depth or comprehensiveness of inquiry.

Intellectual Legacy

Marked by a perceptible receptivity to otherness, and eschewing fanaticism, the *Rasāʾil Ikhwān al-Ṣafāʾ* seems to embody a form of diversity in Islam that can accommodate miscellaneous ancient and monotheistic traditions. Encountering 'veracity in every religion' (*al-ḥaqq fī kull dīn mawjūd*),[25] and seeing knowledge as 'pure nourishment for the soul', the Ikhwān associated the attainment of happiness and the *soteriological* hope for salvation with the scrupulous development of rational pursuits and intellectual quests. They also promoted a 'friendship of virtue' amongst their brotherhood, and venerated the liberal spirit in Islam. Their syncretism, which is not reducible to a mere form of eclecticism, may have been partly influenced by Sabian (Harranian) practices in Mesopotamia; this intellectual impetus may have been a motivating factor behind the Ikhwān's eschatological aspiration to found a spiritual sanctuary that would assist their co-religionists in overcoming the sectarian discord plaguing their era. The Brethren did take heed of the Torah of Judaism and the canonical Gospels of Christianity, besides their pious observance of the Qurʾan and the hadith. This propensity towards ecclesiastic ecumenism is noticeable primarily in their approach to the question of prophecy, and in their effort to accomplish some form of harmony between faith and reason, by reconciling monotheistic revelation with philosophy, in a manner reminiscent of the works of intellectual authorities in Islam of the calibre of al-Fārābī (Alfarabius, d. 950), Ibn Sīnā (Avicenna, d. 1037), and Ibn Rushd (Averroes, d. 1198).

Besides being influenced by ancient Babylonian, Indian, and Persian traditions (Buddhist, Manichaean, Zoroastrian, etc.) the Ikhwān incorporated the fundamentals of Greek science into their epistles.

24 Muḥammad ʿĀbid al-Jābirī, *Takwīn al-ʿaql al-ʿarabī* (6th repr., Beirut: Markaz Dirāsāt al-Wiḥda al-ʿArabiyya, 1994), pp. 202, 204, 212, 265, 268.

25 *Rasāʾil*, Epistle 42, vol. 3, p. 501.

They also attempted to assimilate the range of philosophical thought, including that of Pythagoras, Socrates, Plato, Aristotle, Euclid, Ptolemy, Nicomachus of Gerasa, Galen (Claudius Galenus), Hermes Trismegistus, Plotinus, Porphyry, Iamblichus, and Proclus. Moreover, the compilation of the mathematical and scientific contents of their opus was clearly indebted to the scholarship of experts and polymaths from the generation of al-Kindī (d. 873), along with the unfolding of research associated with the School of Baghdad as it evolved after the founding of the Bayt al-Ḥikma (House of Wisdom).

Oriented by a literal interpretation of the classic analogy of microcosm and macrocosm, and guided by the related fourfold schema (the longstanding belief in a correspondence between the four humours, the four temperaments, the four elements, the four seasons, etc.), the Ikhwān apprehended the human being as a *micro-cosmos* (*al-insān ʿālam ṣaghīr*; Epistle 26) and the universe as a *macro-anthropos* (*al-ʿālam insān kabīr*; Epistle 34). They also enthusiastically attempted to reinstate the equipoise between the psychical directives of the soul and their 'correlative' cosmological impulses. Their analogical thinking was guided by Pythagorean numerological explications of the 'layered order' of the visible universe. Moreover, in the attempt to further ground their effort to reconcile philosophy with revealed religion, they adopted the Neoplatonist account of 'creation by way of emanation' in expression of their onto-theology and its cosmological bearings, whilst imbuing the whole outlook on the universe with Pythagorean and Platonic themes.[26] They were also attracted to the Delphic injunction 'Know thyself!', whilst imparting to it a monotheistic significance in construing it as a pathway to knowing the Creator.

In arithmetic, they relied on Nicomachus; in geometry, on Euclid; in astronomy, on Ptolemy; in music, on Pythagorean theories of harmonics; and in logic, on Aristotle and Porphyry. They also exhibited a taste for occultism and the hermetic arts, following the traditions of Hermes Trismegistus and Iamblichus. It is purported that their fascination with forms of astrological and cosmological symbolism originated in Mesopotamia under the influence of the Sabians of

26 See Baffioni's chapter in the present volume, and also that of Netton.

Harran, in emulation of antique interests in cosmic orderings and the semantics of astral constellations and stellar configurations.

While eruditely synopsising the sciences and wisdoms of the Ancients and the Moderns of their time which, in their eyes, embodied the realities of the world, the Ikhwān were also motivated by their belief in the astrological initiation of a new cycle of time that would result in immense political changes. This conviction may have been a result of talismanic interpretations of stellar conjunctions, which supposedly led the Ikhwān to anticipate the 'radical weakening' of the 'Abbāsid caliphate, as well as pointing to the rise of a competing centre of power in Islam, identified by some as none other than the Fāṭimid dynasty in Egypt. One might here detect the political implications of the Ikhwān's call for the pursuit of happiness in this world and the salvation of the soul in the hereafter, which, in aiming to regain the original purity of the self in its journey back to its source *qua* Creator, would ultimately require the translation of acquired knowledge into the practice of good deeds. By their actions in promoting an earnest companionship of virtue, and by counselling their co-religionists with kindness and affection (Epistle 48), it is clear that the Ikhwān understood proper governance as that which embodies the utopian ideal of the *dawlat al-khayr* ('reign of goodness'), namely, a spiritual polity representing their own version of the 'virtuous city' (*al-madīna al-fāḍila*). While some scholars have argued that the Ikhwān would have been influenced in this regard by the political theory of al-Fārābī (see Epistle 47: 'On the Essence of the Divine Law'), others maintain that it is more likely that they were inspired by adaptations of Plato's *Politeia* (*Republic*), or associated commentaries by Galen. Moreover, in accentuating the political import of their beliefs, the Ikhwān encouraged their followers not to acquiesce under the pressures of power or authority when these accommodate forms of injustice. They avowedly noted that their brothers and friends came from all stations and classes in society, grouping aristocrats and commoners, the sons of 'kings, emirs, viziers, secretaries, tradesmen, and workmen' (Epistle 47). Furthermore, in expression of what may be qualified as a *classical* sense of 'multiculturalism' or 'cosmopolitanism', they pictured the ideal human being as a noble who is graciously 'Persian in breeding, Arab in faith, Ḥanafī in jurisprudence, Iraqi in culture, Hebrew in tradition, Christian in comportment, Syrian in piety, Greek in knowledge, Indian in contemplation, and Sufi in intimation and lifestyle'.[27]

Although many scholars tend to bestow an unquestioning praise upon the Ikhwān's spirit of pluralism, tolerance, moderation, openness, and their expressions of classical forms of liberalism and humanism in Islam, nonetheless, some strict inquirers with traditionalist leanings regard the contents of the *Rasā'il* with theological suspicion.[28] Accordingly, they hold that the Ikhwān's 'excessive' syncretism may have compromised the centrality of the Islamic directives of their thought, and may also have resulted in their judging the articles of faith in Islam as equivalent to those which may be derived from other monotheistic traditions or even older sources.

Contents of the Epistles

In general, fifty-two epistles are enumerated as belonging to the *Rasā'il Ikhwān al-Ṣafā'*, but sometimes they are counted as fifty-one or even as just fifty. Although the actual classification of the sciences in most of the manuscripts of this text displays various discrepancies with that proposed by the Ikhwān themselves in Epistle 7,[29] the compendium of the *Rasā'il* in all its existing textual forms is divided into four parts, which are classed according to the following order: *Riyāḍiyya taʿlīmiyya* (Mathematics), *Jismāniyya ṭabīʿiyya* (Natural Philosophy), *Nafsāniyya ʿaqliyya* (Sciences of the Soul and Intellect), and *Nāmūsiyya ilāhiyya* (Theology).

The first part (*Riyāḍiyya taʿlīmiyya*), which consists of fourteen epistles, deals with the mathematical sciences, treating a variety of topics in arithmetic, geometry, astronomy, geography, and music.[30] It also includes five epistles on elementary logic, which consist of: the

27 *Rasā'il*, Epistle 22, vol. 2, p. 376.

28 For instance, this critique finds an extreme expression in the tradition associated with the works of Ibn Taymiyya.

29 See de Callataÿ's chapter in the present volume.

30 See Wright's chapter on music, and my own chapter on arithmetic and geometry in the present volume.

Isagoge, the *Categories*, *On Interpretation*, the *Prior Analytics*, and the *Posterior Analytics*.

The second part of the *Rasā'il* (*Jismāniyya ṭabīʿiyya*), which groups together seventeen epistles, addresses the physical *qua* natural sciences. It thus treats themes on matter and form, generation and corruption, metallurgy, meteorology, a study of the essence of nature, the classes of plants and animals (the latter being also set as a fable),[31] the composition of the human body and its embryological constitution, a cosmic grasp of the human being as microcosm, and also the investigation of the phonetic and structural properties of languages and their differences.

The third part of the corpus (*Nafsāniyya ʿaqliyya*) comprises ten tracts on the psychical and intellective sciences. Here the Ikhwān set forth, in their own words, the 'opinions of the Pythagoreans and of the Brethren of Purity', and accounted for the world as a 'macranthrope'. They also examined the distinction between the intellect and the intelligible, as well as explicating the symbolic significance of temporal dimensions and epochal cycles, together with a mystical expression of the essence of love (*ʿishq*), an investigation of resurrection, the various types of motion, cause and effect, and definitions and descriptions.

The fourth and last part of the *Rasā'il* (*Nāmūsiyya ilāhiyya*) deals with the *nomic qua* legal and theological sciences in eleven epistles. These address the differences between the varieties of religious opinions and sects, as well as delineating the 'Pathway to God', the virtues of the companionship of the Brethren of Purity, the characteristics of genuine believers, the nature of the divine *nomos* (*al-nāmūs*), the call to God, the actions of spiritualists, of jinn, angels, and recalcitrant demons, the species of politics, the layered ordering of the world, and, finally, the essence of magic and talismanic incantations.

Besides the fifty-two tracts that constitute the *Rasā'il Ikhwān al-Ṣafā'*, this compendium was accompanied by a treatise entitled *al-Risāla al-jāmiʿa* (The Comprehensive Epistle),[32] which acted as the

31 See Goodman's chapter in the present volume.

32 Ikhwān al-Ṣafā', *al-Risāla al-jāmiʿa*, ed. Jamīl Ṣalībā, 2 vols. (Damascus: Maṭbaʿat al-taraqqī, 1949–1951); Ikhwān al-Ṣafā', *al-Risāla al-jāmiʿa*, ed. Muṣṭafā Ghālib (Beirut: Dār Ṣādir, 1974). It is worth noting that Jamīl Ṣalībā refutes the claim

summa summarum for the whole corpus, and was itself supplemented by an abridged appendage known as the *Risālat jāmiʿat al-jāmiʿa* (The Condensed Comprehensive Epistle).

In spite of their commendable erudition and resourcefulness, it is doubtful whether the Brethren of Purity can be impartially ranked amongst the authorities of their age in the realms of science and philosophy. Their inquiries into mathematics, logic, and natural sciences were recorded in the *Epistles* in a synoptic and diluted fashion, whilst sporadically being infused with spiritual, gnostic, symbolic, and occult directives. Nonetheless, their accounts of religiosity, as well as their ecumenical syncretism, together with their praiseworthy efforts to collate the sciences of their time, and to compose a pioneering 'encyclopaedic' compendium, all bear signs of estimable originality.

In terms of the epistemic significance of the *Epistles*, and the intellectual calibre of their authors, it must be stated that, despite being supplemented by oral teachings in seminars (*majālis al-ʿilm*), the heuristics embodied in the *Rasāʾil* were not representative of the most decisive achievements in their epoch in the domains of mathematics, natural sciences, or philosophical reasoning.[33] Moreover, the sciences were not treated with the same level of expertise across the *Rasāʾil*. Consequently, this opus ought to be judged by differential criteria as regards

that *al-Risāla al-jāmiʿa* was authored by the Andalusian mathematician Maslama ibn Aḥmad ibn Qāsim ibn ʿAbd Allāh al-Majrīṭī (d. ca. 1008), as had been suggested in some manuscripts of this epistle which he used in establishing his critical Arabic edition (Ṣalībā, vol. 1, p. 13). As for Muṣṭafā Ghālib, he argues that the *Risāla al-jāmiʿa* was authored by the Imam Aḥmad ibn ʿAbd Allāh, in line also with what we have indicated above.

33 I have discussed this question elsewhere in several publications, including the following studies: Nader El-Bizri, 'The Microcosm/Macrocosm Analogy: A Tentative Encounter between Graeco-Arabic Philosophy and Phenomenology', in *Islamic Philosophy and Occidental Phenomenology on the Perennial Issue of Microcosm and Macrocosm*, ed. Anna-Teresa Tymieniecka (Dordrecht: Kluwer Academic Publishers, 2006), pp. 3–23; Nader El-Bizri, 'The Conceptions of Nature in Arabic Thought', in *Keywords: Nature*, ed. N. Tazi (New York: Other Press, 2005), pp. 63–92; Nader El-Bizri, 'Variations autour de la notion d'expérience dans la pensée arabe', in *L'expérience, collection les mots du monde*, ed. N. Tazi (Paris: Editions la Découverte, 2004), pp. 39–58; Nader El-Bizri, 'Ikhwān al-Ṣafāʾ (Brethren of Sincerity)', in *Encyclopedia of Philosophy*, ed. D. M. Borchert, 2nd ed. (Detroit: Macmillan Reference USA, 2006), pp. 575–577.

the relative merits of each of its epistles. In fairness, there are signs of conceptual inventiveness in the *Epistles*, primarily regarding doctrinal positions in theology and reflections on their ethical-political import, along with signs of an intellectual sophistication in the meditations on spirituality and revelation. Nonetheless, there have been controversies surrounding their philosophical orientations and syncretic approach to *divinalia*, eschatology, prophetology, and angelology.

Despite evident indications of scientific or epistemological shortcomings, the *Rasāʾil* brims with a wealth of ideas, and constitutes a masterpiece of mediaeval literature that presents an erudite and populist adaptation of scientific knowledge.

The corpus is informative in terms of investigating the transmission of knowledge in Islam, the transformative and 'adaptive assimilation' of antique sciences, and the historical evolution of the elements of the *sociology* of learning through the mediaeval forms of the popularisation of the sciences and the systemic attempts to canonise them. By influencing a variety of Islamic schools and doctrines, the Ikhwān's heritage acted as a significant intellectual prompt and catalyst in the unfurling of the history of ideas in Islam, and, as such, it rightfully deserves the station that has been assigned to it among the distinguished Arabic classics and the high literature of Islamic civilisation.

Occasionally verbose, and beset at times by circumlocution and a sense of entrancement with symbolism and occultism, the *Rasāʾil* is nonetheless composed in an eloquent literary style that gives elegant expression to the niceties of classical Arabic. Moreover, the composition of this text displays commendable lexical adaptability, which encompasses the technical idioms of mathematics and logic, the heuristics of natural philosophy, and the diction of religious pronouncements and occult invocations, in addition to poetic verses, didactic parables, and satirical and inspirational fables. Despite the sometimes disproportionate treatment of topics, the occasional hiatus in proofs, irrelevant digressions, or instances of verbosity, such apparent stylistic weaknesses disappear, becoming inconsequential when a complete impression is eventually formed of the architectonic unity of the text as a whole, and of the convergence of its constituent elements as a remarkable *oeuvre des belles lettres*.

Historiography and Hermeneutics: The Dilemmas of Reading the Text

In spite of numerous 'textual' accounts and variegated doctrinal remarks, speculations as to the affiliation of the Ikhwān tend to result in only selective and fragmentary justifications, which ought to be received with a spirit of academic prudence and restraint, whilst adopting cautionary procedures of inquiry in historiography. This matter is further highlighted by the scale of the manuscript tradition of the *Rasāʾil*, which hints at the differential character of the textual evolution, reconstruction, and transmission of its content. Certain manuscripts carry traces of random scribal interventions, or are possibly marked by some deliberate forms of interpolation, along with purely ornamental additions or misinformed omissions. The vagueness surrounding the dating and authorship of this compendium undoubtedly complicates any endeavour to determine the doctrinal affiliation of the compilers of the *Rasāʾil*. Notwithstanding the extensive studies that have been conducted on the genealogy of the Ikhwān al-Ṣafāʾ and the particulars of their credal descent, the specifics of their affiliations do not yet pass without incongruity. The multiplicity of voices that found expression in their tracts may encourage scholars to resist the temptation to confine the Ikhwān's lineage within strict dogma. In this way, some tend to regard their oeuvre, rather, as that of 'free societal agents' or 'liberal thinkers'.[34] While this vagueness might not satisfy many historians of ideas or mediaevalists in Islamic studies, it nonetheless advances a faithful expression of the intellectual syncretism and cognitive openness that pervades the *Rasāʾil* itself. Therefore, one ought, perhaps, to abstain from readily answering the incessant calls to force justifications for any particular *sectarian* assimilation of the Ikhwān's corpus. It may be the case that this tendency in scholarship is due, in part, to the typically biographical-bibliographical methodology of Islamic studies, along with an archivist approach to historiography, in addition to reflecting certain conflicting modern sensitivities or, indeed, *inherited* 'mediaevalist' and 'orientalist' traditions.

34 ʿUmar Farrūkh, 'Ikhwān al-Ṣafāʾ', in *History of Muslim Philosophy*, ed. M. M. Sharif, vol. 1 (4th repr., Delhi: Low Price Publications, 1999), pp. 289–310.

Rethinking a philosophical text like the *Rasā'il* by attempting to retrieve its content through the 'de-structuring' of the history of its transmission conditions our historical consciousness when reading it, and is likewise determined by it. This state of affairs involves more than 'reconstructing the past world to which this work belonged', since our own understanding retains the consciousness that 'we too belong to that world' in attempting to reconstruct it, as well as correlatively, that 'the work belongs to our world'. The distance that separates us from this classical text is not merely historical, but is in essence also hermeneutic.[35] Classics are situated within the dialectics of being preserved as well as being surpassed. And yet, what appears as terminated is sheltered and saved, enfolded within its successors; although, the limited nature of inherited possibilities calls for an opening up of the future in which lies the essence of history and its destining. The classical text, being in essence historical, acquires a future by deepening its past through generational deferrals of the unfolding of its definitive meanings and its resistance to the weathering vicissitudes of history. The archiving, expository, and reproductive efforts ought to be supplemented by productive hermeneutical-critical readings. While the procedures that are followed in identifying, preserving, reconstructing, editing and translating classical texts, and commenting on their transmission do undeniably serve the purposes of establishing successful documents that creditably render the classics accessible, the anguish behind these activities tends to occasionally censor interpretation. From a hermeneutic perspective, the praiseworthy efforts of the custodians of archival recording do not necessarily always encourage original philosophical thinking. The controversies that arise around distortions in the reception or adaptation of classical philosophical works point primarily to concerns in historiography instead of philosophy. After all, a distinction is to be carefully drawn between the philosopher who is motivated by the systemic unfolding of fundamental questions-cum-concepts, and the exegete who is primarily bent on reporting them. Constraints imposed by the custodians of 'archivism' on rethinking the classics rest on an unquestioned metaphysics of time, which construes

35 This aspect of hermeneutics is elaborated by Hans-Georg Gadamer, *Wahrheit und Methode, Gesammelte Werke Band 1* (Tübingen: Mohr Siebeck, 1990).

historicity and temporality in a reductive linearity that is coupled with historicist angst regarding the writing of history. The most troubling of methodological risks in historiography is identified with what falls under the appellation 'anachronism'. Namely, when a classical text is approached by way of de-contextualising its arguments and neglecting the linear sequential nature of its documented transmissions; herein, chronology becomes the prime directive of readability. The charge of 'precursorism' also points to a misreading of history by way of tracing back all the elements of originality in a given tradition to exaggerated influences that it may have had from precedent sources. Moreover, a self-enclosed 'contextualism' studies a tradition within the limits of the unfurling of its own civilisation and the course of development of its local history, while eschewing any appeal to distinct cultural traditions. It is also believed that comparative studies or 'perspectivism' run the risk of being reduced into mere 'metonyms' or sets of selective arguments that suit the intellectual concerns of the interpreter rather than being faithful commentaries that abide closely by the 'original' author or text.[36] Some of these methods attempt to arrest thinking in the name of sustaining a proclaimed textual fidelity in approaching the classical text and sheltering its content, even though such procedures veil the doctrinal leanings that prompt the 'quarrels' amongst mediaevalists. Such controversies in methodology, which may have left their impress on the appropriation of the *Rasā'il*, have been avoided as much as possible through the diversity of the voices that are echoed in this present volume.

36 For an insight into these aspects of methodological prudence that burden mediaevalists, and at times run the risk of 'silently' hindering the flourishing of intellectual originality and conceptual creativeness in Islamic studies, I refer the reader to an expression of this anxious predicament in Robert Wisnovsky, *Avicenna's Metaphysics in Context* (Ithaca, NY: Cornell University Press, 2003), Introduction, pp. 16–18, and Conclusion, pp. 266–267. See also Wisnovsky's references to Muhsin Mahdi, 'Al-Fārābī's Imperfect State', *Journal of the American Oriental Society*, 110 (1990), p. 700; A. I. Sabra, 'The Appropriation and Subsequent Naturalization of Greek Science in Medieval Islam: A Preliminary Statement', *History of Science*, 25 (1987), pp. 223–224; and A. I. Sabra, 'Situating Arabic Science: Locality versus Essence', *Isis*, 87 (1996), p. 664.

The Arabic Critical Edition and Annotated English Translation

Modern academic literature on the *Rasāʾil Ikhwān al-Ṣafāʾ* is reasonably extensive within the field of Islamic studies, and continues to grow, covering works dating from the nineteenth century up to the present, with numerous scholars attempting to solve the riddles surrounding this compendium.[37] The academic rediscovery of the *Rasāʾil Ikhwān al-Ṣafāʾ* in modern times emerged through the monumental editorial and translation efforts of the German scholar Friedrich Dieterici between the years 1861 and 1872. Decades later, this venture was followed by renditions of the *Rasāʾil* into Hindustani and Urdu, with subsequent partial translations into Italian, French, and English;[38] all have been expatiated by extensive studies and analytical-critical commentaries. Several printed editions of the original Arabic have also been established, starting with the *editio princeps* in Calcutta in 1812, which was reprinted in 1846, then a complete edition in Bombay between 1887 and 1889, followed by the Cairo edition of 1928, and the Beirut edition of 1957, and their reprints.[39] Despite the laudable scholarly contribution of these Arabic editions of the *Rasāʾil*, which valuably sustained research on the topic, they are uncritical in character, and they do not reveal their manuscript sources.[40] Consequently, the current printed editions do not constitute definitive primary-source documentations of this classical text. Given this state of affairs, The Institute of Ismaili Studies (IIS) in London has undertaken the publication of a multi-authored, multi-volume, Arabic critical edition and annotated English translation of the fifty-two epistles constituting the *Rasāʾil Ikhwān al-Ṣafāʾ*. The present volume of studies is the first to be published, and serves as an introduction to the series as a whole.

In preparation for the Arabic critical edition of the *Rasāʾil*,

37 As mentioned earlier, controversies arose around the authorship of the compendium, the chronology of its composition, the exact affiliation of the Ikhwān, the ethical-political bearings of their teachings, their place in the history of ideas in Islam, the influences on their thinking, along with investigations centring on the manuscripts of their *Rasāʾil*.

38 Epistle 22: 'On Animals' was translated into Hebrew by the Jewish scholar Kalonymos ibn Kalonymos (d. ca. 1337) around 1316.

39 See note 4 above.

40 See Poonawala's chapter in the present volume.

reproductions of nineteen manuscripts were acquired by the IIS; and the particulars of these acquisitions can be summarised as follows:

I. Bibliothèque Nationale: MS 2303 (1611 CE); MS 2304 (1654 CE); MS 6.647–6.648 (AH 695; Yazd).
II. Bodleian Library, Oxford: MS Hunt 296 (n.d.); MS Laud Or. 255 (n.d.); MS Laud Or. 260 (1560 CE); MS Marsh 189 (n.d.).
III. El Escorial: MS Casiri 895/Derenbourg 900 (1535–1536 CE); MS Casiri 923/Derenbourg 928 (1458 CE).
IV. Königliche Bibliothek zu Berlin: MS 5038 (AH 600/1203 CE).
V. The Istanbul collections (mainly the Süleymaniye and associated libraries): MS Atif Efendi 1681 (1182 CE); MS Esad Efendi 3637 (ca. thirteenth century CE); MS Esad Efendi 3638 (ca. 1287 CE); MS Feyzullah 2130 (AH 704); MS Feyzullah 2131 (AH 704); MS Köprülü 870 (ca. fifteenth century CE); MS Köprülü 871 (1417 CE); MS Köprülü 981 (n.d.).
VI. Tehran: The Mahdavī collection, MS 7437 (AH 640).

It is worth noting that these acquisitions, which consist of the oldest complete manuscripts, along with significant supplementary fragments of an early dating, were each carefully selected from over one hundred extant manuscripts, which are available in public libraries and unrestricted collections.[41]

41 The large number of extant manuscripts and fragments of the *Rasā'il*, the broad range of their geographical sources, and the extended historical timeframe of their reproduction all reveal the extent of the circulation of this influential classical text. I have located over one hundred complete and fragmented manuscripts of the *Rasā'il Ikhwān al-Ṣafā'*, through a variety of catalogues, codices, and secondary sources. These manuscripts are preserved in thirty-nine libraries and collections, noted in alphabetical order by country, as follows: *Egypt*: Dār al-Kutub; Arab League Library (possibly also in the Arab League offices in Tunis); *France*: Bibliothèque Nationale de France; *Germany*: Königliche Bibliothek zu Berlin, Herzogliche Bibliothek zu Gotha, Eberhard–Karlis–Universität (Tübingen), Leipzig (Bibliotheca Orientalis), München Staatsbibliothek; *Iran*: Muṭahharī Library, Tehran University Central Library, Mahdavī (private) Collection; *Ireland*: Chester Beatty Library; *Italy*: Biblioteca Ambrosiana, Biblioteca Vaticana; *Netherlands*: Bibliotheca Universitatis Leidensis; *Russia*: Institut des Langues Orientales (St Petersburg); *Spain*: Biblioteca del Monasterio San Lorenzo de El Escorial; *Turkey*: Süleymaniye, Aya Sofia, Amia Huseyn, Atif Efendi, Esad

Like most critical editions, the IIS edition will be restricted by a relative sense of historical arbitrariness due to what has been preserved or recorded of the extant manuscripts of the text. For instance, the oldest manuscript of the *Rasāʾil*, the MS Atif Efendi 1681, dates back to 1182, and was hence compiled two to three centuries after the period usually associated with the composition of the text. It is also argued by some scholars that it has been either directly affected by scribal interventions, or else indirectly through the transcription of earlier manuscript sources, in terms of 'misleadingly' attributing the *Rasāʾil* to the utterances of the Sufis. Ultimately, the reconstruction of this classical text by way of a critical edition will be undertaken through manuscript reproductions that are significantly distanced in time from the original. The dexterity of the copyists, their deliberate tampering or exercise of restraint and relative impartiality, along with their scribal idioms, would have conditioned the drafting of the manuscripts. Such endeavours would also have been influenced by the intellectual impress of the prevalent geopolitical circumstances in which this text was transcribed, in addition to its channels of transmission. By widening the selection of the oldest manuscripts and fragments, based on the period of the copying, the levels of completeness and clarity, and the recommendations of past and present scholars who have consulted these collections, a suitably grounded critical edition will be produced, and a more reliable textual reconstruction will enable us to reassemble the *Epistles* in a manner that improves access to the contents of the text beyond what is presently available through the printed editions (i.e., those from Bombay, Cairo, and Beirut). It is ultimately hoped that this collective authorial effort will eventually render service to

---

Efendi, Millet Library, Garullah, Köprülü, Kütüphane-i ʿUmūmī Defteri, Manisa (Maghnisa), Rashid Efendi (Qaysari), Topkapi Saray, Yeni Cami, Revan Kishk; *United Kingdom*: Bodleian Library, British Library, British Museum, Cambridge University (Oriental Studies Faculty Library), The Institute of Ismaili Studies (including the Hamdani, Zāhid ʿAlī, and Fyzee Collections), Mingana Collection (Selly Oak Colleges Library, Birmingham), School of Oriental and African Studies (SOAS); *United States*: New York Public Library, Princeton University Library. The findings of this preliminary search will eventually be published in their confirmed final format in a supplement to the present series.

the scholarly community and further complement studies dedicated to the Ikhwān's corpus.

Synopsis of the Chapters of the Present Volume

This volume, which initiates the IIS series of the *Epistles of the Brethren of Purity*, presents selected inquiries by renowned scholars who have dedicated most of their academic efforts over the past decades to the study of the Ikhwān's legacy, and who are also acting as editors, translators, and annotators of the forthcoming critical edition of the *Rasāʾil*. Their diverse and informed scholarly opinions will give expression to the rich academic and intellectual dynamics that surround the interpretation of this compendium, as well as the circumstances determining its reception as a noteworthy classical text of philosophy.

Chapter One
'Why We Need an Arabic Critical Edition with an Annotated English Translation of the Rasāʾil Ikhwān al-Ṣafāʾ'

In the opening chapter, Ismail K. Poonawala offers an analytic survey of modern scholarship on the Brethren of Purity. He probes the sources available on the *Rasāʾil*, and on the Ikhwān al-Ṣafāʾ themselves, with an emphasis on the translations into the European languages, along with a thorough assessment of the relative merits of the existing Arabic publications of this corpus. Ultimately, this effort endorses the production of an Arabic critical edition and annotated English translation of the *Rasāʾil*. In this context, Poonawala reconsiders the questions of the authorship of the *Rasāʾil* and the dating of their composition. He argues in favour of affirming the affiliation of the Ikhwān with the Ismailis, or, at least, that the Brethren were inspired by Ismaili teachings. In the course of his inquiry, he also cautions that 'the search for the authors shows all the signs of a mediaeval polemic against the Ismailis in modern garb'. In addition, his analysis focuses on the specific content of certain epistles in the printed editions of the *Rasāʾil*, while also considering selected manuscripts. He further argues that the transmitted corpus may have been subject to tampering and interpolation, reflecting scribal mischief or doctrinal leanings. He also points out that the oldest extant manuscript is disjoined from 'the original' by about two to three centuries, which confronts scholars with the laborious task of accounting for the history of the compilation of this opus, and similarly of approximating a *stemma* of its manuscript traditions. Furthermore, this state of affairs does indeed confirm the need for an Arabic *critical* edition, and underlines the editorial challenges and authorial obstacles facing those who undertake the academic endeavour to 'reconstruct' a reliable textual model of the *Rasāʾil* that improves access to its content beyond what is currently possible with the available printed editions.

Chapter Two
'The Classification of Knowledge in the Rasāʾil'

In this study, Godefroid de Callataÿ compares the classificatory system devised by the Ikhwān in Epistle 7 of the *Rasāʾil* with the classification that is adopted in the fourfold division of the opus as transmitted in manuscripts and consequently transcribed in the printed editions. He thus reveals the discrepancies between 'the present lists of titles' in the *Rasāʾil* and 'the group of philosophical sciences' as they were described by the Ikhwān themselves in Epistle 7. He also demonstrates how the principal sections of the two systems do not exactly match one another, and that several epistolary headings in the *Rasāʾil* do not fit with some of the subdivisions mentioned in Epistle 7. Based on his analysis, those who favour a longer chronology in situating the composition of the *Rasāʾil* might then confidently claim that the authors of Epistle 7 and the final redactors of the work were not one and the same 'Brethren of Purity'. However, he adds that 'all this is conjectural, and bound to remain so until we obtain a much clearer picture of the social, historical, and epistemological context in which the epistles were produced, collected, and read'. According to de Callataÿ, 'the dating of the *Rasāʾil* is a vexed question', and he adds: 'In spite of a few opinions that remain divergent, there nevertheless seems to be a growing consensus today which accepts that the epistles were composed in or just before the

970s and that they had begun to circulate by the early 980s.' He also notes that 'common sense' would recommend us to take 985/986 as a 'most plausible *terminus ad quem*', given that it was the year of the death of the Būyid vizier with whom Tawḥīdī discussed the *Rasā'il*, as reported in the latter's *Kitāb al-Imtā' wa'l-mu'ānasa*. Nonetheless, de Callataÿ notes that 'these seeming oddities are matters which are best left unsolved for the time being'. His line of research focuses ultimately on the classification of scientific knowledge in its classical forms, while accounting for the compendium as 'a gnostic and philosophical genre of encyclopaedia' that has no equivalent in its era, or in any preceding historical epoch. He also highlights the features determining its syncretism and the unparalleled diversity in its sources. In accounting for the history of ideas in Islam, he seems to be inclined to view authorities as famous as Ibn Sīnā (d. 1037), al-Ghazālī (d. 1111), or Ibn Khaldūn (d. 1406) as being directly indebted to the Brethren's methods and argumentation. And yet, in terms of 'success', he believes that 'it is clear that the Ikhwān's classification cannot compare with al-Fārābī's *Enumeration of the Sciences* (*Iḥṣā' al-'ulūm*), whose impact both in the Orient and the Occident was truly considerable.' De Callataÿ's study provides a contrast to Abbas Hamdani's prominent inquiry in Chapter Three, and it also complements the research that is philosophical in scope, as undertaken by Carmela Baffioni, Ian Richard Netton, and Lenn E. Goodman in their respective chapters.

Chapter Three
'The Arrangement of the Rasā'il Ikhwān al-Ṣafā' *and the Problem of Interpolations'*

Abbas Hamdani's study consists of an updated version of an article that was previously published in the *Journal of Semitic Studies* (29, 1 [1984], pp. 97–110).[42] The revised text that constitutes this chapter has also been supplemented with additional notes and a postscript. It has been selected for the present volume in close consultation with its author, given that it is one of the principal studies embodying his unique thesis

42 We would like to thank the editors of the *Journal of Semitic Studies* and Oxford University Press for permission to include the updated version of Hamdani's article in the present volume.

regarding the dating of the Rasā'il.[43] In his chapter, Hamdani accounts for the difficulties surrounding the authorship and chronology of the composition of the *Epistles*. He argues that determining the identity and affiliation of the Ikhwān, as well as their *madhhab*, all depend on reasonable theses regarding the question concerning the dating of the compilation of the *Epistles*. The common view of the timescale of the composition posits a later chronology for situating the text rather than that espoused by Hamdani. The author argues that this later timescale results in a problematic historiography, which entails that the epistles were composed over a period of almost 'eighty years', and that they may consequently have been drafted by numerous authors working in distinct epochs. Hamdani notes that an earlier chronology would resolve these difficulties; or, at least, that it would indicate a shorter timeframe for the composition, while also restricting the authorship within a smaller circle of contemporaneous compilers who would have drafted the epistles collaboratively. In addition, he suggests that the dividing line between the earlier and the later proposed chronologies is around 297/909; namely, the year of the establishment of the Fāṭimid caliphate in North Africa. Furthermore, Hamdani refers in this context to the findings of Yves Marquet, which situate the composition of the text between 903 and 980, with Epistles 48 and 50, for example, predating the Fāṭimid era (even though Marquet aimed to overcome the ambiguities resulting from the chronological discrepancies related to the dating of individual epistles). Hamdani ultimately rejects the common view that the *Rasā'il* was composed between 961 and 980 (based on Tawḥīdī's story), and he doubts that any epistles could have been written after 909. According to him, a pre-909 dating of the corpus would resolve the internal ambiguities associated with the various dates

43 The dating and authorship questions have also been discussed in some of Abbas Hamdani's other articles. See 'Shades of Shi'ism in the Tracts of the Brethren of Purity', in *Traditions in Contact and Change*, ed. Peter Slater and Donald Wiebe (Waterloo, ON: 1983), pp. 447–460, with notes pp. 726–728; Abbas Hamdani, 'Brethren of Purity, a Secret Society for the Establishment of the Fāṭimid Caliphate: New Evidence for the Early Dating of their Encyclopaedia', in *L'Egypte Fatimide: Son art et son histoire*, ed. Marianne Barrucand (Paris: Presses de l'Université de Paris-Sorbonne, 1999), pp. 73–82. See also related studies by Hamdani in the bibliography of the present volume.

that are suggested for the composition of individual epistles, and that it would not require a rearrangement of the order of the *Rasāʾil* or an interrogation of the coherence of the Ikhwān's classificatory system. Hamdani's proposed timeframe is 260–297/873–909; thus, he considers the internal textual references that are posterior to this dating as interpolations or, possibly, attributable to scribal mischief.

Chapter Four
'The Scope of the Rasā'il Ikhwān al-Ṣafā''

In this chapter, Carmela Baffioni focuses primarily on the Ikhwān's philosophical legacy. She considers the Corpus Aristotelicum (in its Aristotelian and pseudo-Aristotelian forms), whilst exploring the variegated aspects of its influence on the conceptual bearings of the *Rasāʾil*. She also demonstrates how logic is utilised by the Ikhwān as 'a *science* in itself' rather than as 'an *instrument* of science'. Moreover, she shows how Neoplatonism is merged with Neopythagoreanism in the compendium, and convincingly argues that this constituted the Brethren's 'own philosophy' rather than simply having been inspired in its 'synthetic' figuration by foreign sources. She also points to what she identifies as Ismaili leitmotifs or proclivities in the *Rasāʾil*, and attributes the religious orientation of the text to the legacy of the Ismaili *dāʿī* Abū Yaʿqūb al-Sijistānī (d. after 971), with the distinctive mark of his reflections on creationism, prophecy, and resurrection. She also hints at the reception of the *Rasāʾil* within Ismaili intellectual circles, including the impact it had on the burgeoning of the thought of the Ismaili *dāʿī* Ḥamīd al-Dīn Kirmānī. Moreover, she refers to instances of the fusion of Ismaili and Muʿtazilī concepts, while highlighting the accompanying signs of Shiʿi inclinations in thinking, or what she also refers to as ''Alid codes of militancy'. According to her, the *Rasāʾil* expresses a penchant for adapting Greek philosophical notions to fit within a Muslim intellectual context, offering some vivid representations of ninth- and tenth-century attempts to reconcile revelation with reason.

Chapter Five
'The Rasā'il Ikhwān al-Ṣafā' *in the History of Ideas in Islam'*

Ian Richard Netton's chapter complements Carmela Baffioni's study in the preceding chapter in terms of its focus on the philosophical content of the *Epistles*, and in determining its place within the classical history of ideas in Islam. Netton offers an engaging reflection on the classification of this compendium as an 'encyclopaedia', or as being expressive of a Greek-inspired genre of 'encyclopaedism'. He identifies the *Rasā'il* as part of the literary genre of *adab*, or *belles lettres*, and he also celebrates the Ikhwān's enthusiastic passion for classification and narrative, reflecting the intellectual impetus that animated their composition of the *Epistles*. In spite of the historical critique of the style of the *Rasā'il* as at times verbose, repetitive, and complicated, Netton does nonetheless consider this oeuvre a masterpiece of literature. He also presents an analysis of the Ikhwān's consideration of what he refers to as 'the Qur'anic Creator Paradigm', and the way they accounted for it in terms of Neoplatonist principles of emanation, as inspired not only by the foundational *Enneads* of Plotinus (ca. 205–270), but above all by the teachings of Proclus Diadochus (ca. 410–485) in his *Stoikheiôsis theologikê* (*Elements of Theology*). As Netton shows, the Ikhwān surpassed the fundamental Plotinian triad of the One, the Intellect, and the Soul, in their hierarchy of being with its nine levels.[44] Furthermore, Netton investigates the aspects of soteriology and syncretism in the *Rasā'il*, and examines the particularities of the Ikhwān's deeply rooted faith in the promise of salvation and the utopian-cum-optimistic potential for success in the pursuit of happiness. This is evoked in the iconic image of the 'ship of salvation' (*safīnat al-najāt*) sailing the worldly ocean of matter, which is understood as being a gift from God in itself that can be earned through a genuine and sincere fraternal co-operation (*ta'āwun*), and which expresses both the overall ethos of the affiliates of the brotherhood as well as the advantageous consequence of following their teachings.

44 See Netton's chapter in the present volume. Also refer to *Rasā'il*, Epistle 29, vol. 3, p. 56; ibid., Epistle 32, vol. 3, pp. 181–184; ibid., Epistle 38, vol. 3, p. 285.

Chapter Six
'Misled and Misleading . . . Yet, Central in their Influence: Ibn Taymiyya's Views on the Ikhwān al-Ṣafā'"

While the other chapters focus on various hermeneutic, exegetical, explicative, historical, and textual interpretations of the *Rasā'il Ikhwān al-Ṣafā'*, and on contextually situating them in the development of the history of ideas in mediaeval Islamic civilisation, Yahya J. Michot's chapter offers an example of traditionalist credal critiques of the Brethren of Purity. Michot presents an expository account of the refutation of the Ikhwān's philosophy by the Mamlūk-era theologian Taqī al-Dīn Ibn Taymiyya (d. 1328) who construed the Brethren's oeuvre as that of 'esotericist heresiarchs'. This antagonistic turn in accounting for the intellectual legacy of the Ikhwān, as it is encountered in several fragments of Ibn Taymiyya's corpus, complements the studies that are grouped in the other chapters by way of advancing a viewpoint that rejects the philosophy of the Brethren of Purity, rather than celebrating it or espousing its conceptual models. By considering Ibn Taymiyya's vehement reaction against the Ikhwān's teachings, Michot reveals the extent of the influence of the *Rasā'il* in the intellectual *milieu* of the fourteenth century. It seems that the impact of the Ikhwān's legacy had reached a certain 'menacing' degree that ultimately solicited a hostile reaction to the *Rasā'il* from the Damascene theologian, who, in his turbulent era, seems to have regarded their teachings as a 'threat' to the credal tradition that he espoused in expression of his faith. It is in this spirit that Michot's chapter discloses the profound doctrinal antinomies that were at work within the history of ideas in Islam; principally, in terms of showing the adverse responses the *Rasā'il* drew from certain theologians in the context of a wider doctrinal opposition to philosophy, and a disapproval of gnosis or hermetic *bāṭinī* leanings.[45] Moreover, Michot presents aspects of Ibn Taymiyya's understanding of the chronology and affiliation of the Ikhwān, including his speculation that the *Rasā'il* is traceable back to the era of 'the founding of Fāṭimid

45 Ibn Taymiyya's severe critique of the Ikhwān is noted in the context of the acrimony he harboured against thinkers like Ibn Sīnā, Ibn 'Arabī, Ibn Sab'īn, Ibn Rushd, and Ibn Ṭufayl, while also displaying signs of restlessness regarding the 'philosophically-oriented' propositions of al-Ghazālī.

Cairo' in 360/970, and his attribution of the compendium to the Ismailis (whom he referred to disparagingly as 'Ubaydids and Qarmaṭīs). Michot's chapter also contains scholarly translations from Arabic into English of selected fragments from about forty primary sources from Ibn Taymiyya's textual corpus, which principally focus on the theologian's rejection of philosophy and his hypercritical judgement of the Ikhwān. Such material will give scholars the opportunity to inquire further about this fundamental area of investigation in the history of ideas in Islam and the multifaceted nature of its polemics.

Chapter Seven
'Epistolary Prolegomena: On Arithmetic and Geometry'

In my chapter, I focus on mathematics and the way it relates to the epistemology of the Brethren of Purity. My analysis is principally explanatory, and I attempt to elucidate the technical aspects of Epistles 1 and 2 on arithmetic and geometry, while also endeavouring to situate the contents of both tracts within the context of the development of 'mathematical sciences' in the ninth and tenth centuries. I place particular emphasis on offering a detailed exposition of the various mathematical propositions on arithmetic and geometry, as well as indicating their textual sources of transmission, specifically from the Greek heritage, along with their mediation by way of the flourishing of mathematics after the founding of the Bayt al-Ḥikma (House of Wisdom) in Baghdad. I also present some of the most significant aspects regarding the ontological and epistemological entailments of the Ikhwān's philosophical reflections on the intellective bearing of the study of mathematics. This is coupled with an assessment of the pedagogic possibilities and heuristics that the Ikhwān's treatment of the mathematical sciences offers to those seeking to study logic or natural philosophy, and to those who ultimately inquire into the nature of the soul with a view to reflecting on the fundamentals of theology.

Chapter Eight
'Music and Musicology in the Rasā'il Ikhwān al-Ṣafā''

Here, Owen Wright presents a focused analysis and informative explication of the epistle on music (*Risāla fī al-mūsīqā*; Epistle 5), by way of complementing the examination of the 'mathematical sciences' in the *Rasā'il*, as undertaken in the preceding chapter on arithmetic and geometry. Wright endeavours to situate the musical interests of the Ikhwān al-Ṣafā' within the broader context of the history of music and musicology in mediaeval Islamic civilisation. His specialist study offers detailed accounts of the technicalities of the Brethren's inquiries with respect to music, and the particulars of their artistic and intellectual approach to this art, which is part of their mathematical research and related to their broader outlooks on cosmology and natural phenomena. He also elucidates the Ikhwān's belief that the underlying numerical constitution of music provides a profound analogy to the structure of the cosmos, and, furthermore, that music offers the means through which the human soul can potentially 'glimpse the beauties of the higher orders'. Wright examines the nuances of the Ikhwān's grasp of rhythm, melody, and harmony, as well as the therapeutic, aesthetic, moral, and artistic dimensions which can be associated with the production of ordered combinations of musical sounds. Moreover, he accounts for the Ikhwān's views on the expression of the human voice in singing and on the human dexterity in the skilled use of musical instruments. In addition, he presents some insights concerning al-Kindī's legacy in music and the impact that its impetus may have had on the development of the Ikhwān's examination of this art. This study, along with the inquiries that are dealt with in the previous and subsequent chapters respectively, all constitute investigations focused squarely on certain specific arts and sciences within the *Rasā'il*.

Chapter Nine
'*Reading* The Case of the Animals versus Man: *Fable and Philosophy in the Essays of the Ikhwān al-Ṣafā'*'

In the final chapter, Lenn E. Goodman complements the inquiries that were addressed in Chapters 4 and 5 on the philosophical aspects of the Ikhwān's teachings by focusing on the particulars of Epistle 22 (on the animals), with its literary qualities, as well as its ethical-ecological bearings. Besides the significance of this epistle as part of the classical science of 'zoology', primarily in its Aristotelian philosophical genre, this *risāla* also constitutes a wonderful and engaging fable that has momentous ethical and ecological ramifications, in addition to being an example of the Ikhwān's satirical and didactic depiction of the human condition. Goodman shows how the *Rasā'il* corpus was inspired by the epistolary genre of antique narratives, and the way in which its composition gives expression to a taste for 'encyclopaedism'. He also argues that the Ikhwān's style of writing might embody the practices of a secretarial class of the early Islamic *imperium*. Based on Goodman's interpretation, this tract can be classed as an 'Aesopian' satiric fable, which hints to a tradition that may perhaps be traced back to the legacy of the Greek fabulist Aesop (ca. 620–560 BCE), and to a brand of literature that deploys ambiguous and allegorical meanings in order to elude political censorship. According to Goodman, this stylistic 'device' is ultimately used 'to culminate a moral and spiritual critique of human foibles and institutions'. Furthermore, Goodman presents a thorough analysis of the epistle on animals through references to his own reading of Montaigne. He also draws on Judaic sources in the history of narrative in considering his interpretation of this *risāla* as 'an ecological fable'. His readings are further supported by extensive quotations and informative citations from his annotated English translation of Epistle 22. He focuses on issues related to kingship, politics, morality, ecology, and ethics, with reflections on natural religion and on the determinants of rational deliberation. The principal thesis behind his study concerns the question of freedom as the very essence of being human. While the epistle on animals gives more weight to the human condition in the debate with the animals, in terms of the significance that is assigned to the '*immortality* of the human soul', this state of affairs is understood as a potential blessing and, equally, a probable source of eternal wrath. The lesson to be derived from this analysis emphasises the consequences of human comportment within the limits of our freedom, which is granted by means of exercising our rational faculties; this ultimately reveals the burden of responsibility that results from this 'gift', and the seriousness of our consequent accountability.[46]

46 I thank Professor Wilferd Madelung for his remarks on an earlier draft of this prologue.

ONE

Why We Need an Arabic Critical Edition with an Annotated English Translation of the Rasāʾil Ikhwān al-Ṣafāʾ

Ismail K. Poonawala

It is true that we already have three complete editions of all four volumes of the *Rasāʾil Ikhwān al-Ṣafāʾ*,[1] published in Bombay, Cairo, and Beirut in 1887–1889, 1928, and 1957 respectively.[2] However, none of them is a critical edition. The last of these was reprinted in 2004 and 2006. In addition, we have a composite edition by ʿĀrif Tāmir, published in 1995.[3] Hence, it is legitimate to ask the following questions: First, why do we need a critical edition? Second, why do we need an annotated English translation? We will address the second question first because it is simpler and easier to answer. In fact, the *Rasāʾil* has

1 The *Rasāʾil Ikhwān al-Ṣafāʾ* is hereafter cited as *Rasāʾil.*

2 *Kitāb Ikhwān al-Ṣafāʾ wa-Khullān al-Wafāʾ liʾl-imām al-humām quṭb al-aqṭāb mawlānā Aḥmad ibn ʿAbd Allāh*, ed. Wilāyat Ḥusayn, 4 vols. (Bombay: Maṭbaʿat Nukhbat al-Akhbār, 1305–1306/1887–1889); *Rasāʾil Ikhwān al-Ṣafāʾ wa-Khillān al-Wafāʾ*, ed. Khayr al-Dīn al-Ziriklī, 4 vols. (Cairo: al-Maṭbaʿa al-ʿArabiyya bi-Miṣr, al-Maktaba al-Tijāriyya al-Kubrā, 134/1928); *Rasāʾil Ikhwān al-Ṣafāʾ*, ed. Buṭrus al-Bustānī, 4 vols. (Beirut: Dār Ṣādir and Dār Bayrūt, 1377/1957; repr., 2004).

3 *Rasāʾil Ikhwān al-Ṣafāʾ wa-Khullān al-Wafāʾ*, ed. ʿĀrif Tāmir, 4 vols. (Beirut and Paris: Manshūrāt ʿUwaydāt, 1415/1995).

not been translated in its entirety into any Western language. Friedrich Dieterici single-handedly translated most of the epistles into German in summary form, and published them between 1861 and 1872.[4] Hitherto, his translation remains the only one of most of the epistles in a Western language. In 1975, Susanne Diwald published her translation of the third part of the *Rasāʾil* in German, having consulted certain older manuscripts, and she included copious footnotes including collations from other manuscripts.[5] Unfortunately, she did not live long enough to complete translation of the entire work. In light of this, the need for a complete English translation is fully justified. Before we address the second question, it is appropriate to digress a little and briefly review the last two centuries of scholarship concerning the *Rasāʾil*.

From the time of its circulation in public, either towards the end of the third/ninth or the beginning of the fourth/tenth century, the *Rasāʾil* was well known to the educated elite of mediaeval Muslim society. The fact that the epistles circulated and were eagerly read throughout the Muslim world soon after they were issued is attested by numerous witnesses. The anonymous author of *Muntakhab ṣiwān al-ḥikma* not only complements the author(s) of the *Rasāʾil* for their achievement but also quotes approvingly from it.[6] Abū al-Qāsim Maslama al-Majrīṭī (d. 398/1007), a mathematician and astronomer, is generally credited with bringing the *Rasāʾil* to public attention in al-Andalus (Andalusia, Spain), and his student al-Kirmānī, is recognised as having introduced the *Rasāʾil* to Saragossa and areas further north.[7] Two pseudo-Majrīṭī

4 They were published under the general title *Die Philosophie der Araber im X. Jahrhundert* in Leipzig.

5 Susanne Diwald, ed. and trans., *Arabische Philosophie und Wissenschaft in der Enzyklopädie Kitāb Iḫwān aṣ-ṣafāʾ* (III): *Die Lehre von Seele und Intellekte* (Wiesbaden: Otto Harrassowitz, 1975).

6 The original version of the *Ṣiwān al-ḥikma*, probably a compendium of texts studied and discussed in Abū Sulaymān al-Sijistānī al-Manṭiqī's circle, is not extant. Joel L. Kraemer thinks that the *Ṣiwān al-ḥikma* was not compiled by al-Sijistānī in the strict sense of the word, but was probably based upon lecture notes originating from his circle. He has described three existing recensions; see Joel L. Kraemer, *Philosophy in the Renaissance of Islam: Abū Sulaymān al-Sijistānī and his Circle* (Leiden: E. J. Brill, 1986), pp. 119–130, 132. The *Muntakhab* is edited by ʿAbd al-Raḥmān Badawī under the title *Muntakhab ṣiwān al-ḥikma* (Tehran, 1974), pp. 361–362.

7 See J. Vernet, 'Al-Madjrīṭī', *EI2*, vol. 5, pp. 1109–1110.

works, entitled *Ghāyat al-ḥakīm* and *Rutbat al-ḥakīm*, compiled during the first half of the fourth/tenth century, not only refer to the *Rasā'il* but also quote long passages from it.[8] In *Kitāb al-Imtā' wa'l-mu'ānasa*, compiled in Rajab 374/November–December 983, Abū Ḥayyān al-Tawḥīdī also refers to the *Rasā'il* and alleges that the encyclopaedia was put together by a group of Basran thinkers who were his contemporaries but held heretical views.[9] In *Munqidh min al-ḍalāl*, when al-Ghazālī (d. 505/1111) refutes the Ismailis, he refers to the *Rasā'il*.[10] Another Zaydī Mu'tazilī author, Abū Muḥammad al-Yamanī (d. 540/1145–1146), who was well acquainted with Ismaili works, also mentions the *Rasā'il* in his refutation of the beliefs of the Ismailis.[11] In the great compilation entitled *al-Kāmil*, an annalistic history from the beginning of the world up to the year 628/1231, Ibn al-Athīr records that in 555/1160 Ibn al-Murakhkham, the *qāḍī* ('judge') of Baghdad, was arrested by the then newly-installed 'Abbāsid caliph al-Mustanjid. His property was confiscated and his books dealing with philosophy were set on fire. What is worth mentioning in this respect is that Ibn

8 See Abbas Hamdani, 'Brethren of Purity, a Secret Society for the Establishment of the Fāṭimid Caliphate: New Evidence for the Early Dating of their Encyclopaedia,' in *L'Egypte Fatimide: Son art et son histoire*, ed. M. Barrucand (Paris: Presses de l'Université de Paris-Sorbonne, 1999), pp. 73–75. Both treatises were composed in the 330s/940s and 340s/950s, and Hamdani has provided all the relevant information and indicated the applicable sources.

9 Abū Ḥayyān al-Tawḥīdī, *Kitāb al-Imtā' wa'l-mu'ānasa*, 2nd edition, ed. Aḥmad Amīn and Aḥmad Zayn, (Beirut: Manshūrāt Dār Maktabat al-Ḥayāt, 1965), vol. 2, pp. 3–6 (hereafter cited as *Kitāb al-Imtā'*). The relevant passage is translated later in this chapter; see below, note 54.

10 Al-Ghazālī, *al-Munqidh min al-ḍalāl (Erreur et délivrance)*, ed. Farīd Jabre (Beirut: Commission Internationale pour la Traduction des Chefs-d'oeuvre, 1959), Arabic text p. 33, French trans. p. 94; English trans. R. J. McCarty in *Freedom and Fulfillment* (Boston, MA: Twayne Publishers, 1980), p. 89. Ghazālī states that the authors of the *Rasā'il* relied upon 'the feeble philosophy of Pythagoras'. He then adds that Aristotle had already refuted the latter and regarded his teaching weak and contemptible. Hence, Ghazālī describes the *Rasā'il* as a *ḥashw al-falsafa* ('refuse/waste of philosophy').

11 Abū Muḥammad al-Yamanī, *'Aqā'id al-thalāth wa'l-sab'īn firqa*, ed. Muḥammad al-Ghāmidī (Medina: Maktabat al-'Ulūm wa'l-Ḥikam, 1414/1993–1994), vol. 2, p. 513. It is worth noting that he considers the *Rasā'il* an Ismaili work, along with twenty-four others. His heresiography is, therefore, an important source of information about Ismaili works available to him in Yemen during the middle of the sixth/twelfth century.

al-Athīr notes that among those books were *al-Shifāʾ* of Ibn Sīnā and [the *Rasāʾil*] *Ikhwān al-Ṣafāʾ*.[12] These examples clearly show that the *Rasāʾil* was a well-known Ismaili philosophical work.

Renewed interest in this celebrated encyclopaedia began at the dawn of the eighteenth century when a segment of a fascinating fable from one of its epistles was first translated into Urdu, and soon thereafter the original Arabic version was published. The story of modern scholarship on the *Rasāʾil*, therefore, began in Calcutta, India, in 1810, when Mawlawī Ikrām ʿAlī published an Urdu (also called Hindustani)[13] translation of the allegorical dispute between humans and animals before the just king of the jinn from the *Rasāʾil*.[14] On the title page, without giving any details of his source(s), the translator stated that traditionally, the *Rasāʾil* has been ascribed to Abū Sulaymān al-Bustī, known as al-Maqdisī, and others. Readers familiar with the mediaeval bibliographical sources in Arabic will easily discern that Mawlawī Ikrām ʿAlī's ascription of the *Rasāʾil* to Abū Sulaymān al-Maqdisī echoes the most prevalent theory that was put in circulation by Abū Ḥayyān al-Tawḥīdī, as discussed below.

Two years later, in 1812, Shaykh Aḥmad ibn Muḥammad Shurwān al-Yamanī published the same fragment of the fable in its original Arabic, however, under a misleading title, *Tuḥfat Ikhwān al-Ṣafā*[ʾ], with a brief introduction in Arabic and another preface in English by T. Thomason.[15] Based on the manuscript which he consulted, Shaykh

12 Ibn al-Athīr, *al-Kāmil fī al-tārīkh*, ed. C. J. Tornberg (repr., Beirut: Dār Ṣādir, 1979), vol. 11, p. 258.

13 Urdu, the language of Islamic religious and cultural expression in the Indo-Pakistani subcontinent, emerged during Muslim rule in the region. Its vocabulary is extensively derived from Persian and Arabic, but the linguistic base is Indo-Aryan along with its core vocabulary as well as its phonology, morphology, and syntax. 'Hindustani' is used as a synonym for 'Urdu'. See C. Shackle, 'Urdū', *EI2*, vol. 10, p. 874; J. Burton-Page, 'Hindustānī', *EI2*, vol. 3, p. 461.

14 The *Rasāʾil Ikhwān al-Ṣafāʾ* has traditionally been ascribed to Abū Sulaymān (al-Maqdisī) and others; translated from the Arabic into Urdu by Mawlawī Ikrām ʿAlī (Calcutta: Munshī Muḥammad Ṭāha, 1810). It has been reprinted several times.

15 *Tuḥfat Ichwān-oos-Suffa*, in the original Arabic, revised and edited by Schuekh Ahmud-bin-Moohummud Schurwan-ool-Yummunee [Shaykh Aḥmad ibn Muḥammad Shurwān al-Yamanī] with a short preface in English by T. T. Thomason (Calcutta: printed by P. Pereira at the Hindoostanee Press, 1812).

Aḥmad al-Yamanī stated that the author of the book was Ibn al-Jaldī. The editor, however, failed to provide any further information about the author except stating that he was well known. At the end of the introduction, the editor confessed that eventually he stumbled upon a statement by ʿAbd al-ʿAlī ibn Ḥusayn al-Bīrjandī, in his commentary on Ptolemy's (Baṭlamiyūs) *Almagest* (*al-Majisṭī*), which mentioned that the *Rasāʾil* was compiled by a group of *mutakallimīn* ('scholastic theologians'). The word '*tuḥfa*', which was added to the title by the editor without giving any explanation, misled scholars for a while in their search for the title of the book and its authors.

Neither Mawlawī Ikrām ʿAlī nor Shaykh Aḥmad Shurwān al-Yamanī provided any information about the manuscript(s) they relied upon or a detailed list of the contents of the remaining epistles. This meant that scholars who did not have access to the manuscripts of the *Rasāʾil* did not get a clear picture of the number of epistles, their contents or the range of subjects with which they dealt. 'The Dispute Between the Animals and Man',[16] the lengthiest tale within the *Rasāʾil*, is a good example of the authors' socio-political criticism of Muslim society, here couched in animal characters in order to avoid offending the sensibilities of their readers. Publication of this story, first in Urdu translation and then in its original Arabic, was not only received by scholars with enthusiasm but was subsequently translated (either from Urdu or the original Arabic) into various languages, including English. These translations, in turn, aroused new interest among scholars concerning the *Rasāʾil*, their contents, authorship, and their sources. Aloys Sprenger correctly remarked that the publication of the story created considerable sensation due to the novelty of its ideas and its peculiarity of style and language.[17] The subsequent history of modern

Repr., Cairo in 1900 under the title *al-Ḥayawān wa'l-insān: wa-hiya khātimat wa-zubdat Rasāʾil Ikhwān al-Ṣafāʾ*.

16 This section is from the eighth epistle of the second volume, 'On Physical and Natural Sciences'. The epistle itself is entitled '*Fī takwīn al-ḥayawānāt wa-aṣnāfihā*' (On the Genesis of the Animals and their Species).

17 Aloys Sprenger, 'Notices of Some Copies of the Arabic Work Entitled *Rasāyil Ikhwān al-Cafā*'', *Journal of the Asiatic Society of Bengal*, 17 (1848), part 1, pp. 501–507; ibid., part 2, pp. 183–202; repr. in *Islamic Philosophy, Vols. 20–21: Rasāʾil Ikhwān aṣ-Ṣafāʾ wa-Khillān al-Wafāʾ*, ed. Fuat Sezgin et al. (Frankfurt

scholarship in respect to this encyclopaedia revolves around two main issues: First, publishing and translating the table of contents (*fihrist*) of the *Rasā'il* in order to show the reader the full range of the subject matter (sometimes adding more detail about certain epistles in the form of summaries and, at times, discussions of their sources). Second, investigating and discovering the mediaeval Arabic sources that contained information on the purported identity of the authors, in order to lift the veil of mystery that surrounds them. The limitations of this article do not permit further elaboration on this last point, except to state that, in many cases, the search for the authors shows all the signs of a mediaeval polemic against the Ismailis in modern garb. It is hoped that the full version of this research will be published soon.

The Name

'Ikhwān al-Ṣafā' wa-Khullān al-Wafā'' ('Sincere Brethren and Faithful Friends') is a pseudonym assumed by the authors of this celebrated encyclopaedia in order to hide their true identity. The question as to why they had to conceal their identity is closely connected to that of their affiliation with a secretive religio-political organisation whose main objective was to supplant the 'Abbāsid caliphate with a Shi'i imamate.[18] They deliberately concealed their identity not only to avoid detection by 'Abbāsid agents but also to allow their *Rasā'il* wider circulation and appeal to a broad cross-section of Muslim society. In fact, their appeal was not restricted to Muslims alone; rather, the eclectic nature of their philosophy was attractive to non-Muslims as well.

The pen name of the authors of the *Rasā'il Ikhwān al-Ṣafā'*, was derived from the famous animal fables of *Kalīla wa-Dimna* (also known as the *Fables of Bidpai*), translated into Arabic by Ibn al-Muqaffa' (d.

am Main: Institut für Geschichte der Arabisch-Islamischen Wissenshaften, 1999), vol. 1, pp. 201–228.

18 All references, unless otherwise stated, are to the Beirut edition. *Rasā'il Ikhwān al-Ṣafā'*, vol. 4, pp. 18–25, 146–148, 171, 375–381. See also Abdul Latif Tibawi ('Abd al-Laṭīf Ṭībāwī), *'Further Studies on Ikhwān aṣ-Ṣafā''*, *Islamic Quarterly*, *22* (1978), pp. 60–61.

139/756).[19] The authors of the *Rasā'il* were well acquainted with these fables and admired them.[20] The sincere brethren (*ikhwān al-ṣafā'*) in the above-cited fable were bound together by one another's friendship in the story of the ring-dove.[21] Referring to the story, the authors of the *Rasā'il* state:

> And know, O Brother, may God assist you and us with a spirit of His, that you ought to know for certain that you alone cannot redeem yourself from the affliction of this world into which you have fallen because of the [sin] committed by our [fore]father Adam. You need for your redemption and liberation from this world of generation and corruption — and ascension to the world of celestial spheres — and proximity to God's close angels, the assistance of brethren who are sincere advisers and associated with you by bonds of friendship, who are virtuous and have insight into the affairs of religion, so that they can apprise you of the path of the hereafter and how to attain it. So, take an example from the story of the ring-dove mentioned in the book of *Kalīla wa-Dimna* and how it freed itself from the snare.[22]

Furthermore, they narrate a number of anecdotes taken from ancient lore and biographies of the Companions of the Prophet to stress the meaning of sincere and loyal friendship.[23] In order to add an Islamic dimension to the meaning of 'brotherhood' (*ikhwān*), they cite a hadith of the Prophet and a Qur'anic verse. The tradition states:

---

19 *Kalīla wa-Dimna*, the celebrated collection of Indian fables of the *Pančatantra*, was translated into Arabic from the Pahlavi version. F. Gabrieli, 'Ibn al-Muḳaffa'', *EI2*, vol. 3, p. 883.

20 *Kalīla wa-Dimna* is referred to three times, see *Rasā'il Ikhwān al-Ṣafā'*, vol. 2, pp. 124, 474; vol. 4, p. 113. The two main characters, Kalīla and Dimna, are referred to twice; ibid., vol. 2, pp. 244, 330.

21 See *Bāb al-ḥamāma al-muṭawwaqa in Kalīla wa-Dimna*, ed. Ṭaha Ḥusayn and 'Abd al-Wahhāb 'Azzām (Cairo: Dār al-Ma'ārif, 1941), pp. 125–146.

22 *Rasā'il*, vol. 1, p. 100.

23 See the *Rasā'il*, vol. 4, pp. 14–60. In these two epistles, the authors further elaborate the meaning of the specific fraternity of the 'Ikhwān al-Ṣafā'', with examples derived from various traditions showing that its members are ever-ready to sacrifice their own lives for the spiritual welfare of their brethren.

> The faithful are like one person and one soul. Their blood and their possession counterbalance each other and they form one [united] front against the rest.[24]

The Qur'anic verse states:

> Hold fast to God's rope all together; do not split into factions. Remember God's favour to you: you were enemies and then He brought your hearts together and you became brothers [*ikhwān*[an]] by His grace; you were about to fall into a pit of Fire and He saved you from it.[25]

The authors of the *Rasā'il*, having been influenced by Neoplatonism and Neopythagoreanism, gave a fresh nuance to the meaning of the word '*ṣafā*'', moving from the original meaning, 'the purification of one's soul in preparation for being reunited with its divine source', to the already acquired meaning of 'brotherhood in faith'. In the very last epistle, entitled '*Fī māhiyyat al-saḥr wa'l-'azā'im wa'l-'ayn*' (On the Nature of Witchcraft, Incantation, and the Evil-Eye), the authors further define the phrase 'Ikhwān al-Ṣafā'', stating:

> When [our pious brethren] attain the heights of knowledge and loftiness of deeds, they are no longer in need of others with respect to all their worldly needs. When they reach this rank and attain this status, it would be credible to call them Ikhwān al-Ṣafā'. Know, O Brother, that the real meaning of this appellation is an exclusive property, not merely a metaphor, inherent in those who are worthy of it. Know, O Brother, may God the Exalted assist you, that the purity of soul cannot be achieved except after attaining the utmost serenity [with respect to] one's faith and worldly affairs. This means that one should know, according to one's ability and reach, the profession of the unity of God the Most High, knowledge of the true nature of living beings and peculiarities of created beings. One who is not qualified [as described above] cannot be counted among

24 It is transmitted by Ibn Ḥanbal, Abū Dāwūd, Nasā'ī, and Ibn Māja. See A. J. Wensinck et al., *Concordance et Indices de la Tradition Musulmane*, 8 vols., 2nd ed.(Leiden: E. J. Brill, 1992), s.v. k-f-9.

25 Qur'an 3:103.

> the people of purity [*ahl al-ṣafāʾ*] . . . Know, O brother, that it is in the very essence of purity [*ṣafāʾ*] that nothing that is needed by a pure and chaste soul should be absent from it . . . We have already stated that the knowledge of subtle sciences and noble teachings prepare a person in order to improve his physical lot in this world and ameliorate the fate of his soul in the hereafter.[26]

Contents of the Rasāʾil

The *Rasāʾil Ikhwān al-Ṣafāʾ* consists of fifty-one epistles arranged in four groups, preceded by a table of contents (*Fihrist al-rasāʾil*),[27] and followed by a compendium (*al-Risāla al-jāmiʿa*).[28] The four groups/parts are:

1. Mathematical–Philosophical Sciences (*al-Riyāḍiyya al-falsafiyya*): thirteen epistles on numbers (their essence, quantity, and quality), geometry, astronomy, music, geography, theoretical and practical arts, morals, and logic (corresponding to the five treatises of logic: *Isagoge*, *De Categoriae*, *De Interpretatione*, *Analytica Priora*, and *Analytica Posteriora*).
2. Physical and Natural Sciences (*al-Ṭabīʿiyya waʾl-jismāniyya*): seventeen epistles on physics, generation and corruption, mineralogy, botany, the nature of life and death, the nature of pleasure and pain, and the limits of human beings' cognitive ability.

26 *Rasāʾil Ikhwān al-Ṣafāʾ*, vol. 4, pp. 411–412.

27 MS ʿĀṭif Efendi 1681, transcribed in 1182 CE, Istanbul Collection. The ʿĀṭif Efendi manuscript, which is the oldest extant complete copy of the *Rasāʾil* contains fifty-one epistles. This number confirms all of the thirteen internal references in the text. See *Rasāʾil Ikhwān al-Ṣafāʾ*, vol. 1, pp. 282, 327, 361; ibid., vol. 2, p. 152; ibid., vol. 3, pp. 29, 75, 538; ibid., vol. 4, pp. 64, 173, 176, 250, 283, 284. However, in the Bombay, Cairo, and Beirut editions of the *Rasāʾil*, the number of epistles is fifty-two; the source of this discrepancy seems to have stemmed from counting two epistles in the Mathematical–Philosophical section (i.e., vol. 1) either as two separate epistles or as one single epistle. In all three published editions, the twelfth and thirteenth are regarded as two separate epistles, while they are counted as just one in the ʿĀṭif manuscript.

28 *Al-Risāla al-jāmiʿa*, ed. Jamīl Ṣalībā, 2 vols. (Damascus: Maṭbaʿat al-Taraqqī, 1949–1951).

3. Spiritual–Intellectual Sciences (*al-Nafsāniyya al-ʿaqliyya*): ten epistles on the metaphysics of the Pythagoreans and of the Brethren themselves, the intellect, the cycles and epochs, the nature of love, and the nature of resurrection, definitions, and descriptions.
4. Juridical–Theological Sciences (*al-Nāmūsiyya al-ilāhiyya*): eleven epistles on beliefs and creeds, the nature of communion with God, the creed of the Brethren, prophecy and its conditions, actions of the spiritual entities (angels, jinn, and demons), types of political constitutions, providence, magic, and talismans.

The fifty-one epistles in four volumes constitute an encyclopaedia of the philosophical sciences that were current among the Arab Muslims and others in the Middle East at that time. Leaving aside their obvious Shiʿi sympathies and discontent with the established political order, they attempted to reconcile Greek philosophy with Islamic revelation and *sharīʿa* (the 'sacred law'). *Falsafa*, equated with *ḥikma* (literally, 'wisdom'), is defined by them as seeking the path to God. Attainment of that knowledge is their ultimate goal. They further described themselves as a group of fellow seekers after truth, held together by their contempt for the world and its allurements and by their devotion to truth, whatever its origin.[29] They attempted to popularise learning and philosophy, and in appealing to a multiplicity of races and religions they developed a strong strain of interconfessionalism. Their attitude toward other religions was, therefore, strikingly liberal. They argued that religious differences stem from accidental factors such as race, habitat, and time, and do not affect the essential unity and universality of truth. They stated that the Sincere Brethren must not be hostile to any disciple of knowledge nor forsake any book of learning. To this they added unequivocally that they should not cling fanatically to any creed and should not be bigoted against other creeds: 'Our creed comprises all creeds and embraces all sciences.'[30] The compilation of such an encyclopaedia was, therefore, a remarkable achievement, and it occupies a unique place in the history of Islamic thought. In his book

29 *Rasāʾil*, vol. 4, p. 14.

30 Ibid., vol. 1, p. 48; ibid., vol. 4, pp. 41–42, 167–168, 426.

Geschichte der Philosophie im Islam, T. J. De Boer critically reviewed the contents of the *Rasāʾil*, based on the analysis of Friedrich Dieterici.[31] With the publication of De Boer's book, the *Rasāʾil* was assigned its due place in the history of philosophy in Islam.[32]

The authors of the *Rasāʾil* drew on a variety of sources. The Greek element is dominant throughout; for example, Ptolemy in astronomy, Euclid in geometry, Hermes Trismegistus in magic and astrology, Aristotle in logic and physics, and Plato and the Neoplatonists in metaphysics. Of the Neoplatonists, Plotinus and Porphyry exercised the strongest influence on the Brethren. Another pervading influence in the *Rasāʾil* is that of the Neopythagoreans, especially in their arithmetic and music. In their astrology there are traces of Babylonian and Indian elements. Additionally, there are fables of Indian–Buddhist, Persian–Zoroastrian, and Manichaean origin, and quotations from Jewish and Christian scriptures, both canonical and apocryphal. Despite these diverse sources, it is noteworthy that the authors achieved a remarkable overall synthesis. Hence, the question remains: does not such an encyclopaedia deserve to have a critical edition based on the oldest extant manuscripts? Certainly. A critical edition will reveal the extent of interpolation and tampering with the original text. This, in turn, will assist us in navigating the way to establishing the identity of these anonymous authors. Of course, it will not identify the names of the individual authors, but at least it should determine their affiliation to a particular ideological or sectarian group within Islam. Mediaeval and modern scholars have put forward conflicting claims regarding the authorship, with candidates such as the Shiʿi Ismailis, the Qarmaṭī, the Sufis, the Muʿtazila, or some unknown group of thinkers. Let us now examine the three available editions and see whether any of them might already qualify as critical edition.

31 T. J. De Boer, *Geschichte der Philosophie im Islam* (Stuttgart, 1901); English trans. Edward Jones, *The History of Philosophy in Islam* (1903; repr., London: Luzac and Co., 1965), pp. 81–96.

32 Detailed citations are given in my comprehensive study of this question, forthcoming.

The Bombay Edition of the Rasā'il

The full unabridged Arabic text of the *Rasā'il*, comprising four volumes (leaving aside *al-Risāla al-jāmi'a*), was published for the first time in 1305–1306/1887–1889 in Bombay, India. The publisher, al-Shaykh al-Ḥājj Nūr al-Dīn ibn Jīwākhān, a Musta'lī–Ṭayyibī Ismaili–Bohra and a dealer in books, was well acquainted with Ismaili history and literature.[33] Therefore, following the Ismaili tradition, he ascribed the authorship of the *Rasā'il* to the hidden Ismaili Imam Aḥmad ibn 'Abd Allāh ibn Muḥammad ibn Ismā'īl ibn Ja'far al-Ṣādiq, a contemporary of the 'Abbāsid Caliph al-Ma'mūn (d. 218/833).[34]

On the copyright page, the publisher states that the edition is based on an old and accurate copy of the *Rasā'il* ('*bi-nuskhat*[in] *qadīmat*[in] *ṣaḥīḥat*[in] *minhu*'); however, no description of the manuscript is given. Two notes at the end of the fourth volume are worth mentioning in this respect. The first, an encomium written by an Ismaili scholar, Shaykh Muḥammad 'Alī Rāmpūrī (d. ca. 1315/1898), praises the *Rasā'il* and what it contains in the way of philosophical and religious knowledge.[35] He also reproduced a long passage from Idrīs 'Imād al-Dīn's (d. 872/1468) historical work entitled *'Uyūn al-akhbār* concerning the *Rasā'il* and their authorship by the hidden Imam Aḥmad ibn 'Abd Allāh.[36] After endorsing the efforts of the publisher in making this important work available to the reader, Rāmpūrī reveals that despite his efforts, al-Ḥājj Nūr al-Dīn was unable to find an accurate copy of the *Rasā'il*.[37]

The second note, written by Muḥammad Bahā' al-Dīn, the

33 For the publisher, see Ḥusayn al-Hamdānī, *Baḥth tārīkhī fī Rasā'il Ikhwān al-Ṣafā' wa-'aqā'id al-Ismā'īliyya fīhā* (Bombay: The Arabic Library and Co., 1354/1935), p. 20.

34 The title page on all four volumes states: '*Kitāb Ikhwān al-Ṣafā' wa-Khullān al-Wafā' li'l-imām al-humām quṭb al-aqṭāb mawlānā Aḥmad ibn 'Abd Allāh raḥimahu Allāhu ta'āla.*'

35 For Rāmpūrī, see Ismail K. Poonawala, *Biobibliography of Ismā'īlī Literature* (Malibu, CA: Undena Publications, 1977), p. 229.

36 For Idrīs 'Imād al-Dīn, see ibid., pp. 169–172.

37 The Arabic reads, '*nuskha ṣaḥīḥa*'. It is difficult to ascertain what exactly the learned Shaykh meant by that phrase. However, it clearly implies that he was not quite content with the manuscript(s) used for the edition. *Kitāb Ikhwān al-Ṣafā'* (Bombay), vol. 4, p. 411.

proofreader at the press, first affirms that the *Rasāʾil* was compiled by the Imam Aḥmad ibn ʿAbd Allāh, but then adds that it is also said that the *Rasāʾil* was composed by a group of leading intellectuals between the second/eighth and the fourth/tenth century. However, he did not disclose his source for the latter statement.

Before proceeding further, it is worthwhile to draw the reader's attention to the question of interpolation during the transcription and transmission of the text over a long period of time. The question is not unique to the *Rasāʾil*, but applies equally to most texts transmitted over several centuries which lack a history for their textual transmission or known exemplars. A detailed study of this issue will have to wait until a critical edition of the whole encyclopaedia is completed. However, as a preview, I would like to compare certain features of the Bombay edition with the oldest extant manuscript copy of the *Rasāʾil*, transcribed in 578/1182, and preserved in the ʿĀṭif Efendi Library, in Istanbul, Turkey.[38] It should be noted that there is a gap of almost two or three centuries, depending on the theory one subscribes to concerning the authorship and the dating of the *Rasāʾil*, between the time of their compilation and the oldest existing and accessible copy of the entire work. This unaccounted-for interval of two or three centuries leaves us with many unresolved issues concerning additions, subtractions, and modifications introduced into the text. Besides those interpolations, it is obvious that numerous errors might also have been introduced, intentionally or unintentionally, in the process of transcription by the scribes, especially as the book is in several volumes and was widely read and copied.

First, the table of contents given at the beginning of the Bombay edition is much longer than that given in the ʿĀṭif manuscript. In the former, it comprises fifteen pages, while in the latter it is cut down to almost half that size. Second, in the Bombay edition there is an addition after the table of contents that mentions *al-Risāla al-jāmiʿa* and what it contains. No such addition is to be found in the ʿĀṭif manuscript. There is a further addition of three pages to the Bombay edition,

38 The colophon reads: '[Transcription] of the book was completed on Thursday, 13th of Ṣafar in the year 578 by Mawdūd ibn ʿUthmān ibn ʿUmar, the physician, al-Shirwānī.' See MS ʿĀṭif Efendi 1681, fol. 578a.

wherein the epistles are compared metaphorically to a beautiful garden with chirping birds, flowing streams, and a wide variety of flowers and fruits in which one always discovers something that one desires. It ends with an admonition to the reader and those who possess a copy of the *Rasāʾil* that they should exercise their discretion and make sure that the copies of the epistles do not fall in the hands of an unworthy person. Conversely, it adds that those who do possess copies of the *Rasāʾil* should not withhold them from the deserving. The ʿĀṭif manuscript, on the other hand, adds the above counsel very briefly at the end of the table of contents on the very authority of Abū Sulaymān (al-Maqdisī), one of the alleged compilers of the *Rasāʾil*, according to Abū Ḥayyān al-Tawḥīdī.[39] One might argue that such a note defeats the purpose of the authors that their encyclopaedia should circulate widely among the people. Hence, one might be tempted to conclude that this note could have been added later on. However, the internal evidence from the encyclopaedia suggests quite the contrary.[40] In any case, the issue cannot be resolved without examining all existing manuscripts.

Third, in the Bombay edition, at the end of the table of contents, the 'Sincere Brethren and Faithful Friends' are further described as 'People of Justice and Scions of Those Who Extol God and the People Who Possess the Truth and Real Meaning [of Things] concerning the Cleansing of the Soul and Refinement of Character in order to Attain the Ultimate Happiness, the Highest Loftiness, Everlasting Life, and the Final Perfection'.[41] The ʿĀṭif manuscript does not have those extra appellations added to their name. It simply states: 'List of the epistles

39 Al-Tawḥīdī, *Kitāb al-Imtāʿ*, vol. 2, pp. 3–6. Also, see below note 54.

40 The Ikhwān state: 'We do not hide our secrets from the people because of fear from the political authority but we do so as a safeguard for God's gifts to us. Jesus rightly said: "Do not provide wisdom to the undeserving, they might disregard it [literally, 'treat it unjustly'], but do not withhold it from the deserving, [by doing so] you might be unjust to them".' See *Rasāʾil*, vol. 4, p. 166. Elsewhere they also state: 'We should not squander our learning on those who do not deserve it. However, we should not withhold it from those who seek it and deserve it.' Ibid., vol. 4, p. 283.

41 It states: '*Tammat fihrist rasāʾil Ikhwān al-Ṣafā*[ʾ] *wa-Khullān al-Wafā*[ʾ] *wa-ahl al-ʿadl wa-abnāʾ al-ḥamd wa-arbāb al-ḥaqāʾiq wa-aṣḥāb al-maʿānī fī tahdhīb al-nufūs wa-iṣlāḥ al-akhlāq liʾl-bulūgh ilā al-saʿāda al-kubrā waʾl-jalāla al-ʿuẓmā waʾl-baqāʾ al-dāʾim waʾl-kamāl al-akhīr*.'

of Sincere Brethren and Noble Friends' (*Thabat rasā'il Ikhwān al-Ṣafā' wa'l-Aṣdiqā' al-Kirām*).

Fourth, in the table of contents of the Bombay edition there are three peculiar additions: (i) At the beginning, after stating that the fifty-two epistles contain a variety of sciences and wisdom, it adds: *'an kalām al-khulaṣā' al-ṣūfiyya* (literally, 'from the words of faithful Sufis'). (ii) After describing *al-Risāla al-jāmi'a* (the Compendium), it states: '*wa-hiya ithnatāni wa-khamsūna risāla wa-risāla fī tahdhīb al-nufūs wa-iṣlāḥ al-akhlāq*' ('they are fifty-two epistles plus an epistle concerning the cleansing of the soul and refinement of character'). (iii) At the end, that is, following the description of the compendium, it repeats: '*fī tahdhīb al-nufūs wa-iṣlāḥ al-akhlāq li'l-bulūgh ilā al-sa'āda al-kubrā wa'l-jalāla al-'uẓmā wa'l-baqā' al-dā'im wa'l-kamāl al-akhīr*' ('on the cleansing of the soul and refinement of character in order to attain the ultimate happiness, the highest loftiness, everlasting life, and the final perfection'). No such additions are to be found in the 'Āṭif manuscript. On the other hand, in the latter, after the title of each epistle except the first, the following expression is repeated: '*fī tahdhīb al-nafs wa-iṣlāḥ al-akhlāq min kalām al-ṣūfiyya*' ('on the cleansing of the soul and the refinement of the character, from the words of the Sufis'), while such expressions are absent from the Bombay edition. These expressions clearly imply that the *Rasā'il* was compiled by the Sufis. This is another case of obvious interpolation, which suggests that copies of the *Rasā'il* circulated widely among the Sufis who tried to appropriate the epistles for themselves. All the above indicators suggest clearly the extent of interpolation. Although the 'Āṭif manuscript is the oldest complete copy, with very few mistakes due to human error or phonetic corruptions of words, it cannot be considered exempt from interpolation.

The Cairo Edition of the Rasā'il

This version of the *Rasā'il* was published in Cairo in 1928. Neither Khayr al-Dīn al-Ziriklī, who was in charge of proofreading and correction, nor the other redactors were able to collate their text with another manuscript which they claimed was in the National Library in Cairo. At the end of the fourth volume, in the concluding remarks (*kalimat al-khitām*) al-Ziriklī states:

> Praise be to God! Printing of the book *Rasāʾil Ikhwān al-Ṣafāʾ wa-Khullān al-Wafāʾ* has been accomplished. The intention [of the redactors] at the beginning was that I would be in charge of correction and collating the proofs with other manuscripts, especially the manuscript in the National Library in Cairo. However, some [unforeseeable] circumstances prevented the implementation of the original intent. Hence, what I was supposed to do was accomplished by a group of other worthy scholars, such as Amīn Efendī Saʿīd, Shaykh Aḥmad Muṣṭafā, and Shaykh Aḥmad Yūsuf. The care they took upon themselves with [reading] the [proofs of the] book is, therefore, obvious from every [printed] page.[42]

Because al-Ziriklī acknowledged that he was unable to compare multiple manuscripts, the above statement implies that instead of collating the text with other manuscripts, the said group of scholars just read and corrected the proofs grammatically. A. L. Tibawi, who reviewed the Cairo edition, states that when the Cairo editors were asked about their source they refused to disclose the origin of the copy upon which they relied. Tibawi, thus, correctly observes that they have given us something very similar to the Bombay edition, but with some punctuation, divided into paragraphs, and printed on better paper.[43] The question, as to whether the Cairo edition is based on the Bombay one or not, cannot be resolved without collating the two editions. Hence, leaving aside that issue, it should be noted that the Cairo edition was introduced with two forewords, each written by a man of letters: the first by a younger scholar and *adīb*, Ṭāha Ḥusayn (d. 1973), and the second by a senior figure, Aḥmad Zakī Pasha (d. 1934).

In his foreword, Ṭāha Ḥusayn reiterated that the epistles reflect the socio-political conditions and varied intellectual climate that prevailed in the Islamic world during the fourth/tenth century. He stressed the

42 *Rasāʾil Ikhwān al-Ṣafāʾ* (Cairo), vol. 4, p. 479.

43 ʿAbd al-Laṭīf Ṭībāwī, ʻIkhwān aṣ-Ṣafāʾ and their *Rasāʾil*: A Critical Review of a Century and a Half of Research,ʼ *Islamic Quarterly*, 2 (1955), p. 36.

literary aspect of the *Rasāʾil* and lauded the efforts of their authors in making philosophical sciences accessible to the common reader. He expressed the view that the authors, whose goal was to prepare the people for both intellectual as well as political revolution, belonged to the extreme group of the Shiʿa, most probably those Ismailis who dissented from the ruling Fāṭimids. He further opined that when Abū al-ʿAlāʾ al-Maʿarrī (d. 449/1058), the celebrated poet, philosopher, and man of letters, visited Baghdad at the end of the fourth/tenth century he met with a group of the Ikhwān there and attended their meetings.[44] Al-Jundī and ʿAbd al-ʿAzīz al-Maymanī have refuted Ṭāha Ḥusayn's assumption and explained that al-Maʿarrī's expression '*ikhwān al-ṣafā*" is used as a *topos*, a nostalgic remembrance of a happy past in the context of a departure of a friend.[45]

Aḥmad Zakī Pasha's foreword, on the other hand, is a reprint of his previous study entitled *Mawsūʿāt al-ʿulūm al-ʿarabiyya wa-baḥth ʿalā* Rasāʾil Ikhwān al-Ṣafāʾ (Encyclopaedias of Arabic Sciences and a Discussion about the *Rasāʾil Ikhwān al-Ṣafāʾ*), originally published in 1890, wherein he noted and commented on the then–recently-published Bombay edition of the *Rasāʾil*.[46] He reproduced verbatim Ibn al-Qifṭī's (d. 646/1248) account concerning the authorship of the *Rasāʾil*, and rejected the theory that al-Majrīṭī was the author. He also flatly rejected the ascription of the *Rasāʾil* to the hidden Ismaili Imam Aḥmad ibn ʿAbd Allāh, denying that such a person ever existed. He further stated that *ʿUyūn al-akhbār* (quoted by Rāmpūrī in support of his claim that the epistles were compiled by the hidden Imam) was a fictitious work invented by the publisher to secure protection of copy-

44 In a poem in his *Saqṭ al-zand*, al-Maʿarrī mentions '*ikhwān al-ṣafā*". See Abū al-ʿAlāʾ al-Maʿarrī, *Dīwān saqṭ al-zand*, ed. Amīn Hindī (1319/1901), p. 132; English trans. Arthur Wormhoudt, *Saqṭ al-zand: The Spark from the Flint* (Ann Arbor, MI: University Microfilms, 1972).

45 Salīm al-Jundī, '*Abū al-ʿAlāʾ al-Maʿarrī wa-Ikhwān al-Ṣafāʾ*', *Majallat al-Majmaʿ al-ʿIlmī al-ʿArabī*, 16 (1941), pp. 346–351; repr. in *Islamic Philosophy*, ed. Fuat Sezgin et al., vol. 2, pp. 205–210. ʿAbd al-ʿAzīz al-Maymanī, *Abū al-ʿAlāʾ wa-mā ilayh* (Cairo, 1342/1923–1924).

46 Aḥmad Zakī (Pasha), *Mawsūʿāt al-ʿulūm al-ʿarabiyya wa-baḥth ʿalā Rasāʾil Ikhwān al-Ṣafāʾ: Étude bibliographique sur les encyclopédies arabes* (Cairo: Būlāq, 1308/1890).

right for his publication of the *Rasāʾil*. Nevertheless, he opined that the authors of the encyclopaedia may have had Ismaili proclivities.

Before we proceed further, two issues raised by Aḥmad Zakī Pasha need explanation: (i) Ascription of the authorship to the hidden Ismaili Imam or the possibility that the authors had Ismaili proclivities. And (ii) Ibn al-Qifṭī's account. Ḥusayn al-Hamdānī's article '*Rasāʾil Ikhwān al-Ṣafāʾ* in the Literature of the Ismāʿīlī Ṭayyibī *Daʿwat*' marks an important watershed in the history of modern scholarship on the authorship of the *Rasāʾil*.[47] He first pointed out a major distinguishing characteristic of the epistles, which is their outspoken Shiʿi character. Next, he noted the resemblance between certain views expressed in the *Rasāʾil* and in Ismaili doctrines. Lastly, he revealed a long history of the Mustaʿlī–Ṭayyibī Ismaili tradition, which claims that the *Rasāʾil* was issued by the concealed Ismaili Imam Aḥmad ibn ʿAbd Allāh, a contemporary of the ʿAbbāsid Caliph al-Maʾmūn, as a counter-move against the tide of Greek translations that had defiled the Islamic *sharīʿa*. This tradition further asserts that the *Rasāʾil* constituted an attempt to reconcile Greek philosophy with the Islamic *sharīʿa*. Of course, Ḥusayn al-Hamdānī himself expressed doubts as to whether this theory could be verified historically at the current stage of knowledge (i.e., in the 1930s). The point he made, which is still valid, was that the Ismaili tradition 'conclusively proves' a relation between the *Rasāʾil* and the Ismaili movement. This aspect is further elaborated by Abbas Hamdani in several of his articles, as will be seen below.

In 1859, long before the publication of the entire text of the *Rasāʾil*, Gustav Flügel published a list of fifty-one epistles with a German translation of their Arabic titles and a brief discussion about the identity of the authors, the organisation of the Ikhwān's fraternity, and the sources of their philosophy. His attempt to unravel the thorny issue of the authorship of the *Rasāʾil* was based on conflicting accounts reported in later sources, such as Ibn al-ʿIbrī (d. 685/1286), al-Shahrazūrī (d.

47 Ḥusayn al-Hamdānī, '*Rasāʾil Ikhwān al-Ṣafāʾ* in the Literature of the Ismāʿīlī Ṭayyibī *Daʿwat*', *Der Islam*, 20 (1932), pp. 281–300; Ḥusayn al-Hamdānī, *Baḥth tārīkhī fī Rasāʾil Ikhwān al-Ṣafā wa-ʿaqāʾid al-Ismāʿīliyya fīhā*; repr. in *Islamic Philosophy*, ed. Fuat Sezgin et al., vol. 20, p. 21; *Rasāʾil Ikhwān al-Ṣafā wa-Khillān al-Wafāʾ*, vol. 2, pp. 149–180.

after 687/1288), and Ḥājjī Khalīfa (d. 1067/1657). However, he supplemented the afore-cited sources by the then-newly-discovered works of Ibn al-Qifṭī (d. 646/1248) and Ṣafadī (d. 764/1363).

Jamāl al-Dīn Abū al-Ḥasan ʿAlī ibn Yūsuf, known as Ibn al-Qifṭī, a versatile author who held various positions under the Atabaks of Aleppo, is the first author to have resuscitated Abū Ḥayyān al-Tawḥīdī's story about the authorship of the *Rasāʾil*, reproducing the relevant passages from the latter's *Kitāb al-Imtāʿ waʾl-muʾānasa*. In his previously mentioned article, Flügel added Ibn al-Qifṭī's account from a manuscript copy of the latter's *Kitāb Ikhbār al-ʿulamāʾ bi-akhbār al-ḥukamāʾ*, generally known as *Taʾrīkh al-ḥukamāʾ*, which has survived as an epitome by al-Zawzanī (written in 647/1249). The compendium was edited by Julius Lippert in 1903. It contains biographies of physicians, philosophers, and astronomers. In his long entry for Ikhwān al-Ṣafāʾ wa-Khullān al-Wafāʾ, Ibn al-Qifṭī writes:

> They [the Sincere Brethren and Faithful Friends] are a group of people who agreed to compose a book about the [various] types of 'First Philosophy'.[48] They arranged the book into treatises numbering fifty-one. Fifty of them are [devoted] to fifty branches of knowledge, while the fifty-first treatise [called *al-Risāla al-jāmiʿa*] contains the summary of all the [other] treatises in a brief and concise manner. These treatises are fascinating, but they are neither exhaustive nor with obvious [i.e., convincing] proofs and arguments. They [appear] as if they are meant as a warning and as signposts to the goal of one who seeks enlightenment.
>
> As the authors [of the *Rasāʾil*] concealed their names, people differed about who the compilers were. Every group propounded its assertion by way of conjecture and speculation. One group said that [the epistles] contain the words of

48 *Al-Falsafa al-ūlā* as defined by al-Kindī is 'the knowledge of the realities of things, according to human capacity' or, more specifically, he defined metaphysics as the 'knowledge of the First Reality which is the Cause of every reality'. Al-Kindī, *Rasāʾil al-Kindī al-falsafiyya*, ed. M. A. H. Abū Rīda (Cairo, 1950–1953), vol. 1, p. 97; Alfred L. Ivry, 'Al-Kindī on *First Philosophy* and Aristotle's Metaphysics', in *Essays on Islamic Philosophy and Science*, ed. George F. Hourani (Albany, NY: SUNY, 1975), pp. 15–24; Majid Fakhry, *A History of Islamic Philosophy* (New York: Columbia University Press, 1983), p. 70.

> an Imam from the progeny of ʿAlī ibn Abī Ṭālib, but they disagreed widely about the identity of the Imam [to such an extent] that it does not establish the true identity of that Imam. Others maintained that the treatises were authored by some of the early Muʿtazilī theologians.
>
> I [the author, i.e., Ibn al-Qifṭī] continued to look very hard, seeking out information about the author/s until I stumbled across Abū Ḥayyān al-Tawḥīdī's statement in response to a question posed to him by the vizier of Ṣamṣām al-Dawla ibn ʿAḍud al-Dawla around the year 373/983–984.[49]

He then reproduces Abū Ḥayyān al-Tawḥīdī's anecdote from the latter's book *Kitāb al-Imtāʿ wa'l-mu'ānasa*, but without naming the book. Although Ibn al-Qifṭī cites al-Tawḥīdī at great length, he refrains from commenting on it and does not express his own opinion as to whether he accepts the story or not. What is significant in his account is that Ibn al-Qifṭī first acknowledges the prevailing view among certain informed circles at that time, that the *Rasā'il* was compiled by an ʿAlid Imam, irrespective of the latter's identity. All the later sources more or less follow Ibn al-Qifṭī.

Flügel had no access to Abū Ḥayyān al-Tawḥīdī's afore-mentioned work. Hence, neither he nor subsequent scholars were aware of the circumstances under which al-Tawḥīdī had alleged that the authors of the *Rasā'il* were a group of Basran thinkers who, according to him, held heretical views. Tawḥīdī's *Kitāb al-Imtāʿ wa'l-mu'ānasa* was edited by Aḥmad Amīn and Aḥmad Zayn in three volumes and was published between 1939 and 1944.[50]

Two years later, in 1946, Samuel M. Stern published an article and enthusiastically endorsed al-Tawḥīdī's story.[51] What Stern and other scholars have inferred from the latter's report is that the authors of the *Rasā'il* were contemporaries of Abū Ḥayyān al-Tawḥīdī and that

49 Ibn al-Qifṭī, *Ta'rīkh al-ḥukamā'*, ed. Julius Lippert (Leipzig: Dieterich's Verlagsbuchhandlung, 1903), pp. 82–88.

50 Al-Tawḥīdī, *Kitāb al-Imtāʿ*.

51 Samuel M. Stern, 'The Authorship of the Epistles of the Ikhwān aṣ-Ṣafā", *Islamic Culture*, 20 (1946), pp. 367–372; Samuel M. Stern, 'Additional Notes', *Islamic Culture*, 21 (1947), pp. 403–404; repr. in *Islamic Philosophy*, ed. Fuat Sezgin et al., vol. 2, pp. 243–250.

he had met at least one of the authors in addition to being conversant with the text of the *Rasāʾil*. Hence, these scholars conclude that the information given by al-Tawḥīdī is absolutely reliable. In his article 'Abū Ḥayyān al-Tawḥīdī and the Brethren of Purity', Abbas Hamdani for the first time scrutinised al-Tawḥīdī's narration concerning the authorship of the *Rasāʾil* and refuted it.[52]

The *Kitāb al-Imtāʿ waʾl-muʾānasa* (Book of Delight and Conviviality) is a recollection of the author's conversation with Ibn Saʿdān, the vizier of the Būyid emir Ṣamṣām al-Dawla, during their nightly entertainment sessions held at the latter's court in Baghdad.[53] The circumstances surrounding its compilation, given at the beginning of the book, are quite revealing of Abū Ḥayyān's personality. Glancing at the accounts of the first few nights, it becomes obvious to the reader that the vizier used Abū Ḥayyān to find out what gossip circulated among the élite of Baghdad concerning himself and his courtiers. In his account of the seventeenth night, Abū Ḥayyān, reminiscing about what happened, states the following:

> The vizier said: 'Tell me about a more important and a more serious matter that has come to my attention. Verily, I continue to hear from Zayd ibn Rifāʿa statements and teachings that I have never heard before. It has reached me that you visit him frequently, sit with him, copy [books] for him, and you exchange unusual anecdotes with him . . .'
>
> I [Abū Ḥayyān] replied: 'O vizier, you knew him before you knew me, in patronage, experience, and service. He has a long-standing, brotherly, and celebrated relationship with you.'
>
> The vizier replied: 'Leave it aside and describe him to me.'
>
> I responded: 'There is [in Zayd] a preponderance of intelligence, a brilliant mind, a ready wit . . . a vast knowledge in various disciplines of prose and poetry, and an outstanding art of writing in arithmetic and rhetoric . . . '
>
> The vizier said: 'Nevertheless, what is his belief?'

52 Abbas Hamdani, 'Abū Ḥayyān al-Tawḥīdī and the Brethren of Purity', *International Journal of Middle East Studies*, 9 (1978), pp. 345–353.

53 In 373/983–984, Ibn Saʿdān was appointed vizier by the Būyid emir in Iraq, Ṣamṣām al-Dawla (r. 372–376/983–987). The conversation ranged from literature, philology, and philosophy to court and literary gossip.

> I replied: 'He is associated neither with any [specific] ideology nor with any [particular] group, because he himself is simmering with every thing and boiling with every discipline [of knowledge] . . . In fact, he lived in Baṣra for a long time, and there, by chance, he encountered a group that had brought various kinds of sciences and crafts together [in strange harmony with each other]. They are: Abū Sulaymān Muḥammad ibn Maʿshar al-Bustī, known as al-Maqdisī; Abū al-Ḥasan ʿAlī ibn Hārūn al-Zanjānī; Abū Aḥmad al-Nahrajūrī; al-ʿAwfī; and others. Zayd accompanied them and rendered them his services. These cohorts frequented each other on account of their association and became sincere in friendship with one another . . . They further claimed that the *sharīʿa* had been defiled by ignorance and polluted with delusion. There was no way to cleanse and purge it except with philosophy . . . They composed fifty epistles about all branches of philosophy, both theoretical and practical, and set aside a table of contents, and entitled the [entire] compilation *Rasāʾil Ikhwān al-Ṣafāʾ wa-Khullān al-Wafāʾ*. They concealed their names and distributed the epistles among the copyists of manuscripts . . . '[54]

Abū Ḥayyān's account then shifts to another controversial issue connected with the *Rasāʾil* and also debated during this period in Islamic history, namely, the relation of Greek philosophy to Islamic *sharīʿa*, but that is beyond the scope of the present study. The vizier Ibn Saʿdān was arrested and executed in 375/985 — the reason given by the historians for his execution is that the vizier was accused of involvement in the activities of the Qarmaṭīs. Looking at Abū Ḥayyān's report, described briefly above, in hindsight one can surmise that the vizier wanted to draw Abū Ḥayyān into a conversation about what other courtiers thought of him and his close associates. Hamdani correctly observes:

> The *wazīr* Ibn Saʿdān being himself involved in the activities of the Qarmaṭian lobby through Ibn Shāhūyeh must have wanted to find out from Abū Ḥayyān the extent of the public knowledge of his involvement. A question about the philosophy of a known heretical member of his close group such as Zayd would serve the purpose of eliciting such information.

54 Al-Tawḥīdī, *Kitāb al-Imtāʿ*, vol. 2, pp. 3–6.

> There is nothing that Abū Ḥayyān could have added to what Ibn Saʿdān already knew about Zayd. Hence the need of Abū Ḥayyān to bring in the *Rasāʾil* to embellish and give credence to his story.[55]

Abū Ḥayyān, a man of letters, a philosopher and a master of Arabic, did not have much luck in obtaining their patronage from the ruling authorities due to his difficult and resilient character and his tempestuous nature. He spent much of his later life in poverty. Towards the end of his life he burned his books; one of the reasons for this was the neglect in which he had had to live for twenty years. Yāqūt (d. 626/1229), the famous author of an anthology of savants, introduces him as follows:

> He was versatile in all disciplines, such as grammar, language, poetry, literature, jurisprudence, and theology in accord with the Muʿtazila ... He was a Shaykh among the *ṣūfīs*, a philosopher among the men of letters, and a man of letters among the philosophers . . . Nevertheless, he assailed [others] with his tongue ... Making derogatory remarks [to others] was his trait and finding faults [with others] was his trade.

One should bear in mind that Abū Ḥayyān was not exempt from the art of embellishing his tales to give further credence to them, and at times he was accused of outright fabrications.[56] Ibn Abī al-Ḥadīd, the famous commentator of *Nahj al-balāgha*, accused him of manufacturing a treatise in the name of Abū Bakr and ʿUmar ibn al-Khaṭṭāb concerning the right of ʿAlī and ascribing it to the authorship of Abū ʿUbayda.[57] This was probably the reason, when Abū Ḥayyān compiled his *Kitāb al-Imtāʿ* at the urging of his patron al-Būzjānī, to whom he

55 Abbas Hamdani, 'Abū Ḥayyān al-Tawḥīdī and the Brethren of Purity', p. 351.

56 This view is shared by the editors of *Kitāb al-Imtāʿ*. See the Arabic introduction. It is said that none of his contemporaries, including his own teacher and patron, Abū Sulaymān al-Sijistānī, escaped from the scathing criticism of his sharp tongue. See *Kitāb al-Imtāʿ*, vol. 1, p. 33 (criticism of his teacher); Kraemer, *Philosophy in the Renaissance of Islam*, p. 28. Kraemer's translation of the Arabic does not reflect the harsh words used by Abū Ḥayyān.

57 It is called *Risālat al-saqīfa*, in *Trois Épîtres d'Abū Ḥayyān al-Tawḥīdī*, ed. Ibrāhīm al-Kaylānī (Damascus, 1951). For an English summary see Kraemer, *Philosophy in the Renaissance of Islam*, pp. 34–39.

used to send the completed parts of the book, why he requested that his benefactor keep it secret.[58] The book was compiled in Rajab 374/November–December 983, and the vizier, Ibn Saʿdān, was still alive and in power; hence Abū Ḥayyān was afraid of reprisal from the latter should he come to know that the book was embellished with tales and fabrications. Joel Kraemer, who has dealt extensively with Abū Ḥayyān al-Tawḥīdī's life and works, states that al-Tawḥīdī's reliability as a reporter is not beyond reproach and that he was stigmatised as a liar and a forger by mediaeval biographers.[59]

The Beirut Edition

A third complete edition of the *Rasāʾil* was published jointly by Dār Bayrūt and Dār Ṣādir in 1376/1957 in Beirut, with a preface by Buṭrus al-Bustānī. He reiterated what Aḥmad Zakī Pasha had said in the preface to the Cairo edition with some citations translated from De Boer's work. Bustānī is totally silent about the basis for this new edition. Thus, we do no know whether it is based on the Bombay or the Cairo edition, or whether it is a composite edition. What one can say is that it is slightly better than the Cairo edition, in that certain words are vocalised and particularly difficult words and terms are explained in the footnotes.

Manuscripts of the Rasāʾil

Unfortunately, no one has undertaken the task of compiling a comprehensive list of manuscript copies of the *Rasāʾil*, either of the complete or incomplete set of four parts or simply of copies of certain epistles, scattered in private and public libraries around the world. This would be a prerequisite before undertaking any critical edition; however, such a project cannot be accomplished without substantial financial support and time-consuming efforts. Nevertheless, it is encouraging to know that Nader El-Bizri, the Managing Editor of the present project, and the editor of this volume, has established a preliminary register of over one hundred manuscripts, deposited in the libraries of Egypt, various

58 See al-Tawḥīdī, *Kitāb al-Imtāʿ*, vol. 1, p. 12; ibid., vol. 2, p. 1.
59 Kraemer, *Philosophy in the Renaissance of Islam*, pp. 31, 32, 40, 44.

countries in Europe, Iran, Turkey,[60] and the United States, along with additional manuscripts of the *Rasāʾil* that are found and listed in India, Syria, and Yemen.

Based on the above information, the oldest extant complete copy of the *Rasāʾil*, from ʿĀṭif Efendi (MS 1681), was transcribed in 578/1182. Although the exact dating of the encyclopaedia is yet to be determined, most of the scholars who subscribe to Abū Ḥayyān al-Tawḥīdī's story are of the opinion that it was compiled during the fourth/tenth century. Yves Marquet is of the opinion that the compilation of the *Rasāʾil* began much earlier, perhaps during the pre-Fāṭimid period, but that the authors mentioned by al-Tawḥīdī gave the *Rasāʾil* their definite and final form. In several of his articles, Abbas Hamdani, based on early Ismaili sources and certain internal evidence from the *Rasāʾil*, has argued, contrary to all other scholars, for an earlier dating, viz., that they were issued before the establishment of the Fāṭimid dynasty in North Africa in 297/909. Thus, assuming that the *Rasāʾil* was composed during the last decades of the third/ninth century, our oldest extant copy was transcribed almost three centuries later. We can, therefore, state that the text of this manuscript, and, in turn, of the entire encyclopaedia, is not well attested or documented. A number of other manuscripts, such as Esad Efendi (MS 38) transcribed in 686/1287; Feyzullah (MS 310) transcribed in 704/1304–1305; Bibliothèque Nationale, in Paris (MS 48) transcribed in 709/1309–1310, just to mention a few, are as valuable to us as the ʿĀṭif manuscript. It should be noted that in the Mustaʿlī–Ṭayyibī Ismaili literature of the Yemeni period there are extensive citations from the *Rasāʾil*. Hence, collation of those sources with the extant manuscripts might reveal whether the Ismailis had access to a different exemplar or even to a distinct family of manuscript tradition. Given the situation, we should expect to find numerous variant readings and interpolations. Thus, a critical edition is likely to reveal many interesting things and to assist us in determining the period during which it was compiled.

60 See 'The Arabic Critical Edition and Annotated English Translation' section of Nader El-Bizri's prologue of the present volume.

TWO

The Classification of Knowledge in the Rasā'il*

Godefroid de Callataÿ

The work most commonly known as the *Rasā'il Ikhwān al-Ṣafā'* (*Epistles of the Brethren of Purity*) is a gnostic and philosophical encyclopaedia which was written in Arabic during the classical age of Islam and whose nature, contents, and purposes have no equivalent of any kind either inside or outside the Muslim world. Scholarship devoted specifically to this work has only started to develop in recent times, so that large parts of the encyclopaedia remain unexplored, or at least insufficiently investigated. To this day, only one section out of the four that form the whole corpus has been edited on a scholarly basis,[1] and

* This is the revised version of a paper given in May 2003 at The Institute of Ismaili Studies, London, as part of the Occasional Lectures Series, organised that year by Leonard Lewisohn. I have much pleasure in thanking him for his kind invitation to give this talk, an electronic version of which is available on the IIS website. I also wish to express my warmest gratitude to Farhad Daftary and Nader El-Bizri for offering to publish this study in the present volume. A shorter and handier version of this paper may be found in my recent *Ikhwan al-Safa': A Brotherhood of Idealists on the Fringe of Orthodox Islam* (Oxford: Oneworld, 2005), where it appears as Chapter 4 ('Encyclopaedism'). I should like to thank Patricia Crone for her valuable comments and suggestions.

1 Susanne Diwald, ed. and trans., *Arabische Philosophie und Wissenschaft in der Enzyklopädie Kitāb Iḫwān aṣ-ṣafā' (III): Die Lehre von Seele und Intellekt* (Wiesbaden: Otto Harrassowitz, 1975).

there remains a majority of epistles which have never been properly translated into English or any other language.[2]

It is now generally agreed that the authors of the *Epistles* were high-ranking men of learning from the Shiʿi community, that they lived in Basra (Southern Iraq) during the fourth century of Islam (tenth century CE), and that they had at least some connections with the Ismaʿili tradition. The encyclopaedia as we know it consists of fifty-one or fifty-two epistles, each dealing with roughly one particular topic of human knowledge, to which is added a 'Concluding' or 'Comprehensive Epistle' (*al-Risāla al-jāmiʿa*) at the end of the corpus. The epistles are visibly classified according to an order designed as a step-by-step progression towards the most difficult aspects of human wisdom. The esoteric nature of certain parts of the encyclopaedia, especially in the last epistles, is a remarkable peculiarity of the *Rasā'il.* Another very conspicuous feature of the corpus is the great diversity and considerable eclecticism of its sources, along with the almost unparalleled scope of the matters involved.

In recent times, several studies have been devoted to the sources and contents of the *Rasā'il Ikhwān al-Ṣafā'*, most notably by Yves Marquet,[3]

2 It is in view of this that the present volume introduces the series of the Arabic critical editions and annotated English translations of the fifty-two epistles of the *Rasā'il Ikhwān al-Ṣafā'*.

3 Yves Marquet, especially *La philosophie des Iḫwān al-Ṣafā': Thèse présentée devant l'Université de Paris IV, 1971* (Algiers: Société nationale d'édition et de diffusion, 1973); a revised, yet substantially unchanged, version of this book was published in Paris and Milan (Archè, 1999). See also Yves Marquet, '910 en Ifrīqiya: Une épître des Ikhwān al-Ṣafā'', *Bulletin d'Études Orientales*, 30 (1978), pp. 61–73; Yves Marquet, 'Les Ikhwān al-Ṣafā' et l'ismaélisme', in *Convegno sugli Ikhwân al-Ṣafā', Roma, 1979* (Rome: Accademia Nazionale dei Lincei, 1981), pp. 69–96; Yves Marquet, 'Les Ikhwān al-Ṣafā' et le Christianisme', *Islamochristiana*, 8 (1982), pp. 129–158; Yves Marquet, 'Les Épîtres des Ikhwān al-Ṣafā', œuvre ismaïlienne', *Studia Islamica*, 61 (1985), pp. 57–79; Yves Marquet, *La philosophie des alchimistes et l'alchimie des philosophes. Jābir ibn Ḥayyān et les 'Frères de la Pureté'* (Paris: Maisonneuve et Larose, 1988); Yves Marquet, 'La détermination astrale de l'évolution selon les Frères de la Pureté', *Bulletin d'Études Orientales*, 44 (1992), pp. 127–146; Yves Marquet, 'Ibn al-Rūmī et les Ikhwān al-Ṣafā'', *Arabica*, 47 (2000), pp. 121–123.

Ian Richard Netton,[4] Carmela Baffioni,[5] and the author of the present survey.[6] We also find a few studies in which the Ikhwān's way of classifying the sciences is briefly discussed for itself,[7] or else compared with other famous Islamic systems, such as those of al-Kindī (d. ca. 870), al-Fārābī (d. 950), Ibn Sīnā (d. 1037), or Ibn Khaldūn (d. 1406)[8]. Yet, to the best of my

---

4 Ian Richard Netton, 'Brotherhood versus Imamate: Ikhwān al-Ṣafā' and the Ismā'īlīs', *Jerusalem Studies in Arabic and Islam*, 2 (1980), pp. 253–262; Ian Richard Netton, 'Foreign Influences and Recurring Ismā'īlī Motifs in the *Rasā'il of the Brethren of Purity*', in *Convegno sugli Ikhwān al-Ṣafā', Roma, 1979* (Rome: Accademia Nazionale dei Lincei, 1981), pp. 49–67; Ian Richard Netton, *Muslim Neoplatonists: An Introduction to the Thought of the Brethren of Purity (Ikhwān al-Ṣafā')* (London: Routledge Curzon, 2002).

5 Carmela Baffioni, 'Traces of "secret sects" in the *Rasā'il* of the Ikhwān al-Ṣafā'', in *Shī'a Islām, Sects and Sufism: Historical dimensions, religious practice and methodological considerations*, ed. Frederick De Jong (Utrecht: M. Th. Houtsma, 1992), pp. 10–25; Carmela Baffioni, *Frammenti e testimonianze di autori antichi nelle epistole degli Ikhwān al-Ṣafā'* (Rome: Istituto Nazionale per la Storia antica, 1994).

6 Godefroid de Callataÿ, 'Astrology and Prophecy: The Ikhwān al-Ṣafā' and the Legend of the Seven Sleepers', in *Studies in the History of the Exact Sciences in Honour of David Pingree*, ed. Charles Burnett et al. (Leiden: Brill, 2003), pp. 758–785; Godefroid de Callataÿ, 'Sacredness and Esotericism in the *Rasā'il Ikhwān al-Ṣafā'*', in *Al-kitāb: La sacralité du texte dans le monde de l'Islam*, Actes du Symposium International tenu à Leuven et Louvain-la-Neuve, 2002, *Acta Orientalia Belgica: Subsidia*, 3, ed. Daniel De Smet, Godefroid de Callataÿ, and Jan van Reeth (Brussels, Louvain-la-Neuve, and Leuven: Société Belge d'Études Orientales, 2004), pp. 389–401.

7 Seyyed Hossein Nasr, *Islamic Science: An Illustrated Study* (Westerham: Westerham Press, 1976); Shlomo Pinès, 'Une encyclopédie arabe du 10e siècle: Les Épîtres des Frères de la Pureté, *Rasā'il Ikhwān al-Ṣafā'*', *Rivista di storia della filosofia*, 40 (1985), pp. 131–136.

8 Louis Gardet and Georges Chehata Anawati, *Introduction à la théologie musulmane: Essai de théologie comparée* (Paris: Vrin, 1948), pp. 101–124 ('Le *kalām* et les sciences musulmanes'); Marc Bergé, 'Épître sur les sciences d'Abū Ḥayyān at-Tawḥīdī: Introduction, traduction, glossaire technique, manuscrit et édition critique', *Bulletin d'Études Orientales*, 18 (1964), pp. 241–300 (esp. pp. 248–254: 'Place de la *Risāla fī al-'ulūm* dans les classifications arabes des sciences'); Franz Rosenthal, *Das Fortleben der Antike in Islam* (Zurich and Stuttgart: Artemis Verlag, 1965), pp. 77–101; Charles Pellat, 'Les encyclopédies dans le monde arabe', *Cahiers d'histoire mondiale*, 9 (1966), pp. 631–658; Charles Pellat, Živa Vesel, and E. van Donzel, 'Mawsū'a', *EI2*, vol. 6, pp. 903–911; Živa Vesel, *Les Encyclopédies persanes: Essai de typologie et de classification des sciences* (Paris: Editions Recherche sur les Civilisations, 1986); Carmela Baffioni, 'Oggetti e caratteristiche del curriculum delle scienze nell'Enciclopedia dei Fratelli della Purità', in *Studi arabo-islamici in memoria di Umberto Rizzitano*, ed. Gianni di

knowledge, no significant attempt has been made so far to appraise the originality of the Brethren's own system. It is the aim of this contribution to present some results of my current exploration of this topic.

* * *

First of all, one must clarify which kind of classification we are talking about. For, on the one hand, there are those fifty-one or fifty-two epistles in the arrangement that has come down to us through the manuscript tradition and whose sequence may indeed qualify as a hierarchy of sciences in its own right. And then we have, on the other hand, the properly called classification of sciences that the Brethren set forth in Epistle 7, namely the one entitled 'On the Scientific Arts and their Aim'. Indeed, the two lists differ from each other in several places, and certain discrepancies are even so marked that, in themselves, they would seem to bear witness to a historical process of re-elaboration.

It seems appropriate to begin with the classification of sciences which the authors themselves outlined in the second half of Epistle 7.[9]

Stefano (Trapani: Istituto di Studi Arabo Islamici 'Michele Amari' di Mazara del Vallo, 1991), pp. 25–31; Wolfhart Heinrichs, 'The Classification of the Sciences and the Consolidation of Philology in Classical Islam', in *Centres of Learning: Learning and Location in Pre-modern Europe and the Near East*, ed. Jan Willem Drijvers and Alastair A. MacDonald (Leiden: Brill, 1995), pp. 119–139; Hans Hinrich Biesterfeldt, 'Medieval Arabic Encyclopedias of Science and Philosophy', in *The Medieval Hebrew Encyclopedias of Science and Philosophy: Proceedings of the Bar-Ilan University Conference*, ed. Steven Harvey (Dordrecht, Boston, and London: Kluwer, 2000), pp. 77–98; Hans Hinrich Biesterfeldt, 'Arabisch-islamische Enzyklopädien: Formen und Funktionen', *Die Enzyklopädie im Wandel vom Hochmittelalter bis zur Frühen Neuzeit*, Akten des Kolloquiums des Projekts D im Sonderforschungsbereich 231 (1996), ed. Christel Meier (Munich: Wilhelm Fink Verlag, 2002), pp. 43–83.

9 I have used the four-volume Beirut edition (ed. Buṭrus al-Bustānī, Dār Ṣādir, 1957), which (pending the completion of the IIS Arabic critical edition) remains the more commonly used throughout the world, in spite of its numerous deficiencies. The text of 'Epistle 7' is to be found in vol. 1, pp. 258–275. The classification itself starts on p. 266. For a translation (into French) of the whole epistle, see 'Ikhwān al-Ṣafā': des arts scientifiques et de leurs objectifs (*Épître VII* des Frères de la Pureté)', ed. and trans. Godefroid de Callataÿ, *Le Muséon*, 116 (2003), pp. 231–358.

For us, the most important part of this text is the overall presentation of the system, which begins with the following lines:

> We should like to mention the types of science and their subdivisions, so as to direct the seekers of knowledge to their aims and guide them to their objectives. For the wishes of souls for different sciences and branches of culture are like the desires of bodies for foods that differ in taste, colour, and smell. (*Rasāʾil*, vol. 1, p. 266)

These preliminary words look like an invitation merely to single out from the entire corpus of sciences one or two particular fields according to one's tastes. They do not seem to presuppose, as such, any logical or rational sequence of the fields of knowledge that are mentioned next. In other terms, they could as well have been part of a typical piece of *adab* literature like, say, the *Epistle on the Sciences* of Abū Ḥayyān al-Tawḥīdī (d. 1023), which is neither a systematic nor an exhaustive enumeration of sciences.[10] What comes next in Epistle 7 makes it clear, however, that the Ikhwān had a much more elaborate system in mind.

A Tripartite Division of Sciences

The passage reads:

> Know, my brother, that there are three kinds of science with which people busy themselves; namely, the propaedeutic [i.e.,

10 Al-Tawḥīdī, *Risāla fī al-ʿulūm*, ed. Marc Bergé as 'Épître sur les sciences', *Bulletin d'Études Orientales*, 18 (1964), pp. 286–298. See Bergé's observation (p. 248): 'À vrai dire cet auteur nous pose, par cette oeuvre, le problème général du développement des sciences chez les Arabes, en abordant, de façon plus précise, la question de leur classification. Mais la simple lecture du plan et de l'analyse de la *Risāla* nous a montré que cet auteur a eu des visées plus modestes et ne se pose pas devant nous en docteur légiférant sur cette classification ou voulant, de façon prolixe, défendre la cause des Arabes et de leur prédication à l'essor des sciences.' The sequence according to which the 'sciences' are dealt with by Tawḥīdī is: jurisprudence, the Book, prophetic tradition, analogical reasoning, theology, grammar, language, logic, medicine, astronomy/astrology, arithmetic, geometry, stylistics, and mysticism.

introductory] sciences, the sciences pertaining to revealed law, and the sciences of true philosophy.

- The propaedeutic [sciences] are those concerning the proper rules established, for the most part, for the pursuit of livelihoods and the improvement of life in this world. They are of nine kinds: (1) writing and reading; (2) language and grammar; (3) calculation and operations; (4) poetry and prosody; (5) auguries and auspices, and the like; (6) magic, talismans, alchemy, mechanical devices and the like; (7) professions and crafts; (8) sale and purchase, trades, cultivation, and breeding; (9) the study of campaigns and history.
- The types of religious sciences established for the healing of souls and the pursuit of the hereafter are of six kinds: (1) the science of revelation; (2) the science of interpretation [*ta'wīl*]; (3) the science to do with transmissions and reports (from past religious authorities); (4) the science of jurisprudence, norms, and laws; (5) the science relating to remembrance, exhortations, asceticism, and mysticism; (6) the science of the interpretation of dreams. The learned in the science of revelation are those who recite the Qur'an and know it by heart. The learned in the science of interpretation are the Imams and the successors of the prophets. Those who know about transmissions are the specialists of the Tradition. Those who know about the laws and norms are the jurists. Those who know about remembrance and exhortation are the worshippers, the ascetics, the monks, and the like. The learned in the interpretation of dreams are the dream interpreters.
- The philosophical sciences are of four kinds: (1) mathematical; (2) logical; (3) physical; (4) divine. [Further divisions of the philosophical sciences will be discussed below.] (*Rasā'il*, vol. 1, pp. 266–267)

The Propaedeutic Sciences

Firstly, there are the sciences which the Ikhwān call the propaedeutic (or, disciplinary or training) sciences and which they define as 'the sciences of education [*ādāb*] which have been established mainly for the pursuit of livelihoods and the improvement of life in this world'. The Brethren do not look down on them, as all these sciences have proved useful in the terrestrial accomplishment of man, yet their very segregation from the rest clearly suggests that they were felt to be inferior to the sciences of the two other groups, whose purpose is not restricted to life here below.

The Ikhwān were not the first thinkers to speak of propaedeutic or 'training' sciences (*'ilm al-riyāḍiyyāt*). In his *Epistle on the Number of Books by Aristotle*,[11] al-Kindī used exactly the same words, yet in his writing the expression unambiguously referred to the four mathematical sciences that make up the so-called 'Pythagorean quadrivium', namely, arithmetic, geometry, astronomy, and music. At least from the time of Plato (fourth century BCE), the importance of these four mathematical sciences as a prerequisite to any other studies had been endorsed in the West by such great authorities as Nicomachus of Gerasa (fl. between 50 and 150), Boethius (d. 524), and Isidore of Seville (d. 636), such that it became a commonplace of any discussion about philosophy and its divisions in the mediaeval schools of Western Europe in the twelfth and thirteenth centuries.[12] This tradition of four liberal arts also found its way into Islamic thinking, as we can see from al-Kindī's treatise on the number of Aristotle's books but also from countless other forms of evidence.[13] The Pythagorean quadri-

11 Al-Kindī, *Risāla fī kutub Arisṭūṭālīs*, in *Rasā'il al-Kindī al-falsafiyya*, ed. Abū Rīda (Cairo, 1369/1950), pp. 363–384. For a detailed discussion of this work as a classification of knowledge in its own right, see Angel Cortabarria Beitia, 'La classification des sciences chez al-Kindī', *Mélanges de l'Institut Dominicain d'Études Orientales*, 11 (1972), pp. 49–76, esp. pp. 61–70.

12 Let us single out, among countless other studies in this field, Guy Beaujouan, 'L'enseignement du *quadrivium*', in *La scuola nell'Occidente latino dell'alto medioevo*, Settimane, 19 (Spoleto: Centro italiano di studi sull'Alto Medioevo, 1972), vol. 2, pp. 639–723; Guy Beaujouan, 'The Transformation of the Quadrivium', in *Renaissance and Renewal in the Twelfth Century*, ed. Robert Louis Benson and Giles Constable (Cambridge, MA: Harvard University Press, 1982) pp. 463–487; Ilsetraut Hadot, *Arts libéraux et philosophie dans la pensée antique* (Paris: Études Augustiniennes, 1984); Brigitte Englisch, *Die Artes liberales im frühen Mittelalter (5.–9. Jh.): Das Quadrivium und der Komputus als Indikatoren für Kontinuität und Erneuerung der exakten Wissenschaften zwischen Antike und Mittelalter* (Stuttgart: Franz Steiner Verlag, 1994).

13 On the tradition of the four liberal arts in Islam, see Majid Fakhry, 'The Liberal Arts in the Mediaeval Arabic Tradition from the Seventh to the Twelfth Centuries', in *Arts Libéraux et Philosophie au Moyen Âge*, Actes du quatrième

vium was sometimes enlarged so as to include engineering and other 'educational sciences' (*ʿilm al-taʿālīm*), as al-Fārābī calls them in his famous *Enumeration of the Sciences.*[14] Very often, though, it held its original structure without alteration, as for instance in the *Epistle on the Parts of Intellectual Sciences* by Ibn Sīnā.[15] In any case, what matters most here is the observation that the Ikhwān al-Ṣafāʾ did not range the 'science of number' (i.e., applications of mystical number theory) among their disciplinary or training sciences. Rather, they chose to range the whole block of mathematics as a specific section of their ultimate group of sciences — the philosophical sciences — to which we shall return later in greater detail. As for the training sciences, their list does indeed include a section headed 'Calculations and operations';

congrès international de philosophie médiéval, Montreal, 1967 (Montreal and Paris: Institut d'études médiévales, 1969), pp. 91–97; Charles E. Butterworth, 'Paris est et sagesse ouest. Du *Trivium* et *Quadrivium* dans le monde arabe médiéval', in *L'enseignement des disciplines à la Faculté des arts (Paris et Oxford, XIIIe–XVe siècles)*, Actes du colloque international, Studia Artistarum, 4, ed. Olga Weijers and Louis Holtz (Turnhout: Brepols, 1997), pp. 477–493; Godefroid de Callataÿ, 'Les arts libéraux en Islam', in *Une lumière venue d'ailleurs. Héritages d'Orient et d'Occident au Moyen Age*, Actes du colloque international tenu à Louvain-la-Neuve, 2005, ed. Godefroid de Callataÿ and Baudouin Van den Abeele (Turnout: Brepols, forthcoming).

14 Al-Fārābī, *Iḥṣāʾ al-ʿulūm*, ed. and trans. Angel González Palencia in *Alfarabi. Catálogo de las Ciencias* (Madrid: Consejo Superior de Investigaciones Científicas, 1932). This enlarged quadrivium, which consists of seven disciplines (arithmetic, geometry, optics, astronomy, music, science of weights, and mechanics) made its way into the West most notably via Gundissalinus's translations and adaptations of al-Fārābī's treatise, where the three new sciences appear as *scientia de aspectibus*, *scientia de ponderibus*, and *scientia de ingeniis* respectively. On al-Fārābī's classification, see Muhsin Mahdi, 'Science, Philosophy, and Religion in Alfarabi's *Enumeration of the Sciences*', in *The Cultural Context of Medieval Learning*, Proceedings of the First Colloquium on Philosophy, Science, and Technology in the Middle Ages, ed. John E. Murdoch and Edith D. Sylla (Dordrecht and Boston: D. Reidel Publishing Company, 1973), pp. 113–147.

15 Ibn Sīnā, *Risāla fī aqṣām al-ʿulūm al-ʿaqliyya*, in *Tisʿ rasāʾil fī al-ḥikma wa'l-ṭabīʿiyyāt* (Cairo, 1326/1908). A French translation of the work was produced by Georges Chehata Anawati, 'Les divisions des sciences intellectuelles d'Avicenne', *Mélanges de l'Institut Dominicain d'Études Orientales*, 13 (1977), pp. 323–335. See also Edouard Weber, 'La classification des sciences selon Avicenne à Paris vers 1250', in *Études sur Avicenne*, ed. Jean Jolivet and Roshdi Rashed (Paris: Belles Lettres, 1984), pp. 77–101; Henry Hugonnard-Roche, 'La classification des sciences de Gundissalinus et l'influence d'Avicenne', in *Études sur Avicenne*, ed. Jean Jolivet and Roshdi Rashed (Paris: Belles Lettres, 1984), pp. 41–75.

however, doubtless the Brethren used it to refer to a very practical and strictly mundane use of numbers.

Let us now briefly consider the other sections of this first group. Writing and reading, grammar, and poetry and prosody, all these being parts of what we would call the science of language, could easily be justified here as other kinds of foundational learning. 'In the beginning was the Word': so does it seem to be the case with several Islamic classifications of sciences also. The first chapter of al-Fārābī's *Enumeration of the Sciences* is the one devoted to the 'science of the language' (*ʿilm al-lisān*). In a similar way, Ibn al-Nadīm's monumental *Fihrist* (dated 987/988), which could stand as a catalogue of sciences in its own right, starts with a section which 'describes the languages of the peoples, Arab and foreign, the characteristics of their methods of writing, their types of script and forms of calligraphy'. To see that the sciences of language, in the *Rasāʾil* as well, are placed in the beginning is no surprise, then. What is more significant, once again, is to find that all those fields are contemplated in their everyday applications only. There is no need, I think to justify the presence of disciplines like crafts, trades, cultivation, breeding, and the like, which are all clear examples of subjects — should we say 'arts' or 'sciences'? — whose interest does not overstep the bounds of this world. Yet the same must be said, we note, of the science of 'campaigns and history', and even of magic, alchemy, and the like, which are thus all regarded here as exclusively profane activities. In all, the group of propaedeutic sciences leaves us with the impression that it has been set up primarily so as to serve as a kind of lumber-room of mundane practices. But this could be regarded, after all, as a typical feature of *adab* literature.

The Religious and Conventional Sciences

Moving to the second group of sciences, we first have to take notice of its heading and definition. The Brethren called this group 'the religious and conventional sciences' (*al-ʿulūm al-sharʿiyya al-waḍʿiyya*),[16] explaining

16 This still corresponds to the Ikhwān's tripartite division of the sciences (as mentioned above), rather than to the sections of the corpus, which each bear the title of a type of science.

that these are the sciences that 'have been set up for the healing of souls and the pursuit of the hereafter'. The notion that has to be emphasised here is certainly that of conventionality. The Ikhwān speak of sciences that 'have been set up', in exactly the same manner as with the training sciences. Obviously, the religious sciences differ radically from those, in that they concern not this world but the other. Yet they do share with them the remarkable character of being conventional, that is, purposefully invented or created. The Ikhwān identify six categories of religious sciences and the category of people connected with each. There is no call for lengthy discussion about the science of revelation in its Qur'anic form (*tanzīl*) or the 'stories and traditions' (*riwāyāt wa-akhbār*) or 'jurisprudence' (*fiqh*), for all these branches are to be expected in this context. Worthier of note, perhaps, is that theology (*kalām*) which is frequently associated with jurisprudence in many Muslim systems of classification, is not even named here. Like everyone else in the mediaeval Islamic world, the Brethren seem to have turned to legal scholars for knowledge of the religious law, but for an understanding of ultimate reality they obviously relied on the divine knowledge of their imams, on the one hand, and their own philosophical endeavours on the other; an account of 'a science of interpretation' (*ta'wīl*), as a prerogative of the Imams and the successors of the Prophets, also confirms this matter plainly. True to their eclecticism, the Ikhwān did not hesitate to include 'Sufism' (*taṣawwuf*) and various types of ascetic practices — whether Muslim or not — as religious sciences, too. The last science in this group is another science of interpretation, namely, the interpretation of dreams. This is a science, or we might say an 'art', which is legitimised in Islam by certain traditions of the Prophet (*ḥadīth*), and even by so famous a Qur'anic passage as the *Sūrat Yūsuf*. As such, it is often enumerated in Islamic classifications of the sciences, as for instance in Ibn Khaldūn's *Muqaddima*, where it is also ranged among the religious sciences.

The Philosophical and Real Sciences

This takes us to the third and last group, which the Brethren call the 'philosophical and real sciences' (*al-falsafiyya al-ḥaqīqa*) and which consists of four main disciplines, namely: the mathematical (*al-riyāḍiyyāt*), the logical (*al-manṭiqiyyāt*), the physical (*al-ṭabīʿiyyāt*), and the divine (*al-ilāhiyyāt*) sciences. For the present inquiry, this is also the most interesting part of the classification since, as the authors themselves point out in various places, the philosophical sciences are those for which they have composed individual epistles. In this respect, it also seems worthwhile to quote a few lines from the passage in which the Ikhwān effect the transition between the development of philosophical sciences and the final exhortation of Epistle 7:

> We have produced an epistle for each branch of the abovementioned sciences and in them adduced some of those meanings [*maʿānī*],[17] and we have completed them with a general epistle to awaken the negligent and guide the beginners, excite the interest of students and serve as a path for those who learn. Be happy by means of it, my brother, and show this epistle to your brethren and friends; make them desirous of science, urge them to renounce this world, and show them the way to the final abode! (*Rasāʾil*, Epistle 7, vol. 1, p. 274)

With this last group of philosophical sciences we come to more familiar ground. So familiar, it seems, that the Brethren did not even bother to define this last group nor to tell their readers how these sciences came to be. What we are to infer is that the sciences of that group have not been devised, but that they exist *per se*. It is in that sense, I would argue, that the Ikhwān can claim that they are 'real'. Just a bit of common sense would be enough to make up what is lacking: that despite their different natures, the philosophical sciences and the religious sciences have the same purpose or objective, which is the happiness of the soul in the world hereafter. In a passage from Epistle 28 (*Rasāʾil*, vol. 3, pp. 30–31), which is dedicated to the limits of human knowledge,[18] the Ikhwān compare these different ways of reaching the same goal to the various places from which pilgrims converge on the Sacred House of God.

Now let us proceed to consider the Ikhwān's division of philosophy.

17 Namely, the particulars of each branch of the sciences as delineated in Epistle 7.

18 For a French translation of this epistle, see 'Ikhwān al-Ṣafāʾ: Sur les limites du savoir humain (*Épître XXVIII* des Frères de la Pureté)', ed. and trans. Godefroid de Callataÿ, *Le Muséon*, 116 (2003), pp. 479–503.

As is well known, Aristotle distinguished physics, mathematics, and metaphysics as the three parts of what he called 'theoretic philosophy', whose purpose is the study of the intelligible beings (*al-maʿqūlāt*). Physics, he said, deals with entities that cannot exist or be conceived as separate from matter and motion. At a higher level of abstraction, mathematics is concerned with entities which can be rationally isolated from matter and motion, but which nevertheless require both so as to exist. The highest level of abstraction falls to metaphysics, which deals with those intelligible beings that are not only conceivable as separate from matter and motion, but which can also exist without them. The Aristotelian division of speculative philosophy was transmitted to the mediaeval West by Boethius, who in his *De Trinitate* spoke of those three parts as '*philosophia naturalis*', '*mathematica*', and '*theologica*'.[19] In Islam, the threefold scheme was taken up by al-Kindī and all his successors in the science of philosophy, the only point of discussion being the places in the sequence ascribed to physics and mathematics respectively. From an ontological point of view, the sequence just mentioned should evidently be preferred. Yet from what has been said earlier we can also understand why the mathematical sciences, that is, ultimately, the Pythagorean quadrivium, might be regarded as a type of propaedeutic learning of its own.

This, we may note, seems to be the case with our text, where mathematics come before physics and metaphysics. Now, with the Ikhwān that other rational — and quite common — sequence is broken up by the incorporation of logic into the whole system. This, however, is nothing to be amazed at. Following in the footsteps of the Alexandrian commentators of Late Antiquity, the Arabs had long been accustomed to regard the whole set of Aristotle's logical sciences as a prerequisite tool (*organon* is the Greek word) for the study of every rational science.[20] As a result, logic and mathematics could both be viewed as necessary preliminaries to the general study of philosophy.

19 Boethius, *De Trinitate*, II, 3.

20 On the crucial role of Alexandrian commentators in the transmission of classification schemata, especially via the tradition of *prolegomena* to philosophy (such as those of Ammonius, Elias, David, or Paul the Persian), see Christel Hein, *Definition und Einteilung der Philosophie: Von der spätantiken Einleintungsliteratur zur arabischen Enzyklopädie* (Frankfurt am Main: P. Lang, 1985);

We may now focus on the way the Ikhwān further divide the group of philosophical sciences. It would be interesting to quote verbatim the passage from Epistle 7 in which the Ikhwān explain and comment on each one of these subdivisions. In order to save some space, I shall here restrict myself to present that part of the text in the form of the table below:

Table 1

(1) Mathematical sciences	*(2) Logical sciences*	*(3) Natural sciences*	*(4) Divine sciences*
Arithmetic	Poetics	Science of corporeal principles	Knowledge of the Creator
Geometry	Rhetoric	Science of the heaven and the world	Science of spiritual beings
Astronomy	Topics	Science of coming-to-be and passing-away	Science of psychic beings
Music	Analytics	Science of atmospheric events	Science of governance (with five subdivisions: prophetic, royal, public, domestic, private)
	Sophistry	Science of minerals	Science of the Return
		Science of plants	
		Science of animals	

This table calls for some explanation. Aristotle's legacy is, of course, paramount. Not only the general structure but even each part of entire sections like logic or physics is purely Aristotelian in its very appellation; they will not retain our attention here. Nor should we dwell for any longer on the mathematical quadrivium of the first section, as I think enough has been said on that before.[21] Definitely the most original

Dimitri Gutas, 'Paul the Persian on the Classification of the Parts of Aristotle's Philosophy: A Milestone Between Alexandria and Bagdad', *Der Islam*, 60, 1 (1983), pp. 231–267.

21 As Baffioni notes in 'Oggetti e caratteristiche' (see note 8 above), p. 28: 'Le "propedeutiche" del primo gruppo [i.e., of the philosophical sciences], evidentemente, sono tali in un senso ben diverso dale scienze poste al primo livello del *curriculum*: la loro esplicitazione rende infatti chiaro che si tratta delle

section — and therefore the most interesting to look at — is the last one, which immediately strikes the reader with its non-Aristotelian elements. First of all, we learn that there is no such thing as a single divine science, a fact which can be legitimately compared with Aristotle's 'science of being as being' or with the '*philosophia prima*' of mediaeval scholasticism. Instead, there are five divine sciences, including the 'science of governance' and the 'science of the Return'. The last three divisions of the art of governance, namely the 'public', the 'domestic', and the 'private' fit nicely with Aristotle's subdivision of practical philosophy into politics, economics, and ethics, respectively. Yet the 'prophetic' and the 'royal' parts stand wholly apart from any Aristotelian scheme, and should be understood, rather, in relation to the Brethren's highly sophisticated, and decisively Shiʿi, conception of prophetic cycles of history.[22] As for the eschatological 'science of the Return', by which the Brethren mean the return to the heavenly abode from which the human soul has fallen, it is probably best explained in the broader context of the Neoplatonic theory of emanation. In spite of these manifold borrowings, one cannot help being impressed by the internal coherence of the Brethren's own scheme of divine sciences, for this scheme mirrors a kind of double journey between God and the ultimate goal of His creation, namely the man enabled 'to know', in other words, to become an accomplished philosopher. On the one hand, we notice the phase of descent, whereby the human soul gradually descends, or falls, from the Creator to the individual (necessitating the 'private' art of governance, i.e., ethics, at the end of the sequence). On the other hand, we find that to this descending phase corresponds an ascending phase, whereby the soul of the true philosopher is able to rise up again towards its point of origin. It is no wonder, then, that the 'science of the Return' is given final place in the classification of sciences as a whole. The Brethren, who regularly refer to science as the food of the soul, would no doubt credit this particular food with having the best taste, colour, and smell. In

scienze aventi ruolo propedeutico *all'interno* delle *Rasā'il*: aritmetica, geometria, astronomia, musica.'

22 Marquet, 'La détermination astrale de l'évolution selon les Frères de la Pureté', *Bulletin d'Études Orientales*, 44 (1992), pp. 127–146. See also Ikhwān al-Ṣafāʾ, *Les révolutions et les cycles (Épîtres des Frères de la Pureté, XXXVI)*, ed. and trans. Godefroid de Callataÿ (Beirut and Louvain: Academia-Bruylant, 1996).

passing, it is interesting to note that the Brethren speak of 'knowledge' (*maʿrifa*, suggestive of recognition) rather than 'science' (*ʿilm*) when they come to deal with the Creator.

Comparing the Two Systems

At this stage, I think it will be appropriate to consider the list of the fifty-one or fifty-two epistles that make up the corpus of the Ikhwān as it has come down to us. The following table shows the titles of sections and of individual epistles as they have been preserved in the manuscript tradition, some of which exhibit an embellished tone.

Table 2

Section I: Mathematical Sciences (14 Epistles)	
Epistle 1 (vol. 1, p. 48)	On Numbers
Epistle 2 (vol. 1, p. 78)	The Epistle Entitled *Jūmaṭriyā*, Dealing with Geometry [*handasa*] and an Account of its Quiddity
Epistle 3 (vol. 1, p. 114)	The Epistle Entitled *Asṭrūnūmiyā*, Dealing with the Science of the Stars and the Composition of the Spheres
Epistle 4 (vol. 1, p. 158)	On Geography [*al-Jughrāfiyā*]
Epistle 5 (vol. 1, p. 183)	On Music [*al-Mūsīqā*]
Epistle 6 (vol. 1, p. 242)	On the Arithmetical and Geometrical Proportions with respect to the Refinement of the Soul and the Reforming of Characters
Epistle 7 (vol. 1, p. 258)	On the Scientific Arts and their Objects
Epistle 8 (vol. 1, p. 276)	On the Practical Arts and their Objects
Epistle 9 (vol. 1, p. 296)	On the Explanation of Characters, the Causes of their Differences, their Types of Disease, and Anecdotes Drawn from the Refined Manners of the Prophets and the Cream of the Morals of the Sages
Epistle 10 (vol. 1, p. 390)	On the *Isagoge* [*Īsāghūjī*]
Epistle 11 (vol. 1, p. 404)	On the Ten Categories [*Qāṭīghūriyās*]
Epistle 13 (vol. 1, p. 420)	On the Meaning of the Analytics [*al-Anūlūṭīqā al-ūlā*]

Epistle 14 (vol. 1, p. 429)	On the Meaning of the Posterior Analytics [*al-Anūlūṭīqiyā al-thāniya*]
Section II: The Corporeal and Natural Sciences (17 Epistles)	
Epistle 15 (vol. 2, p. 5)	Where One Accounts for Matter, Form, Motion, Time, and Place, together with the Meanings of These [Things] When They are Linked to Each Other
Epistle 16 (vol. 2, p. 24)	The Epistle Entitled 'The Heavens and the World', on the Reforming of the Soul and the Refinement of the Characters
Epistle 17 (vol. 2, p. 52)	Where One Accounts for the Coming-To-Be and the Passing-Away
Epistle 18 (vol. 2, p. 62)	On Meteors
Epistle 19 (vol. 2, p. 87)	Where One Accounts for the Coming-To-Be of the Minerals
Epistle 20 (vol. 2, p. 132)	On the Quiddity of Nature
Epistle 21 (vol. 2, p. 150)	On the Kinds of Plants
Epistle 22 (vol. 2, 178)	On the Modalities of the Coming-To-Be of the Animals and of their Kinds
Epistle 23 (vol. 2, p. 378)	On the Composition of the Corporeal System
Epistle 24 (vol. 2, p. 396)	On the Sense and the Sensible, with respect to the Refinement of the Soul and the Reforming of the Characters
Epistle 25 (vol. 2, p. 417)	On the Place Where Drops of Sperm Fall
Epistle 26 (vol. 2, p. 456)	On the Claim of the Sages that Man is a *Microcosm*
Epistle 27 (vol. 3, p. 5)	On the Modalities of the Birth of Particular Souls in the Natural Corporeal System of Man
Epistle 28 (vol. 3, p. 18)	Where One Accounts for the Capacity of Man to Know, which Limit He [can] Arrive at, what He [can] Grasp of the Sciences, which End He Arrives At and which Degree of Nobility He Rises To
Epistle 29 (vol. 3, p. 34)	On the Point of Death and Birth
Epistle 30 (vol. 3, p. 52)	On What is Particular to Pleasure; On the Wisdom of Birth and Death and the Quiddity of Them Both
Epistle 31 (vol. 3, p. 84)	On the Reasons for the Differences Between Languages, Graphic Figures, and Expressions

Section III: The Sciences of the Soul and of the Intellect (10 Epistles)	
Epistle 32 (vol. 3, p. 178)	On the Intellectual Principles of the Existing Beings According to the Pythagoreans
Epistle 33 (vol. 3, p. 199)	On the Intellectual Principles According to the Brethren of Purity
Epistle 34 (vol. 3, p. 212)	On the Meaning of the Claim of the Sages that the World is a *Macranthrope* [*Insān Kabīr*]
Epistle 35 (vol. 3, p. 231)	On the Intellect and the Intelligible
Epistle 36 (vol. 3, p. 249)	On Revolutions and Cycles
Epistle 37 (vol. 3, p. 269)	On the Quiddity of Love
Epistle 38 (vol. 3, p. 287)	On Rebirth and Resurrection
Epistle 39 (vol. 3, p. 321)	On the Quantity of the Kinds of Motions
Epistle 40 (vol. 3, p. 344)	On Causes and Effects
Epistle 41 (vol. 3, p. 384)	On Definitions and Descriptions
Section IV: The Nomic, Divine, and Legal Sciences (11 Epistles)	
Epistle 42 (vol. 3, p. 401)	On Views and Religions
Epistle 43 (vol. 4, p. 5)	On the Quiddity of the Way [Leading] to God — How Powerful and Lofty is He!
Epistle 44 (vol. 4, p. 14)	Where One Accounts for the Beliefs of the Brethren of Purity and the Doctrine of the Divine Men
Epistle 45 (vol. 4, p. 41)	On the Modalities of the Relations of the Brethren of Purity, their Mutual Help and the Authenticity of the Sympathy and Affection [They Have for Each Other], Whether It Be for the Religion or for What Pertains to This World
Epistle 46 (vol. 4, p. 61)	On the Quiddity of Faith and the Characteristics of the Believers who Realise [Those Things]
Epistle 47 (vol. 4, p. 124)	On the Quiddity of the Divine '*Nomos*', the Conditions of Prophecy and the Quantity of the Characteristics [of the Prophets]; On the Doctrines of the Divine Men and of the Men of God
Epistle 48 (vol. 4, p. 145)	On the Modalities of the Call to God
Epistle 49 (vol. 4, p. 198)	On the Modalities of the States of Spiritual Beings
Epistle 50 (vol. 4, p. 250)	On the Modalities of Types of Governance and of their Quantity
Epistle 51 (vol. 4, p. 273)	On the Modalities of the Arrangement of the World as a Whole
Epistle 52 (vol. 4, p. 283)	On the Quiddity of Magic, Incantations, and the Evil Eye

Let us now juxtapose the two systems which the text would seem to invite us to compare, namely, the present list of titles and the group of philosophical and true sciences as described in Epistle 7 (i.e., Tables 1 and 2 above). In the same way as this group of philosophical sciences (as described in Epistle 7), the whole corpus of the *Rasā'il*, as we have it, is divided into four main sections; so far, so good. But here already come the first discrepancies, since it is readily apparent that the main sections of the two systems do not correspond exactly. Despite its title, the first section of the *Rasā'il* incorporates the logical sciences, giving the appearance of having combined the first two groups of the system of classification in Epistle 7. As a consequence of this conflation, the group of physical sciences has shifted to the second section. As for the last group of the classification, that of divine sciences, it appears to be split into two different sections, dealing respectively with the 'Sciences of the Soul and of the Intellect' and the 'Nomic, Divine, and Legal Sciences'. These are, to be sure, significant changes. But we may immediately note other differences, such as, for instance, the great number of *rasā'il* whose titles do not coincide with any of the subdivisions mentioned in Epistle 7.

In the introduction to his *La philosophie des Iḫwān al-Ṣafā'*, Yves Marquet attempted to determine the evidence for the view that, in various passages throughout the encyclopaedia, 'our *Epistles* keep the traces of a certain vagueness, both in the order of chapters, and in the number of treatises in each section'.[23] Adducing a certain number of indisputable indications from the text itself, Marquet arrived at the following conclusions: (1) At the time when the first epistle of the group of physical sciences was written — that is, the one concerning matter, form, etc. — only five epistles of Section I, and seven of Section II had already been compiled. (2) Some epistles from the first and second sections were later modified, whether it be by amplification or by division of their contents. In an earlier draft, there was, for instance, only one epistle on logic. (3) Each one of the four sections was subsequently extended, or completed with the incorporation of new epistles. Needless to say, the comparison of our two systems confirms each of these points. The changes, already evident

23 Marquet, *La philosophie* (as in note 3 above), p. 13.

in the mathematical and the physical sections, tend to become ever more frequent as we approach the end of the corpus.

That said, the Ikhwān's assertion that they have dedicated a specific epistle to each one of the subdivisions remains, to a very large extent, valid. The encyclopaedia opens with the four sciences of the quadrivium (arithmetic in Epistle 1, geometry in 2, astronomy in 3, and music in 5). The only peculiarity is that a *risāla* on geography has now been intercalated between those on astronomy and music, but this is hardly surprising since geography may indeed be conceived of as a sort of natural appendix to astronomy. The titles of the five *rasā'il* on logic correspond, not to the five sciences mentioned in Epistle 7 (that is, poetics, rhetoric, topics, analytics, and sophistry), but rather to the famous *Kitāb al-Burhan* (Book of Demonstration) — in other words, the *Posterior Analytics* (treated in Epistle 14) and to its four indispensable preliminaries, namely: the *Isagoge* (Epistle 10), the *Categories* (Epistle 11), the *Peri Hermeneias* or *De Interpretatione* (Epistle 12), and the *First Analytics* (Epistle 13). The section on natural sciences is, as we have said, the one in which the original sequence has been best preserved. Indeed, each of the seven parts of physics is given a specific *risāla* (from 15 to 22), with only one noteworthy addition, namely the one on the quiddity of nature, in Epistle 20.

Clearly the most interesting point brought out by this comparison concerns the last section, where the variations can no longer be regarded as negligible. Thus, apart from the epistle on spiritual beings, which duly figures as the subject of Epistle 49, the only other science to be found as such in the encyclopaedia is the last one, the science of the Return, but we note immediately that this *risāla*, that is, Epistle 38, has been placed in the third, not the fourth, section. As for the science of governance and its own subdivisions, it would certainly be a mistake to assimilate it too readily with what the Ikhwān report in Epistle 50, on the species of governance.

How are such apparent oddities to be accounted for? At the risk of disappointing the reader, I would argue that these are matters which are best left unsolved for the time being. Certainly, one could put forward chronological reasons, and assume, for instance, that a certain hiatus separated the writing of Epistle 7 (with its systematic and carefully considered classification of the sciences) and the overall compilation of the *Rasā'il*. Those who, like Marquet, favour a longer chronology could certainly claim that the authors of Epistle 7 and the final redactors of the work were not the same 'Brethren of Purity'. In our present state of knowledge, one could even surmise that the arrangement of the *Rasāʾil* as we know it should not be ascribed to the authors themselves, but rather to later scribes or scholars. But all this is conjectural, and bound to remain so until we obtain a much clearer picture of the social, historical, and epistemological context in which the epistles were produced, collected, and read. As with so many other vexed questions concerning the Ikhwān, this kind of speculation will profit greatly from the forthcoming edition, established on a truly scholarly basis, of the entire corpus of epistles.

At any rate, it is unrealistic to expect perfect correspondence between the classification in Epistle 7 and the actual sequence of epistles comprising the collection, as one soon realises if one is willing to concede that the *Rasāʾil* is merely the most visible part of the Brethren's undertaking. In many places, the Brethren refer or allude to their secret meetings, known as 'sessions of science' (*majālis al-ʿilm*), and they make it very clear that the highest grade of their teaching programme has not been committed to writing. As Marquet rightly summarised in the book referred to above, 'the *Epistles* are at the same time the master's book and the student's handbook, yet a handbook which must be completed with some oral teaching'.[24] The section of the encyclopaedia in which the discrepancies with the classification in Epistle 7 are especially numerous is precisely the last one, which pertains to the highest level of esotericism.

Conclusions

These observations should not, however, prevent us from trying to draw a certain number of conclusions about the scheme devised by the Ikhwān as a whole. What I should especially like to emphasise is that this scheme in itself is profoundly original, even if many of its constitutive parts are not.

Regarding the general division of philosophy, the Ikhwān were satisfied with taking up the Aristotelian division of speculative philosophy, to which they simply added logic. This way of proceeding was surely not

24 Ibid., p. 20.

original. We find it already in Islam with al-Kindī's treatise on the number of books by Aristotle, and we have no reason to doubt that he used the same scheme in his other works specifically dedicated to the classification of knowledge. Al-Fārābī expresses a similar view in his *Enumeration of Sciences* (*Iḥṣā' al-ʿulūm*), where logic, the 'doctrinal' sciences (*taʿālīm*), and the binomial physics/metaphysics form the substance of chapters 2, 3, and 4 respectively. In the same way as al-Kindī and al-Fārābī, the Brethren must have felt that the Aristotelian corpus could itself serve as the backbone of the entire system of classification, at least for logic, the natural sciences, and the divine sciences, but at the same time it could not have escaped their notice that his corpus was deficient with respect to the group of mathematical sciences, with the result that Aristotle's system clearly needed to be supplemented by those of other classical authorities. But in this respect also, the Ikhwān proved not to be great innovators. What they did was to keep the original fourfold structure of mathematical sciences as such, claiming to derive their knowledge in those fields directly from Pythagoras, Euclid, Nicomachus, and Ptolemy.[25] Out of the three new doctrinal sciences of al-Fārābī's *Enumeration*, i.e., optics, the science of weights, and mechanics, the Ikhwān mention only the last, which they range among the propaedeutic sciences, along with magic, talismans, and alchemy, as we have seen. Did they know of the other two? It is difficult to say. That they were more reluctant than al-Fārābī to break up the original scheme is easier to determine, for they were clearly much more indebted to the Pythagorean tradition than he was.[26]

The originality of the Brethren's classification is to be found in other, markedly more important aspects. First of all, the ontological partition between the disciplines of everyday life (the group of propaedeutic or

25 See *Rasā'il*, Epistle 1, vol. 1, p. 49, where the four disciplines of the quadrivium are mentioned side by side, along with the names of the four Greek authorities just mentioned (or else the title of his work, in the case of Ptolemy). The passage associates arithmetic with Pythagoras and Nicomachus, geometry with Euclid, and astronomy with the *Almagest*. Curiously enough, no name is cited in relation to music, but perhaps we may conjecture that a reordering of certain words occurred at some point in the manuscript tradition, since Pythagoras may well have originally been associated with music.

26 See Antonella Straface, 'Testimonianze pitagoriche alla luce di una filosofia profetica: la numerologia pitagorica negli Ikhwān al-Ṣafā'', *Annali dell'Istituto Universitario Orientale*, 47 (1987), pp. 225–241.

educational sciences) and the sciences which the soul requires in order to find happiness in the world hereafter (the two other groups) is, as far as I can see, a motif not found in any other Islamic classification of knowledge. Needless to say, this partition is consistent with, and owes a great deal to, the general purpose of our authors. As we have already noted, throughout the *Rasā'il* the Brethren emphasise the point that science is the food of the soul and that only by improving his knowledge may a man hope to liberate his soul from this corrupted world. In any event, it was decisively innovative on the Ikhwān's part to identify the *ādāb* sciences with nothing more than a group of mundane practices.

Let us now tackle the point I should like to single out as the most crucial to this discussion. When considering Islamic systems of classification for the sciences, it is customary to refer to another, undoubtedly more famous, system of bipolarity, namely, that of the conventional, religious, and properly Islamic sciences, and the rational, philosophical, and 'foreign' sciences. This bipolarity is possibly nowhere better evidenced than in al-Khwārizmī's *Keys to the Sciences* (*Mafātīḥ al-ʿulūm*), where the whole spectrum of scientific disciplines is ranged under two different headings, given respectively as 'The Religious Sciences and the Arabic Sciences Connected with Them' and 'The Non-Arabic Sciences, from the Greeks as well as from Other Nations'.[27] The date of composition of al-Khwārizmī's treatise may be fixed in the last quarter of the tenth century, and most probably shortly after the year 977.[28] From

27 Al-Khwārizmī, *Mafātīḥ al-ʿulūm*, ed. Gerlof van Vloten (Leiden: Brill, 1895), p. 5. Actually, Book 1 ('Arabic Sciences') consists of six chapters: (1) Jurisprudence, (2) Theology, (3) Grammar, (4) Art of Secretary, (5) Poetry and Prosody, (6) Historical Accounts; Book 2 ('Non-Arab Sciences') consists of nine chapters: (1) Philosophy, (2) Logic, (3) Medicine, (4) Arithmetic, (5) Geometry, (6) Astronomy, (7) Music, (8) Mechanics, (9) Alchemy. As Hans Hinrich Biesterfeldt notes in 'Medieval Arabic Encyclopedias', pp. 86–87 (see note 8 above): 'The six chapters of the first book constitute three pairs: first the central sciences of the divine law, jurisprudence and theology, then the instrumental sciences, linked to the Arabic language, grammar and writing, and finally the metiers central to the office of the secretary. 1.3–1.6 may also be summarised as *ʿulūm adabiyya* ('literary disciplines'). The nine chapters of the second book can be grouped as: (1) Philosophy and its Instrument, Logic, (2) The Quadrivium of the Four Mathematical Sciences, and (3) The Three Particular Disciplines of Medicine, Mechanics, and Alchemy.'

28 Charles E. Bosworth, 'A Pioneer Arabic Encyclopedia of the Sciences', *Isis*, 54,

about the same period, we have al-ʿĀmirī's *Exposition of the Merits of Islam* (*al-Iʿlām bi-manāqib al-Islām*), a work of a more polemical nature also known to offer a bifurcate scheme of sciences, in which the philosophical sciences of physics, metaphysics, and mathematics are said to correspond with the religious sciences of *ḥadīth*, *kalām*, and *fiqh* respectively.[29] One way or another, as we said, almost every subsequent discussion of the sciences in Islam retains some trace of this fracture between *al-ʿulūm al-naqliyya* ('the transmitted sciences') and *al-ʿulūm al-ʿaqliyya* ('the intellectual sciences'), from al-Ghazālī's brief *Risālat al-ladunniyya* in the eleventh century,[30] to much more extended works such as al-Āmulī's *Nafāyis al-funūn* in fourteenth-century Persia,[31] or the bibliographical nomenclature compiled by al-Qalqashandī in fifteenth-century Egypt.[32]

Looking back, it is a pity that we do not possess more of the works that al-Kindī and al-Fārābī are said to have written about the classification of the sciences. Nonetheless, from their extant writings on the subject, it appears that neither of these two great masters based their

---

1 (1963), p. 100: 'The *Mafātīḥ al-ʿulūm* is dedicated to a vizier who served the Persian dynasty of the Samanids, which ruled from Bukhara over Transoxiana and Khurasan during the ninth and tenth centuries, and this dedication and certain other internal evidence enables us to fix the date of the book's composition at some time shortly after 977 AD.'

29 Al-ʿĀmirī, *Kitāb al-Iʿlām bi-manāqib al-islām*, ed. Aḥmad Ghurāb (Cairo, 1967). See also Mohammed Arkoun, 'Logocentrisme et vérité religieuse dans la pensée islamique, d'après *al-Iʿlām bi-manāqib al-Islām* d'al-ʿĀmirī', *Studia Islamica*, 35 (1972), pp. 5–51; Hans Hinrich Biesterfeldt, 'Abū al-Ḥasan al-ʿĀmirī und die Wissenschaften', *Zeitschriften der Deutschen Morgenländischen Gesellschaft*, Suppl. 3 (1977), pp. 335–341.

30 Al-Ghazālī, *Risālat al-ladunniyya* (Cairo, 1353/1934), in *al-Jawāhir al-ghawālī*.

31 Āmulī, *Nafāyis al-funūn wa-ʿarāyis al-ʿuyūn*, ed. A. Shaʿrānī (Tehran, 1377–1379/1957–1959).

32 On this, see Gaston Wiet, 'Les classiques du scribe égyptien', *Studia Islamica*, 18 (1963), pp. 41–80, where a French translation of al-Qalqashandi's nomenclature is given. On late Arabic classifications, see Régis Blachère, 'Quelques réflexions sur les formes de l'encyclopédisme en Égypte et en Syrie du VIIIème/XIVème à la fin du IXème/XVème siècle', *Bulletin des Études Orientales*, 23 (1970) pp. 7–19; Ulrich Marzolph, 'Medieval Knowledge in Modern Reading: A Fifteenth-Century Arabic Encyclopaedia of *omni re scibili*', in *Pre-modern Encyclopaedic Texts: Proceeding of the Second COMERS Congress, Groningen, 1996* , ed. Peter Binkley (Leiden: Brill, 1997), pp. 407–419.

own classifications on this partition. In both the *Epistle on the Number of Books by Aristotle* (*Risāla fī kammiyyāt kutub Arisṭūṭālis*) and the *Prime Philosophy* (*al-Falsafa al-ūla*), al-Kindī speaks of the human and the divine 'sciences' as two different ways to acquire knowledge. The former, he says, is the science of the philosophers: it is based on human reasoning and therefore implies long and painful efforts. The latter is the exclusive property of the prophets, who do not have to labour for it since they receive it directly by divine inspiration.[33] This is, needless to say, quite different from the division between the philosophical and the religious sciences that is found in later classifications. Nor does al-Fārābī's *Iḥṣā' al-ʿulūm* qualify in any respect as providing some kind of model precursory to our scheme. For it is a striking feature of al-Fārābī's work that it contains a chapter in which the science of politics (*al-ʿilm al-madanī*), with its Greek influence explicitly acknowledged, is combined with, or rather simply appended to, the two typically Islamic sciences of *fiqh* and *kalām*.[34]

The dating of the *Rasā'il* is another vexed question, which I shall not take up again here. In spite of a few opinions that remain divergent, there nevertheless seems to be a growing consensus today which accepts that the epistles were composed in or just before the 970s and that they had begun to circulate by the early 980s. At all events, common sense would recommend that we take 985 or 986 as the most plausible *terminus ad quem*, since it was the year of death of the vizier Ibn Saʿdān with whom the litterateur al-Tawḥīdī — the same as the above-mentioned author of the *Epistle on the Sciences* — discussed the *Epistles* in this well-known passage from the *Kitāb al-Imtā' wa'l- mu'ānasa* (Book of Pleasure and Conviviality).[35] If this chronology is correct, it means that

33 See al-Kindī, *Risāla fī kammiyyāt kutub Arisṭūṭālis*, ed. Abū Rīda (Cairo: Dār al-Fikr al-ʿArabī, 1369/1950), p. 373: 'For that science [i.e., the divine science] is a privilege of the prophets (God pray for them!) inaccessible to mankind. It is one of their wonderful properties, by which I mean, one of the specific marks through which they are distinct from the rest of mankind.'

34 Jean Jolivet, 'Classifications des sciences', in *Histoire des sciences arabes*, ed. Roshdi Rashed, vol. 3 (Paris: Seuil, 1997), p. 261: 'Elle [*political science*] comprend deux parties: l'une fait connaître ce qu'est le bonheur, l'autre traite des conduites et des mœurs bonnes ou mauvaises, du gouvernement, etc.'

35 Abū Ḥayyān al-Tawḥīdī, *Kitāb al-Imtā' wa'l-mu'ānasa* (Cairo, 1953), part 2, pp. 4–5.

the Brethren were indeed among the first in Islam, if not the first, to make the distinction between the rational and the traditional sciences. Were it for this alone, their classification deserves more attention than it has received so far.

These considerations lead naturally to another, equally tricky issue, namely that of the Ikhwān's impact on subsequent generations of thinkers in Islam. For quite obvious reasons, the influence of the Ikhwān was usually not duly acknowledged. But there seems to be a broad agreement today among scholars admitting that the Brethren's impact was very significant indeed, and that it clearly overstepped the limits of the Dār al-Islām, since at least some parts of the encyclopaedia did make their way to the Latin world. Unfortunately, no general investigation of this topic has been conducted so far, so that we must content ourselves with mentioning a few indisputable examples, such as the works by al-Idrīsī (d. 1165), al-Qazwīnī (d. 1283), or the famous Pseudo-Aristotelian *Secretum Secretorum*, in which entire passages from the *Rasā'il* appear almost word for word. Regarding more specifically the problem of the classification of sciences, it would also be premature, I think, to attempt tipping the scales either way, although I would personally be inclined to view authorities as famous as Ibn Sīnā (d. 1037), al-Ghazālī, (d. 1111), or Ibn Khaldūn (d. 1406) as directly indebted to the Brethren's methods and argumentation. In terms of success, it is clear that the Ikhwān's classification cannot compare with al-Fārābī's *Enumeration*, whose impact both in the East and the West was considerable. Nevertheless, in spite of the provisional character, it is hoped that the present enquiry will have convinced the reader that the significance of the Brethren's own system — which is intrinsically more elaborate than al-Fārābī's — deserves re-evaluation.

THREE

The Arrangement of the Rasāʾil Ikhwān al-Ṣafāʾ *and the Problem of Interpolations*[*]

[with Postscript]

Abbas Hamdani

One of the mysteries surrounding the mediaeval Islamic encyclopaedia known as the *Rasāʾil Ikhwān al-Ṣafāʾ* relates to the identification of the exact religious persuasion of its authors.[1] The question is complicated not only by the anonymous nature of the work but also by its ambiguous style. Scholars have differed on the related questions of the date and

* This chapter is an updated version of an article that was previously published in the *Journal of Semitic Studies*, 29, 1 (1984), pp. 97–110, and it is supplemented with a Postscript. Minor alterations to the text have been made, and where there has been an addition to the notes, this has been indicated by square brackets. The author of this chapter (Abbas Hamdani) and the editor of this volume (Nader El-Bizri) thank the editors at the *Journal of Semitic Studies* and Oxford University Press for granting permission to republish it.

1 See Abbas Hamdani, 'Shades of Shīʿism in the Tracts of the Brethren of Purity', in *Traditions in Contact and Change*, ed. Peter Slater and Donald Wiebe (Waterloo, ON: Wilfrid Laurier University Press, 1983), pp. 447–460, notes pp. 726–728[; and Abbas Hamdani, 'Brethren of Purity, a Secret Society for the Establishment of the Fāṭimid Caliphate: New Evidence for the Early Dating of their Encyclopaedia' in *L'Ègypte Fatimide: Son art et son histoire*, ed. Marianne Barrucand (Paris: Presses de l'Université de Paris-Sorbonne, 1999), pp. 73–82].

authorship of the *Rasāʾil*.[2] The passages in the work that give rise to these differences of opinions cannot be properly interpreted until several early *Rasāʾil* manuscripts have been consulted. Unfortunately, the three main printed editions of the complete work are based on a single, late Indian manuscript.[3] Similarly, the three printed editions of the concluding *al-Risāla al-jāmiʿa* are based on late manuscripts.[4] Susanne Diwald's edition of the *Rasāʾil*, though incomplete, benefits from the inclusion of a critical apparatus, being based on several early manuscripts; however, it comprises just the third part of the text, and only in German translation.[5] The task of reference to diverse topics in such a large work is rendered all the more difficult by the absence of indices. There exists only a small concordance of the titles of the fifty-two *rasāʾil* and their page numbers in the Bombay, Cairo and

2 Summaries of this controversy can be found in ʿAbd al-Laṭīf Ṭībāwī, 'Ikhwān aṣ-Ṣafāʾ and their *Rasāʾil*: A Critical Review of a Century and a Half of Research', *Islamic Quarterly*, 2 (1955), pp. 28–46; Samuel M. Stern, 'The Authorship of the Epistles of the Ikhwān aṣ-Ṣafāʾ', *Islamic Culture*, 20, 4 (1946), pp. 367–372, with addendum in ibid., 21, 4 (1947), pp. 403–404; Samuel M. Stern, 'New Information about the Authors of the "Epistles of the Sincere Brethren"', *Islamic Studies*, 3 (1964), pp. 405–428; Abbas Hamdani, 'Abū Ḥayyān al-Tawḥīdī and the Brethren of Purity', *International Journal of Middle Eastern Studies*, 9 (1978), 345–353; Abbas Hamdani, 'An Early Fāṭimid Source on the Time and Authorship of the *Rasāʾil Ikhwān al-Ṣafāʾ*', *Arabica*, 26, 1 (1979), pp. 62–75.

3 (a) *Kitāb Ikhwān al-Ṣafāʾ*, ed. Wilāyat Ḥusayn, 4 vols. (Bombay: Maṭbaʿat Nukhbat al-Akhbār, 1305–1306/ ca. 1888); (b) *Rasāʾil Ikhwān al-Ṣafāʾ*, ed. Khayr al-Dīn al-Ziriklī, 4 vols. (Cairo: al-Maṭbaʿa al-ʿArabiyya bi-Miṣr, 1928), with two separate introductions by Ṭāhā Ḥusayn and Aḥmad Zakī Pasha; (c) *Rasāʾil Ikhwān al-Ṣafāʾ*, ed. Buṭrus al-Bustānī, 4 vols. (Beirut: Dār Ṣādir, 1957) — page numbers in the present article refer to this edition [; (d) *Rasāʾil Ikhwān al-Ṣafāʾ*, ed. ʿĀrif Tāmir, 5 vols. (Beirut and Paris: Manshūrāt ʿUwaydāt, 1995)].

4 (a) *al- Risāla al-jāmiʿa*, ed. Jamīl Ṣalībā, 2 vols. (Damascus: Maṭbaʿat al-Taraqqī, 1949–1951); (b) *al-Risāla al-jāmiʿa*, ed. Muṣṭafā Ghālib (Beirut: Dār Ṣādir, 1974) — page numbers in the present article refer to this edition of the *Jāmiʿa* [; (c) the fifth and last volume of *Rasāʾil Ikhwān al-Ṣafāʾ*, ed. ʿĀrif Tāmir (Beirut and Paris: Manshūrāt ʿUwaydāt, 1995) referred to above; *Risālat jāmiʿat al-jāmiʿa*, ed. ʿĀrif Tāmir (Beirut: Dār al-Nashr liʾl-Jāmiʿīyyīn, 1959) is not part of the original manuscript, but from a much later volume found in the Nizārī collections of Syria].

5 Susanne Diwald, trans. and ed., *Arabische Philosophie und Wissenschaft in der Enzyklopädie Kitāb Iḫwān aṣ-ṣafāʾ (III): Die Lehre von Seele und Intellekt* (Wiesbaden: Otto Harrassowitz, 1975).

Beirut editions, along with an index of poetical references prepared by David R. Blumenthal.[6]

The questions relating to the identity of the authors of the *Rasāʾil* and their *madhhab* are largely dependent on the question of the date of the composition of the *Rasāʾil*. A later chronology may suggest a composition of the encyclopaedia over a long period, by several authors living at different times, and several revisions and rearrangements of the component tracts. It can also accommodate the internal allusions to a much later period. An earlier chronology would presuppose a shorter period of composition, the minimum of rearrangement and revision, and a committee of authors writing at the same time under a coordinator or an editor and working on a planned sequence of composition. Internal references to a later period would have to be proved as interpolations. To my mind, the dividing line between these two chronologies is represented by the year 297/909, that is, the establishment of the Fāṭimid caliphate. In fact, in much of the scholarly argument about the dating of the *Rasāʾil*, this has proved to be the battle line. Having taken their stand on the date of composition of the *Rasāʾil*, scholars have argued whether its authors were Sunnis or Shiʿis; if Sunnis, whether they were Muʿtazilī or Sufi; if Shiʿis, whether they were Zaydī, Ithnāʿasharī, Fāṭimid, or Qarmaṭī. The later chronology is the easier and safer of the two and is generally accepted. I have, however, preferred the earlier and more difficult and have argued the case for it in previous articles.[7] The following two sections relate strictly to questions of arrangement and interpolation. The question of religious affiliation is treated here as subsidiary.

Number and Arrangement of the Rasāʾil

Ṭībawī has pointed out that among the many contradictions contained in the *Rasāʾil Ikhwān al-Ṣafāʾ*, there is one that relates to the number of *rasāʾil*.[8] The total number of these epistles that have come down to us

6 David R. Blumenthal, 'A Comparative Table of the Bombay, Cairo, and Beirut Editions of the *Rasāʾil Ikhwān al-Ṣafāʾ*', *Arabica*, 21, 2 (1974), pp. 186–203.

7 See notes 1 and 2 above.

8 Ṭībāwī, 'Ikhwān aṣ-Ṣafā' and their *Rasāʾil*', p. 38.

is fifty-two. There is also the *Fihrist* at the beginning which describes the contents of the *Rasā'il*. To these should be added *al-Risāla al-jāmiʿa* at the end. Although *al-Jāmiʿa* is considered a separate work, in fact it forms an integral part of the *Rasā'il*. It is not necessarily a summary of the encyclopaedia, but is intended to make comparatively more explicit the meanings only implied in many of the themes of the main work. At several places in the work itself, however, the epistles are mentioned either as fifty-one or fifty-two, excluding *al-Jāmiʿa* and the *Fihrist*.

Abū Sulaymān al-Manṭiqī (ca. 300–375/912–985), philosopher-scientist of the court of the Būyid Emir Aḍud al-Dawla, mentions fifty-one *rasā'il*,[9] whereas his pupil, the celebrated man of letters Abū Ḥayyān al-Tawḥīdī (ca. 320–414/933–1023), counts fifty.[10] Neither of them include *al-Jāmiʿa* or the *Fihrist*. Ibn al-Qifṭī (d. 642/1244)[11] follows Abū Ḥayyān's count, whereas Ḥājjī Khalīfa (d. 1067/1656)[12] follows that of al-Manṭiqī. The Yemeni Ṭayyibī *dāʿī* Idrīs ʿImād al-Dīn (d. 832/1438) mentions fifty-two and *al-Jāmiʿa*.[13] Moreover, he gives a summary of all the epistles based on the *fihrist*.

The problem of the number is not as confusing as that of the arrangement of the epistles. Friedrich Dieterici, writing as early as 1858, suggested that the epistles were written over a long period and completed about 370/980.[14] He based himself on Abū Ḥayyān al-Tawḥīdī's account[15] as well as on certain internal allusions. Lately, Yves Marquet

9 Abū Sulaymān al-Manṭiqī (al-Sijistānī), *Kitāb Ṣiwān al-ḥikma*, ed. ʿAbd al-Raḥmān Badawī (Tehran, 1974), pp. 361–362.

10 Abū Ḥayyān al-Tawḥīdī, *Kitāb al-Imtāʿ wa'l-mu'ānasa*, ed. Aḥmad Amīn and Aḥmad al-Zayn, 3 vols. (Beirut: Manshūrāt Dār Maktabat al-Ḥayāt, 1939–1944; repr., 1953), vol. 1, pp. 20–22.

11 Ibn al-Qifṭī, *Ta'rīkh al-ḥukamā'*, ed. Julius Lippert (Leipzig: Dieterich's Verlagsbuchhandlung, 1903), p. 82.

12 Ḥājjī Khalīfa, *Kashf al-ẓunūn*, ed. Gustav Flügel, 7 vols. (Leipzig, 1835–1858), vol. 3, p. 460.

13 Idrīs ʿImād al-Dīn, *ʿUyūn al-akhbār*, ed. Muṣṭafā Ghālib, vol. 4 (Beirut, 1973), pp. 367–390.

14 Friedrich Dieterici, *Die Philosophie der Araber im X. Jahrhundert n. Chr. aus den Schriften der Lauteren Brüder*, 8 vols. (Leipzig, 1858–1872), vol. 1, pp. 141–142.

15 Contained in Abū Ḥayyān's *al-Imtāʿ wa'l-mu'ānasa* (see note 10 above), and studied in detail in A. Hamdani, 'Abū Ḥayyān al-Tawḥīdī and the Brethren of Purity' (see note 2 above).

has endorsed this view by stating that the epistles were composed between 350/961 (that is, shortly before the Fāṭimid conquest of Egypt) and 370/980 (that is, shortly before Abū Ḥayyān's report).[16] He also considers some tracts to have been written before the establishment of the Fāṭimid caliphate in North Africa (297/909) and some of them shortly after that, thus increasing the range of composition considerably, from about 290/903 to 370/980, that is, about 80 years.[17] He feels that the forty-eighth epistle, which is couched in the words of the expected Imam himself and which gives the *bashāra* (good tidings) of his appearance, is the oldest and that it was composed long before 297/909. Also, he considers the fiftieth epistle, in which the Imam speaks of his *du'āt* ('missionaries') having discovered him and of the work that is in progress for the establishment of the new state, to have been written about the same time but shortly after the forty-eighth *risāla*. Then, according to Marquet, comes the forty-fourth *risāla*, which he thinks must have been composed in 298/910, shortly after the establishment of the Fāṭimid caliphate. He characterises it as a *cris de triomphe*. He thinks that, on the whole, the fourth part of the *Rasā'il Ikhwān al-Ṣafā'* is earlier and that the first, second and third parts were added later. There is an exception, however. He thinks that in the fourth epistle of the first part, the section on the *ta'āqub al-duwal* belonged to an earlier period and was inserted here later. Finally, he considers the epistle on animals as the latest because it ends with words that suggest that it was composed with the help of the ruling Imams. The original positions of the ninth, fifteenth and forty-first epistles are also said by Marquet to have been different.[18]

In a subsequent note, Marquet analyses a poem contained in the forty-seventh epistle and concludes that it was composed at the time of the Fāṭimid caliph al-Qā'im (r. 322–334/934–946) on account of certain references, which he interprets as follows: the reference to 'the sun rising from the west', as the establishment of the Fāṭimid caliphate

16 Yves Marquet, 'Iḫwān al-Ṣafā'', *EI2*, vol. 3, pp. 1071–1076.

17 Yves Marquet, *La philosophie des Iḫwān al-Ṣafā'* (Algiers, 1973), p. 8; Yves Marquet, '910 en Ifrīqiya: Une épître des Ikhwān al-Ṣafā'', *Bulletin d'Études Orientales*, 30 (1978), pp. 61–73.

18 See Yves Marquet, 'Ikhwān al-Ṣafā', Ismaïliens et Qarmates', *Arabica*, 24, 3 (1977), pp. 233–257.

already achieved; the reference to the 'rebel' (*mārid*) as the last Aghlabid ruler; and the reference to the 'Dajjāl' (meaning 'falsifier'; sometimes translated as 'Antichrist'), as the Khārijī leader Abū Yazīd al-Ifranī, who opposed the Fāṭimid caliphs al-Qā'im and al-Manṣūr.[19]

Much of this internal evidence is vague and could be interpreted otherwise. There is nothing in the forty-fourth epistle to prove that it was written in 298/910 and in a mood of celebrating the advent of the Fāṭimid caliphate. All it suggests is the impending completion of the mission of the past prophets and philosophers by the establishment of the rule of the Ahl al-Bayt. No doubt, there is an air of imminence about this *risāla*, but it certainly does not suggest a *fait accompli*. As for the forty-seventh epistle, the reference to 'the sun rising from the west' should be taken in connection with the exploitation of this Shiʿi tradition by the *dāʿī* Abū ʿAbd Allāh al-Shīʿī, who began his work in North Africa in 280/893 on behalf of the Fāṭimids much before the establishment of their caliphate. Again, 'Dajjāl' could refer to the arch-enemy of the Fāṭimids, the ʿAbbāsid caliph himself, and not necessarily to Abū Yazīd, and the 'rebel' (*mārid*) could possibly be the last Aghlabid or any one of the many opponents of Abū ʿAbd Allāh. Again, to say that the twenty-first epistle is the last because the concluding benediction states that the *risāla* is completed 'with the help of the Creator of the worlds and of Muḥammad and his progeny, the rightly-guiding Imams' (*Rasā'il*, vol. 2, p. 377) would be wrong. The 'Imams' need not necessarily mean the Fāṭimid caliphs but could instead mean the Imams of the 'period of concealment' (*satr*).

Although Marquet's thesis that the *Rasā'il Ikhwān al-Ṣafā'* represents a movement for the Fāṭimid caliphate is quite correct, his arguments for the post-297/909 composition of most of the epistles are not. There is, therefore, no need to suggest a wholesale rearrangement and revision of the epistles. I see no reason to dispute the original order of the epistles except for minor changes of arrangement. Instead of considering the forty-eighth and fiftieth epistles as the beginning of the encyclopaedia, I would think that they were written at the end. I agree with Marquet in placing these epistles before the establishment

19 Yves Marquet, 'Note annexe: à propos d'un poème Ismaïlien dans les épîtres des Iḫwān aṣ-Ṣafā'', *Studia Islamica*, 55 (1982), pp. 137–142.

of the Fāṭimid caliphate, but I think that the other epistles were written before them roughly in the order in which they appear and in accordance with a previously established plan. There seems to have been a hurried assembly of the epistles by the compiler and a quick and incomplete revision by him to correct cross-references and the numbering of the tracts before issuing them to the public.

Originally, the *rasā'il* numbered fifty-one, excluding *al-Risāla al-jāmi'a* at the end of the encyclopaedia and a *fihrist* at the beginning. Thus we have a reference to fifty-one *rasā'il* in *Rasā'il*, vol.1, p. 327; *Rasā'il*, vol. 4, pp. 173, 250, 282, 284.[20] The *Fihrist* seems to have been first composed after the completion of Part III, as it is first mentioned at the beginning of Part IV, i.e., at the beginning of the forty-second epistle (*Rasā'il*, vol. 3, p. 401). It is also mentioned in the last epistle (*Rasā'il*, vol. 4, p. 284) and in *al-Jāmi'a*, p. 537.[21] The compilation of *al-Risāla al-jāmi'a* seems also to have begun when the *Fihrist* was written; so *al-Jāmi'a* is mentioned in the *Fihrist* (*Rasā'il*, vol. 1, p. 43) and in Part IV (*Rasā'il*, vol. 4, pp. 250, 262, 340).

The *rasā'il* as they stand today are not fifty-one, but fifty-two. The fifty-second was originally the fifty-first, as is admitted in it (*Rasā'il*, vol. 4, p. 284), and at the beginning of it fifty preceding epistles are mentioned (*Rasā'il*, vol. 4, p. 283). This means that the one that is now listed as the fifty-first was actually the fifty-second or the last. Thus the fifty-first and fifty-second were each intended as fifty-first and then were both retained, one of them having to become the fifty-second. Marquet has pointed out that five pages (*Rasā'il*, vol. 4, pp. 276–281) of the present fifty-first *risāla* from its total of nine (*Rasā'il*, vol. 4, pp. 273–282) are a paraphrase of a section (*Rasā'il*, vol. 2, pp. 167–172) of the 'twenty-first' epistle.[22]

After having completed the fifty-two epistles, the *Fihrist* and *al-Jāmi'a*, the compiler of the encyclopaedia seems to have amended the *Fihrist* and the first epistle by stating that the total of the epistles

20 All the page numbers here and in what follows refer to the Beirut edition (1957).

21 Page numbers of *al-Risāla al-jāmi'a*, refer to the Beirut edition by Muṣṭafā Ghālib (Dār Ṣādir, 1974).

22 Marquet, *La philosophie des Ikhwān al-Ṣafā'* (1973), p. 11.

was fifty-two (*Rasāʾil*, vol. 1, pp. 43, 77). He specifically mentions the titles and describes the contents of both the fifty-first and fifty-second epistles. He does not, however, go beyond the first epistle in correcting this number in references in the subsequent epistles — thus fifty-one, the original but incorrect total, remains uncorrected in *Rasāʾil*, vol. 1, p. 327; *Rasāʾil*, vol. 4, pp. 173, 250; and in the fifty-first epistle itself (*Rasāʾil*, vol. 4, p. 282). This is evidence of the fact that before its dissemination there was only one quick and hurried revision of the encyclopaedia, along with the *Fihrist* and *al-Jāmiʿa*, and that there was no continuous process of revision over a long period of time.

Marquet thinks that the seventh and eighth epistles on crafts, which were supposed to be in Parts III and II respectively, were shifted to Part I by way of a later revision.[23] The change of place for the eighth only is alluded to in the epistle itself in two passages (*Rasāʾil*, vol. 1, pp. 276, 286). Since the *Fihrist* mentions these two epistles in the order in which they have come down to us, even if there was a change of place, it could have been effected before the composition of the *Fihrist* and been connected with the stage of planning the order rather than with the stage of revision.

Concerning *al-Risāla al-jāmiʿa*, and its relation to the corpus of the entire encyclopaedia, the *Fihrist* states:

> All these epistles are like introductions to it [*al-Jāmiʿa*], arguments towards it and samples for it . . . *al-Risāla al-jāmiʿa* from our epistles is the sum and substance of what we have presented . . . This is the *Fihrist* of the Epistles of the Pure Brethren and Sincere Friends, of the Just People and the Praiseworthy Children. They [i.e., the epistles] are fifty-two and an epistle about the disciplining of souls and the rectitude of manners. (*Rasāʾil*, vol. 1, p. 43)

This shows that *al-Jāmiʿa* was an integral part of the *Rasāʾil*; also, that it was meant to elucidate the meaning of the main work and to make explicit what was implicit in the latter. It is not referred to as the fifty-third epistle but as a *risāla* in addition to the fifty-two. This proves it was a work for separate distribution among a special élite.

23 Ibid., pp. 11–12.

Thus later writers such as Abū Ḥayyān al-Tawḥīdī would have had no knowledge of it.

Let me now summarise my differences with Marquet's interpretations. According to him, the composition of the *Rasāʾil* continued from around 290/903 to 370/980, that is, a little short of eighty years. This would take care very neatly of the internal references, such as a verse of the poet Abu al-Fatḥ al-Bustī (360–401/970–1010) (*Rasāʾil*, vol. 2, p. 59; *Rasāʾil*, vol. 3, p. 248);[24] verses of al-Mutanabbī composed in 347/958 (*Rasāʾil*, vol. 4, pp. 48, 76); a reference to *al-ashāʿira* (al-Ashʿarī died in 324/934) (*Rasāʾil*, vol. 3, p. 161); and also, the twelve qualifications of an ideal ruler that appear both in the *Rasāʾil* (vol. 4, pp. 129–130) and in *al-Madīna al-fāḍila* of al-Fārābī (d. 339/950).[25] It could also be reconciled with Abū Ḥayyān al-Tawḥīdī's so-called contemporary report about the alleged authors of the *Rasāʾil* by making them the last in a chain of authors and responsible for introducing into the *Rasāʾil* references of a later period. Marquet also thinks in terms of an elaborate rearrangement and continuous revision of the *Rasāʾil* over a long period of time by several successive authors.

In my opinion, the authors must have been a compact group, writing not consecutively but simultaneously with a well-laid out plan both of the content as well as the arrangement of the *Rasāʾil* and also of its sequel *al-Jāmiʿa*. One of them must have served as a co-ordinator–editor (in the main, the authors usually refer to themselves in the plural; sometimes only in the singular, as in *Rasāʾil*, vol. 1, p. 372). The style is the same throughout, with the same clichés of affectionate address, the same simple and eloquent diction, the same building up of the expectations for what is to come and the same catholic appeal to all faiths and philosophies. The language is also similar, with Persian words and verses slipping in, suggesting the eastern origin of the authors (*Rasāʾil*, vol. 1, pp. 139, 209, 235). I believe that the composition must have been planned and achieved by a group of authors sometime between 260/873, when the last Imam of the Ithnāʿashariyya disappeared, and 297/909,

24 The identification of al-Bustī as the author of the verse in question was made by Ṭībāwī, in 'Ikhwan aṣ-Ṣafāʾ and their *Rasāʾil*', p. 37.

25 Al-Fārābī, *Kitāb Ārāʾ ahl al-madīna al-fāḍila*, ed. Albert Nader, 2nd ed. (Beirut: Dār al-Mashriq, 1968), pp. 127–128 (henceforth cited as *al-Madīna al-fāḍila*).

when the Fāṭimid caliphate was established; but closer to the later date than the former. This period was characterised in mediaeval Islamic history by strong messianic expectations and a really revolutionary preparation for the establishment of a caliphate to rival the ʿAbbāsids. As we have noted before, minor editorial changes must have been made just before issuing the *Rasāʾil* to the public. Regarding the later references found in the *Rasāʾil*, I should think that they were scribal interpolations, having nothing to do with the original text.

Interpolations in the Rasāʾil

The ninth-/fifteenth-century Yemeni Ṭayyibī *dāʿī*, Idrīs ʿImād al-Dīn, after describing the contents of the *Rasāʾil* and *al-Jāmiʿa*, states:

> One of the later writers who were separated from the pure friends [*awliyāʾ*] of God and who deviated from His guidance and became transgressors said that the epistles were not compiled by one of the hidden Imams and argued [his point by the presence of] a verse in the *Rasāʾil* that belongs to Aḥmad ibn al-Ḥusayn al-Mutanabbī; and it is: 'In the body, there is a soul that does not age with its [body's] aging, although the body deteriorates with its [soul's decay]. These epistles were composed by the above-mentioned Imam, Aḥmad ibn ʿAbd Allāh ibn Muḥammad ibn Ismāʿīl ibn Jaʿfar — may the blessings of God be on him, on his ancestors and on the pure among his descendants, without doubt and contradiction, without falsehood and deceit. No doubt, this verse has been inserted by one of the later copyists and this is not hidden from the thoughtful investigators.[26]

Here we need not go into the discussion about the Ismaʿili position on the *Rasāʾil*'s authorship, which has already been studied at length by me in a previous article.[27] What is relevant is the fact that as early as the ninth/fifteenth century there was the realisation that the text of the *Rasāʾil* had been tampered with by copyists. The above-mentioned verse of al-Mutanabbī is from a poem written for the Ikhshīdid regent

26 Idrīs ʿImād al-Dīn, *ʿUyūn al-akhbār*, vol. 4, pp. 392–393.
27 Abbas Hamdani, 'An Early Fāṭimid Source' (see note 2 above).

al-Kāfūr in 347/958.[28] This verse, along with two others following it, are introduced with the expression *kamā qāla al-qā'il* ('as said by the sayer') in *Rasā'il*, vol. 4, p. 48. Many of al-Mutanabbī's verses are highly memorable and have become proverbial. A copyist could have written some such verses in the margin in support of an argument contained in the text, and a later copyist could have neatly introduced the marginal note into the text just because it blended well with that text. The problem can properly be solved by the comparison of older manuscripts, but in the case of the *Rasā'il*, the oldest known manuscript is the Istanbul, Atif [Efendi] 1681 written in 587/1182.[29] Even if we accept that the encyclopaedia was composed in Abū Ḥayyān's time, the oldest available manuscript represents a lapse of two centuries. If we accept an earlier dating for the *Rasā'il*, the lapse is even greater. Much scribal mischief could have happened during that period.

To illustrate the above point, let us take another verse which is introduced, in *Rasā'il*, vol. 2, p. 59, with '*kamā qīla*' ('as it is said'), and again, in *Rasā'il*, vol. 3, p. 248, with the phrase, '*wa-kadhālika ishārātu al-ḥukamā'i shi'ran*' ('and such are the indications of the philosophers in verse'). The verse itself can be translated: 'Strive in your soul and [try] to perfect its qualities for you are a human being by soul and not by body'. It has been identified by Ṭībawī as belonging to Abu'l-Fatḥ al-Bustī.[30] Like those of al-Mutanabbī, this one of al-Bustī is also memorable and has become a proverbial verse, but here we have an interesting contradiction. Although al-Bustī's verse appears in all the three printed editions in two places, it does not appear in one Istanbul manuscript in either place, although there is everything else from the text surrounding the verse.[31] It is also omitted from a Tehran manuscript dated 686/1287.[32]

The suspicion of interpolation in the above two instances is further strengthened in the case of three verses in *Rasā'il*', vol. 4, p. 76 which are

28 Al-Mutanabbī, *Dīwān*, ed. Nāsīf Yāzijī, vol. 2 (Cairo, n.d.), p. 353.

29 See Diwald, *Arabische Philosophie und Wissenschaft*, p. 17.

30 Ṭībāwī, 'Ikhwan aṣ-Ṣafā' and their *Rasā'il*', p. 37.

31 MS Istanbul, Ahmet III 2128 dated ca. 800/1400 (Arab League, Cairo, microfilm no. 210), no folio number.

32 MS Majlis-i Shūrā-yi Millī 4707 (Arab League, Cairo, microfilm no. 172), no folio number.

introduced by the expression, '*kamā qāla al-muḥaqqiq shiʿran*' ('as the editor has said in verse'). Here the editorial interpolation is quite obvious, as rightly pointed out by Ṭībawī, who further adds: 'A clearer example is perhaps provided in the line after "*kamā qāla al-qā'il*" (*ar-Risālah al-Jāmiʿah*, vol. 1 [Damascus, 1948], p. 464), which is not found in either the Taimuria or the Tehran manuscripts of the tract.'[33]

There is one verse in *Rasā'il*, vol. 3, p. 272, that is introduced by the phrase '*qāla al-shāʿir*' ('the poet has said'). Louis Massignon has identified the poet as Ibn al-Rūmī (221–283/835–896) and his time as constituting the *terminus a quo*, which accords well with the theory that the *rasā'il* were composed between 260/873 and 297/909. [Ibn al-Rūmī in his turn mentions the Ikhwān al-Ṣafā' twice, for which more details are given in the postscript to this article.][34]

As for the Persian verses in the *Rasā'il*, there are seven cited in *Rasā'il*, vol. 1, p. 139; two in *Rasā'il*, vol. 1, p. 209; and six in *Rasā'il*, vol. 1, p. 235. Although they are cited anonymously, they are not likely to be interpolations. They are traceable [in the fragments preserved from the very early pre-Firdawsī poets Rūdakī (d. 329/940) and Shuhayd al-Balkhī (d. 315/927).[35] For Rūdakī, reference could be made to Saʿīd Nafīsī's work,[36] and for Shuhayd, to Dhabīḥ Allāh Ṣafā''s *Ganj-i-sukhan*.[37] (A few more details about these Persian verses are given in the postscript to this article.)] This would also agree with Massignon's *terminus a quo*, and would not militate against the pre-909 composition of the *Rasā'il*.

All other poets cited by name in the *Rasā'il* are very early ones

33 Ṭībāwī, 'Ikhwan aṣ-Ṣafā' and their *Rasā'il*', p. 35, note 5.

34 Louis Massignon, 'Sur la date de la composition des *Rasā'il Ikhwān al-Ṣafā*'', *Der Islam*, 4 (1913), p. 324.

35 Reference could be made to Gilbert Lazard, ed. and trans., *Les premiers poètes persans (IX–Xe siècles): Fragments rassemblés* (*Ashʿār-i parakanda-yi qadīmītarīn shuʿarā-yi fārsī-zabān*)[, 2 vols. (Tehran and Paris: Département d'Iranologie de l'Institut Franco-Iranien and Adrien-Maisonneuve, 1964)].

36 Saʿīd Nafīsī, *Muḥīṭ-i zindagī-u aḥwāl u ashʿār-i Rūdakī*, 3rd ed. (Tehran: Sepehr, 1958), p. 509 (see also Nafīsī's introduction); cf. Saʿīd Nafīsī, *Fihrist al-Majdūʿ*, ed. ʿAlī Naqī Munzawī (Tehran, 1966), editorial note at p. 157. I am grateful to Dr. Jalāl Badakhshānī for pointing out these references to me.

37 [Dhabīḥ Allāh Ṣafā', *Ganj-i-sukhan*, vol. 1, *From Rūdakī to Anwarī*, (Tehran: Ibn Sīnā Press, n.d.). I am grateful to Dr. Farhad Daftary for tracing for me the references to all the Persian verses cited.]

such as Imruʾ al-Qays, Basūs, al-Nābigha and ʿUrwa ibn Ḥizām (of the Umayyad period). There is similarly no name of any Islamic philosopher or scientist or theologian except that of Abū Maʿshar Jaʿfar ibn Muḥammad, the astrologer, whose report is quoted (*Rasāʾil*, vol. 4, pp. 288–289).[38] Abū Maʿshar died in 272/885, that is, within the *Rasāʾil*'s period of composition suggested by me, namely 260/837 to 297/909.

Several religious groups are mentioned, but they are all very early, except two, *al-Kayyāliyya*, (*Rasāʾil*, vol. 1, p. 217; *Rasāʾil*, vol. 2, p. 206) and *al-Ashāʿira* (*Rasāʾil*, vol. 3, p. 161). According to the Ismaʿili *dāʿī* Idrīs ʿImād al-dīn, Aḥmad ibn al-Kayyāl was one of the *duʿāt* (ʻmissionaries') of the hidden Imam ʿAbd Allāh, the successor of Muḥammad ibn Ismāʿīl ibn Jaʿfar al-Ṣādiq; he revolted against the Imam and formed a faction called *al-Kayyāliyya*, whom the Ikhwān also call the Niners.[39] Wladimir Ivanow thinks al-Kayyāl died in 207/822,[40] whereas Massignon says the rebel *dāʿī* died ca. 270/883.[41] Wilferd Madelung determines al-Kayyāl's activity around ʻthe turn of the 3rd/9th century'.[42] In any case, the mention of *al-Kayyāliyya* in the *Rasāʾil* does not conflict with our *terminus ad quem*, i.e. 297/909, for the composition of the encyclopaedia.

While describing the futile conflicts of opinion between several Islamic and pre-Islamic religious groups, the authors of the encyclopaedia have this to say: ʻThus the people of different *sharīʿa*s kill and curse each other as do the *Nawāṣib* and the *Rawāfiḍ*; the *Jabariyya* and the *Qadariyya*; the *Khawārij* and the *Ashāʿira*, etc.' (*Rasāʾil*, vol. 3, p.

38 A discussion on this report and Abū Maʿshar's time is contained in Zāhid ʿAlī, *Tārīkhe Fāṭimiyyīn* (in Urdu), pp. 521–522.

39 *Rasāʾil*, vol. 4, pp. 357–358.

40 Wladimir Ivanow, ʻIsmailis and Qarmatians', *Journal of the Bombay Branch of the Royal Asiatic Society*, n.s., 16 (1940), p. 64.

41 Louis Massignon, ʻEsquisse d'une bibliographie Qarmate', in *A Volume of Oriental Studies Presented to Edward G. Browne on his 60th Birthday (1922)*, ed. T. W. Arnold and Reynold A. Nicholson (Cambridge: [Cambridge] University Press, 1922), p. 331.

42 Wilferd Madelung, ʻAl-Kayyāl', *EI2*, vol. 4, p. 847. The later dating of al-Kayyāl's time in Madelung's article is based on his preference for Abū al-Maʿālī's account (ca. 485/1092) over that of the *dāʿī* Idrīs. Al-Shahrastanī (d. 548/1153) seems to have had a better knowledge of al-Kayyāl's status and ideas (*al-Milal*, vol. 1, pp. 138–141); no time is given by him, but a period prior to the beginning of the Fāṭimid caliphate is implied.

161). The inclusion of '*al-Ashāʿira*' was used by Marquet to determine the date of composition as after 324/936 (i.e., the date of al-Ashʿarī's death), at least for the epistle in which it occurs.[43] Since *al-Ashāʿira* as a group would have been formed only long after the death of al-Ashʿarī, a later post-dating of the epistle would seem necessary. In a previous article, I argued that since the authors seem to be against all the groups mentioned in the above quotation, this would suggest that they themselves advocated a very 'traditionalist' ideology; however, the generally liberal thrust of their writing makes that rather unlikely. On this assumption, I was inclined to consider this passage as an editorial interpolation.[44] But it seemed that I would have to modify my opinion, because the passage occurs in the midst of the description of other conflicting religious groups. There was, however, another consideration. The *Nawāṣib* and the *Rawafiḍ* seemed legitimately paired as conflicting schools; so did the *Jabariyya* and the *Qadariyya*; but somehow the *Ashāʿira* did not contrast properly with the *Khawārij*. The *Muʿtazila* would have made a better pair to the *Ashāʿira*, and the *Murjiʾa* to the *Khawārij*. It was necessary, therefore, to compare the printed texts of the *Rasāʾil* with earlier manuscripts of the work to determine the proper reading. I am grateful to my friend François de Blois for checking the manuscripts of the *Rasāʾil* in the British Library and the Bibliothèque Nationale (*fonds arabe*). In both the British Library manuscript Or. 2359 (n.d., fol. 43b) and the Paris manuscript 2305 (AH 1153, fol. 273a), the relevant expression is '*al-Murjiʾa waʾl-Khawārij*'. In the Paris manuscript 2304 (AH 1065, fol. 274b), the reading is '*al-Khawārij*', without its corresponding rival group, and in the Paris manuscript 2303 (AH 1020, fol. 310b), the reading is '*al-Murjiʾa*', also without its corresponding rival group. Nowhere is there '*al-Ashāʿira*', not even in Dieterici's translation, where it is '*al-Murjiʾa waʾl-Khawārij*'.[45] So it seems that '*al-Ashāʿira*' was a mistaken reading in the manuscript used for the Bombay edition of the *Rasāʾil* (1888) which was then repeated in the Cairo edition (1928) and again in the Beirut edition (1957).

43 Marquet, 'Rasāʾil Ikhwān al-Ṣafāʾ'.
44 A. Hamdani, 'Shades of Shīʿism', p. 448.
45 Dieterici, *Die Philosophie der Araber*, vol. 7, p. 213 (see note 14 above).

Thus, the inclusion of '*al-Ashāʿira*' is of no relevance for dating the composition of the *Rasāʾil* after all.

The *Rasāʾil* contains another passage (*Rasāʾil*, vol. 4, pp. 129–130) which, although it is not in the nature of an interpolation, has, nevertheless, been described by Shlomo Pinès as 'plagiarism' from al-Fārābī's *al-Madīna al-fāḍila.*[46] This passage enumerates the twelve qualifications of an ideal ruler. Pinès assumes that the authors of the *Rasāʾil* wrote after al-Fārābī and, therefore, that they took these qualifications wholesale from the latter. In a separate study,[47] I have argued that Pinès' assumption is not correct, as both al-Fārābī and the authors of the *Rasāʾil* were probably using Ḥunayn ibn Isḥāq's (194–264/809–877) Arabic commentary on Plato's *Republic,*[48] which would explain the similar language and enumeration in both. The time of Ḥunayn accords well with the pre-Fāṭimid caliphate composition of the *Rasāʾil.*

Since the famous fourth-/tenth-century man of letters Abū Ḥayyān al-Tawḥīdī named four specific contemporary writers as authors of the otherwise anonymous *Rasāʾil,*[49] it has become almost traditional for both mediaeval and modern scholars to accept his account as fact. In a detailed study published recently, I have shown that Abū Ḥayyān was not really concerned with ascertaining the authorship of the *Rasāʾil* but with maligning a contemporary, Zayd ibn Rifāʿa, whom he accused of heresy by associating him with the supposed, 'heretical' authors of the 'heretical' work, the *Rasāʾil.*[50] Abū Ḥayyān was neither aware of the correct number of the epistles nor of *al-Risāla al-jāmiʿa.* There is only one valuable aspect of Abū Ḥayyān's report, and that is that it shows

46 Shlomo Pinès, 'Some Problems of Islamic Philosophy', *Islamic Culture*, 11, 1 (1937), p. 71; cf. al-Fārābī's *al-Madīna al-fāḍila* , pp. 105–108.

47 Abbas Hamdani, 'Al-Fārābī and the Brethren of Purity', paper presented to the Ninth Annual Meeting of the Middle East Studies Association of North America, Louisville, 1975.

48 Plato, *Republic*, in *The Dialogues of Plato*, ed. and trans. Benjamin Jowett, 4th ed. (Oxford: Clarendon Press, 1953), vol. 2, pp. 1–62 (translator's introduction), 163–499 (translation). Book VI of the *Republic*, 484a–511d (pp. 342–375 of Jowett, vol. 2), contains the qualifications of the Philosopher-King. Ḥunayn's Arabic commentary has been lost but is referred to in Ibn al-Nadīm, *Fihrist*, (Tehran, 1971), p. 306.

49 Al-Tawḥīdī, *al-Imtāʿ waʾl-muʾānasa*, vol. 2, pp. 3–6 (see note 10 above).

50 A. Hamdani, 'Abū Ḥayyān al-Tawḥīdī and the Brethren of Purity' (see note 2 above).

that the encyclopaedia was known in his time, which would preclude any later dating, although his report cannot be relied on to establish whether the work was written contemporaneously or earlier.

This brings us to a particular dating suggested by Paul Casanova, based on an astrological prediction in the forty-eighth epistle of the encyclopaedia (*Rasāʾil*, vol. 4, p. 142) along with other collateral passages.[51] My critique of Casanova's theory [as mentioned in the Postscript to this chapter] argues that Casanova confuses the actual astronomical occurrence with the ability of the *Rasāʾil*'s authors to predict a date correctly. Again, according to the astronomical calculations, several dates are possible; reliance on one, therefore, has no particular significance. Moreover, the time fixed by Casanova, 418/1027 to 439/1047, is obviously incorrect as we know that around 373/983 Abū Ḥayyān was already reading the *Rasāʾil*. In fact, the encyclopaedia was in existence much before the time of Abū Ḥayyān.

For the 'proper' dating of the *Rasāʾil*, I have relied on both internal and external evidence. The former is contained in my article 'Shades of Shīʿism in the Tracts of the Brethren of Purity',[52] and the latter, in my article 'An Early Fāṭimid Source on the Time and Authorship of the *Rasāʾil Ikhwān al-Ṣafāʾ*'.[53] Accordingly, one can date the composition as between 260/873 and 297/909. All internal references posterior to this period can be proved to be interpolations.[54]

51 Paul Casanova, 'Une date astronomique dans les Épîtres des Ikhwān aṣ-Ṣafāʾ', *Journal Asiatique*, 11, 5 (1915), pp. 5–17.

52 See note 1 above.

53 See note 2 above.

54 The original article, that was published in the *Journal of Semitic Studies*, 29, 1 (1984), pp. 97–110, consisted of a revised edition of an earlier paper given at the Annual Meeting of the American Research Centre in Egypt held in San Francisco on 14 April 1980. I am grateful to the University of Wisconsin-Milwaukee for the travel grant which enabled me to attend this meeting and to Professor C. F. Beckingham and Mr François de Blois for making several suggestions and corrections.

POSTSCRIPT

This article was originally published in 1984, and it needed to be corrected in some minor respects, and updated with subsequent research. All additions have been put in square brackets. The rest of the *addenda* are given in this postscript.

Notes 35–37
In my article 'Brethren of Purity, a Secret Society for the Establishment of the Fāṭimid Caliphate' (see note 1 above), I have cited the verses of Ibn al-Rūmī (d. 283/896) referring to 'Ikhwān al-Ṣafā'' twice. He says he is contemporary with them ('*'āshartuhu*') and that they have lived long ('*bi-tafāḍul al-a'amārī*'). Four verses and their translation are given with comments (pp. 76–78).

Note 37
Rūdakī's (d. 329/940) and Shuhayd al-Balkhī's (d. 315/927) Persian verses are cited in the *Rasā'il* in the epistles on astronomy and music (*Rasā'il*, vol. 1, pp. 139, 209, 235). As they are distorted in Arabic print, they are cited again correctly and translated in my paper 'The Relation Between the Persian Authorship of the *Rasā'il Ikhwān al-Ṣafā'* and the Time of its Composition', given at the Annual Meeting of the Middle East Studies Association of North America at Washington DC in November 2005 (awaiting publication). I am thankful to Dr. Farhad Daftary of The Institute of Ismaili Studies, London, for his help in extracting the references to these verses.

Note 48
Plato (see note 48 above for the full reference) does not have twelve qualifications of an ideal ruler (he has only ten), but the Ikhwān (*Rasā'il*, vol. 4, pp. 129–130), al-Fārābī (*al-Madīna al-fāḍila*, pp. 127–128) and Ibn Rushd (d. 595/1198; commentary on Plato's *Republic* in Hebrew translation, ed. Ralph Lerner [Ithaca, NY: Cornell University Press, 1974]) each have twelve qualifications, although the arrangement and style are different. In substance, however, they are the qualifications as given by Plato. The Ikhwān have a more expansive style, whereas

al-Fārābī's is concise and pithy. I do not think any one of these three has taken from the other two, but the number twelve suggests that they all derive from a common early source. Had Ḥunayn ibn Isḥāq's (d. 264/871) commentary, which was referred to by Ibn al-Nadīm (d. 376/987) (*Fihrist*, Beirut ed., p. 246) as stated above, been available, we might have known that he had probably grouped Plato's qualifications under twelve headings — that number seems to have had some religious significance and was not abandoned by any of our three Muslim sources. It is also possible that Ḥunayn's commentary was based on Galen's summary of Plato's *Republic* (as suggested by Richard Walzer in *Plato Arabus* [1951], vol. 1, p. 2) and that it was, in fact, Galen who grouped the qualifications under twelve headings.

Note 51
My comments on Paul Casanova's article (see note 51 above) have now been published in my article 'A Critique of Paul Casanova's Dating of the *Rasā'il Ikhwān al-Ṣafā*", in *Mediaeval Isma'ili History and Thought*, ed. Farhad Daftary (Cambridge: Cambridge University Press, 1996), pp. 145–152, in which the passage on the astrological prediction in the *Rasā'il* (vol. 4, p. 142) is translated and commented upon. It is interesting that the whole of Epistle 48, in which this prediction occurs, is placed in the mouth of the Imam himself.

Note 52
Besides the quotations given in my article 'Shades of Shī'ism', other quotations given in my later article 'Brethren of Purity, a Secret Society' (for both, see note 1 above) should be referred to, since they provide some internal evidence. What is more important is to study the time-layer of the *Rasā'il*'s ideas in *all* fields of knowledge, and that would, I think, provide a better understanding in support of an earlier (i.e., ninth-century) dating of its composition.

FOUR

The Scope of the Rasāʾil Ikhwān al-Ṣafāʾ

Carmela Baffioni

The Work and its Authors

The richness of the contents of the *Rasāʾil Ikhwān al-Ṣafāʾ* makes this corpus the most complete of mediaeval encyclopaedias of science. It precedes by at least two centuries the best-known examples in the Latin world — those by Alexander Neckham, Thomas de Cantimpré, Vincent de Beauvais, and Bartholomaeus Anglicus respectively, which date back to the thirteenth century — and it holds appeal for scholars and readers for a variety of reasons. It consists of extremely heterogeneous material, usually identified as the *summa* of foreign knowledge in ninth-century Islam, reworked in such a way as to represent the insights needed for an elite training for salvation.

Its technical content makes the corpus particularly interesting for philologists and historians of culture and science, yet it also addresses religious matters of central relevance to Islam: God, His unity and uniqueness, His attributes, the Creation, angels, human destiny, good and evil, and resurrection.

A blending of scientific and religious issues is evident from the beginning, though the sciences are the main topic in the first and the second sections and feature prominently in the third; theology is the primary subject of the rest of the corpus. The common distinction between philosophy/science and theology is not evident in the *Rasāʾil*, despite the headings of the four sections, which are misleading insofar as the subject of the treatises is at times inconsistent with the section heading.

This has been helpfully interpreted as being the result of a 'stratified' compilation of the encyclopaedia between the ninth and the tenth centuries. The supposed reworking of the whole corpus by members of the Brotherhood may have caused some epistles to be shifted from their original places to where we find them now; the issues of chronology and authorship of the *Rasāʾil* are connected to this.[1]

The structure and content of the work deserve further consideration, even though it is among the relatively well-explored topics of Islamic philosophy. Its broad scope and apparently contradictory aspects cannot be explained solely by formal reasons. We should ask, rather, whether the authors pursued aims other than those stated; were those aims evident, they might clarify the true nature of the corpus.

The current state of research plainly indicates that the encyclopaedia was assembled within the course of at least a century and a half, at some point between about 840 and 980. The present writer's opinion is that such a collective enterprise must have been conducted with a special use in mind for the wealth of foreign philosophical and

---

1 The *Rasāʾil* is usually dated to the tenth century, though a few scholars differ. See Paul Casanova, 'Une date astronomique dans les Épîtres des Ikhwān aṣ-Ṣafā'', *Journal Asiatique*, 11, 5 (1915), pp. 5–17; and Abbas Hamdani, 'A Critique of Paul Casanova's Dating of the *Rasāʾil Ikhwān al-Ṣafāʾ*', in *Mediaeval Ismaʿili History and Thought*, ed. Farhad Daftary (Cambridge: Cambridge University Press, 1996), pp. 145–152. Yves Marquet's opinion, that the corpus was begun in the late ninth century and reworked during the tenth (see e.g., 'Ikhwān al-Ṣafā'', *EI2*, vol. 3, pp. 1072–1073), deserves careful consideration. See also Abbas Hamdani's chapter in the present volume.

Regarding the authorship, I see no reason to reject the Arabic sources that mention the supposed names of its authors. See Samuel M. Stern, 'The Authorship of the Epistles of the Ikhwān aṣ-Ṣafā'', *Islamic Culture*, 20 (1946), pp. 367–372; and Samuel M. Stern, 'New Information about the Authors of the "Epistles of the Sincere Brethren"', *Islamic Studies*, 3 (1964), pp. 405–428; Abbas Hamdani, 'Abū Ḥayyān al-Tawḥīdī and the Brethren of Purity', *International Journal of Middle Eastern Studies*, 9 (1978), pp. 345–353.

scientific knowledge employed, in view of the doctrinal conflicts and political changes of those times. In this sense, the current view of the *Rasāʾil* as a 'popular' work should be rejected.[2]

Attempts to define the ideological commitment of the Ikhwān have already been made, of course: their link to the Shiʿa is widely recognised, though scattered references in negative terms to the 'hidden imams' (*al-aʾimma al-mastūrīn*) appear contrary to Ithnāʿasharī Shiʿism. Other elements such as references to septenary cycles and a distinction between the elect and the masses have been alleged as proof of the Ismaili inspiration of the Ikhwān — initially, by Henry Corbin.[3] Their political (and cosmological) conceptions have been understood as placing them among the representatives of the 'official' Ismaili rulers, the Fāṭimids, at an early stage, or even under the influence of the Qarmaṭīs.[4] In general, evidence supporting such theories is mostly external, in the form of doctrines or elements of doctrines discussed or mentioned in the corpus. But the extreme variety (and often inconsistency) of the ideas it contains has prevented scholars from agreeing on the authors' religious or ideological commitment and, consequently, on their goals.[5]

Philology, History, and Doctrine versus Ideological Commitment

Let us first consider the reworking of the multi-faceted 'scientific' material of the *Rasāʾil*.

Indian, Persian, and Christian references are brought together

2 See, most recently, U. Rudolph, *Islamische Philosophie. Von den Anfängen bis zur Gegenwart* (Munich: Beck, 2004), p. 39.

3 For septenary cycles, see, e.g., *Rasāʾil*, Epistle 34, vol. 3, pp. 219.18–24. All references in this essay are to the Beirut edition (Dār Ṣādir, 1957).

4 See e.g., Yves Marquet, 'Imāmat, resurrection et hiérarchie selon les Ikhwān aṣ-Ṣafāʾ', *Revue des Études Islamiques*, 30 (1962), pp. 49–142; or Abbas Hamdani, 'An Early Fāṭimid Source on the Time and Authorship of the *Rasāʾil Ikhwān al-Ṣafāʾ*', *Arabica*, 26 (1979), pp. 62–75; Yves Marquet, 'Ikhwān al-Ṣafāʾ, Ismaïliens et Qarmaṭes', *Arabica*, 24, 3 (1977), pp. 233–257.

5 Susanne Diwald shows the matter as not at all unquestionable. See *Arabische Philosophie und Wissenschaft in der Enzyklopädie Kitāb Iḫwān aṣ-ṣafāʾ (III): Die Lehre von Seele und Intellekt* (Wiesbaden: Harrassowitz, 1975), p. 26 ff.

in the *Rasā'il*,[6] but attention is here confined to the Greek heritage. Comparisons between the Ikhwān's quotations and the extant Arabic versions of classical works often show that they may well have used the translations now at our disposal. Where such comparisons are not possible, their extracts usually correspond to the extant original texts. We should conclude, then, that in quoting, reporting, and perhaps in translating ancient sources, the Ikhwān displayed the same philological and philosophical expertise usually associated with the scholars to whom we owe the flourishing translation movement in 'Abbāsid Baghdad. In some cases, they have even preserved the only Arabic fragments of Greek authors known to us: the story of Gyges from Plato's *Republic* in the epistle on magic is a well-known case.[7]

In spite of this, and of a mention of the theory of reminiscence,[8] Platonic references in the encyclopaedia are scarce, and when they do occur they are religiously rather than philosophically oriented.[9] This can perhaps be explained by the influence of the Hellenistic *curricula scientiarum*, which began with a reflection on the Stoic *bios theôrêtikos*, continued with Aristotle's works corresponding to the 'scientific' side of knowledge, and ended with Platonic writings corresponding to 'theological' knowledge.[10]

---

6 A good survey of these can be found in Ian R. Netton, *Allah Transcendent: Studies in the Structure and Semiotics of Islamic Philosophy, Theology and Cosmology* (Richmond: Curzon Press, 1989).

7 It is the last in the corpus and does not feature at all in some editions. Alessandro Bausani considers it spurious and not of great worth. See Alessandro Bausani, *L'enciclopedia dei Fratelli della Purità. Riassunto, con Introduzione e breve commento, dei 52 Trattati o Epistole degli Ikhwan as-safa'* (Naples: Istituto Universitario Orientale [Dipartimento di Studi Asiatici], 1978), pp. 12, 279. It is clearly distinct from the other epistles in style rather than content, but internal evidence against such a hypothesis may be found in the reference to the whole corpus and to the Ancients in *Rasā'il*, Epistle 52, vol. 4, p. 284.4–7, and the 'ending' in vol. 4, p. 312.6–12.

8 In this author's experience, the theory of reminiscence is seldom referred to in Muslim sources. See *Rasā'il*, Epistle 42, vol. 3, p. 424.13–20.

9 The majority of Platonic references are to be found in the fourth section. See Carmela Baffioni, 'Frammenti e testimonianze platoniche nelle *Rasā'il* degli Ikhwān al-Ṣafā'', in *Autori classici in lingue del Vicino e Medio Oriente*, ed. G. Fiaccadori (Rome: Istituto Poligrafico e Zecca dello Stato, Libreria dello Stato, 2001), pp. 163–178.

10 The question is fully discussed by Cristina D'Ancona Costa, *La casa della*

References relating to the sciences are generally unquestionable: for instance, in the first section, the works of Euclid, Nicomachus, and even Archimedes are the basis of mathematical tenets (Epistles 1, 2, and 6); Ptolemy supports astronomy and geography (Epistles 3 and 4); and the Pythagoreans and Nicomachus are the sources for musical theory (Epistles 5 and 6).

The situation is different for philosophy: the wealth of available sources led the Ikhwān to support their doctrines with a variety of theoretical models, often merged together. They thoroughly reworked the Greek texts in line with their aims, which differed completely from those of the Ancients; other than in the special use of ancient philosophy, this is generally evident from the structure of the encyclopaedia.

Different Philosophical Approaches in the Reworking of Ancient Sources

The religious approach to ancient heritage is a consequence of the fact that philosophy is a way towards salvation. This is why the Ikhwān introduced it as an 'imitation of God according to human capacity'. This definition immediately indicates that Hellenism was the basis for the authors' first theoretical and historical reference. This is confirmed by the fact that their source is once again Plato — or, more accurately, a Hellenistic reading of him.[11]

The Hellenistic influence is remarkable, even when the Ikhwān rely on Aristotelian sources. This is first evident in their treatment of logic: Epistles 10–14 describe the contents of the first five books of the *Organon* according to the Neoplatonic order.[12] At the beginning of the second section, when they present Aristotelian material, the Ikhwān

sapienza. La trasmissione della metafisica greca e la formazione della filosofia araba (Milan: Guerini e Associati, 1996), pp. 30–31.

11 See Carmela Baffioni, 'Sulla ricezione di due luoghi di Platone e Aristotele negli Ikhwān al-Ṣafā'', *Documenti e studi sulla tradizione filosofica medievale*, 8 (1997), pp. 479–492. Cf. Paul Kraus, *Jābir ibn Ḥayyān. Contribution à l'histoire des idées scientifiques dans l'Islam. Jābir et la science grecque* (Paris: Les Belles Lettres, 1986), p. 99, note 4.

12 Namely, Porphyry's *Isagoge* and Aristotle's *Categories, On Interpretation*, and *Prior* and *Posterior Analytics*. Two more Aristotelian books were part of the *Organon* in late antiquity (*Rhetoric* and *Poetics*), but they are not dealt with in

again adhere to their Hellenistic arrangement. Epistle 15 echoes themes addressed in the *Physics*; Epistle 16: 'On Heaven and the World' finds the authors in line with the Muslim tradition that transformed the ancient *De Caelo* into a new book, *De Caelo et Mundo*; Epistle 17 deals with *De Generatione et Corruptione*; whilst Epistle 18: 'On Meteorological Phenomena' and Epistle 19: 'On Minerals' recall the contents of *Meteorology* I–III and of *Meteorology* IV respectively.

Indebtedness to Aristotle is confirmed in other treatises such as the 'twin' epistles — Epistle 24: 'On Sense and Sensation' and Epistle 35: 'On Intellect and Intellection' — or Epistle 39: 'On the Kinds of Movement', but this considers the movement of the whole cosmos, from the stars to the natural and voluntary movements of human beings, and the aim is to demonstrate the existence of a God who created the world and who will bring it to an end. Ancient theories on the movements of the stars finally evolve, in Epistle 36: 'On Cycles and Revolutions', into what Godefroid de Callataÿ, following Cumont, has recently defined as a '*fatalisme astral*'.[13]

But the Ikhwān's ancient sources are not only Aristotelian: there are echoes in Epistle 19 of the Hellenistic theory of the sympathies and antipathies of beings and of the correspondence between super-lunar and sub-lunar worlds, stars, and minerals — one of the main tenets of alchemy, the Greek foundation for which was *Meteorology* IV.[14]

The Aristotelian succession is discontinued at Epistle 20, which introduces a Neoplatonic conception of Nature that also recalls certain Ismaili philosophical positions.[15] An order of angels is described

---

the encyclopaedia. As to the omission of the other Aristotelian works of logic (*Topics* and *Sophistical Refutations*), see below in the present chapter.

13 See Ikhwān al-Ṣafā', *Les révolutions et les cycles (Épîtres des Frères de la Pureté, XXXVI)*, ed. and trans. Godefroid de Callataÿ (Beirut and Louvain-la-Neuve: Al-Burāq and Academia-Bruylant, 1996), pp. 36–37.

14 This is sometimes considered spurious. See the bibliography in Carmela Baffioni, *Il IV libro dei 'Meteorologica' di Aristotele* (Naples: Bibliopolis, 1981), pp. 386–392.

15 For instance, Abū Ya'qūb al-Sijistānī's representation of Nature as 'agent of power' and 'principle of movement'. See Carmela Baffioni, 'Antecedenti greci nel concetto di "natura" negli Iḫwān al-Ṣafā", in *Enôsis kai Philia (Unione e amicizia): Omaggio a Francesco Romano*, ed. M. Barbanti, G. R. Giardina, and P. Manganaro (Catania: CUECM, 2002), pp. 545–556.

that precedes the more famous elaborations by Ibn Sīnā (Avicenna, d. 1037) and Shihāb al-Dīn Suhrawardī (d. 1191).[16]

One might wonder why Epistle 20 itself does not open the second section. There may be a historical and perhaps philological reason for this choice: the authors were obliged to base themselves on traditions other than Aristotle's.

Epistle 21 speaks of plants, Epistle 22 of animals, and Epistle 23 and subsequent epistles, of human beings: subjects for which they had to rely mainly on pseudo-Aristotelian or wholly non-Aristotelian sources. Although no work by Aristotle on plants was known to the Arabs, a rich botanical and agricultural corpus was nonetheless available.[17] The parts of the Corpus Aristotelicum on animals — excepting the *De Partibus* and the *De Generatione Animalium*, both translated fully by Yaḥyā ibn al-Biṭrīq — reached them in abridged form under the title *Kitāb al-Ḥayawān*; and, in terms of content and aims, Epistle 22 differs significantly from Aristotle's work because the whole subject is placed within a well-known 'ontological' dispute. Finally, Aristotelian works on the various aspects of human beings were not always available.[18]

In the arrangement of their encyclopaedia, the Ikhwān might have maintained the Hellenistic form of the *curricula scientiarum*, mentioned above. Among the philosophical schools and doctrines reflected in the encyclopaedia, hermetism has an important role, as is evident specifically in Epistle 3 and, in a wider perspective, in Epistle 51.[19] But the Ismaili influence is especially remarkable, particularly in

16 This might, then, demonstrate a possible link between the Ikhwān and the 'illuminative' philosophy, a theme I will consider in the near future. On the 'angelology' of the Ikhwān, cf. also Epistles 25 and 49.

17 The Aristotelian heritage reached Islam via Theophrastus and Nicolaus Damascenus. See Nicolaus Damascenus *De Plantis: Five Translations*, ed. and introduced by H. J. Drossaart-Lulofs and E. L. J. Poortman (Amsterdam, Oxford, and New York: North-Holland Publishing Company, 1989); see also Muṣṭafā al-Shihabī, 'Filāḥa [Middle East]', *EI2*, vol. 2, pp. 899–901.

18 See Francis E. Peters, *Aristoteles Arabus: The Oriental Translations and Commentaries on the Aristotelian Corpus* (Leiden: Brill, 1968).

19 See Yves Marquet, *La philosophie des Ikhwān al-Ṣafā de Dieu à l'homme* (Lille: Service de reproduction des thèses, 1973); Yves Marquet, *La philosophie des alchimistes et l'alchimie des philosophes: Jâbir ibn Ḥayyân et les 'Frères de la Pureté'* (Paris: Maisonneuve et Larose, 1988); Yves Marquet, 'Les références à Aristote dans les Épîtres des *Ikhwān aṣ-Ṣafā*", in *Individu et société: L'influence*

accounting for the hierarchical structure of the universe and of the three natural realms.

In their discussion of embryology in Epistle 25, the Ikhwān oscillate between the Aristotelian and Ismaili heritage, adopting the Aristotelian thesis of the origin of the embryo in the sperm and the menstrual blood,[20] but emphasising a new astronomical and astrological approach absent in the Greek tradition and thought to have originated with Muslim scientists, from whom it passed to the Latin world.[21]

The Structure of the Encyclopaedia

The first topic to be addressed is the quadrivium: arithmetic is treated in Epistle 1, geometry in Epistle 2, astronomy in Epistle 3, and music in Epistle 5. Between these last two is inserted an epistle on geography (Epistle 4), evidently considered as part of the science of astronomy.[22]

With regard to the trivium sciences, there is no place for rhetoric in a corpus oriented on absolute truth: as will be stated later, dialectics

d'Aristote dans le monde méditerranéen, ed. T. Zarcone (Istanbul, Paris, Rome, and Trieste: Isis, 1988), pp. 159–164.

20 This means that they did not accept the ancient Hippocratic theory, later developed by Galen, of a female sperm coexisting with the male sperm in the formation of the human being.

21 See Charles S. F. Burnett, 'The Planets and the Development of the Embryo', in *The Human Embryo: Aristotle and the Arabic and European Traditions*, ed. G. R. Dunstan (Exeter: Exeter University Press, 1990), pp. 95–112. Such formulations are found, for instance, in *Sirr al-khalīqa* by Balīnūs (the Arabic name for Apollonius of Tyana) but also in physicians such as ʿAlī ibn al-ʿAbbās al-Majūsī and ʿArib ibn Saʿid al-Qurṭubī (both tenth century). See Carmela Baffioni, 'L'influenza degli astri sul feto nell'Enciclopedia degli Ikhwān al-Ṣafā'', *Medioevo. Rivista di storia della filosofia medievale*, 23 (1997), pp. 409–439; and Carmela Baffioni, 'L'embriologia araba fra astrologia e medicina. Abu Maʿšar al-Balkhī e Muḥammad ibn Zakarīyā' al-Rāzī', in *La diffusione dell'eredità classica nell'età tardo-antica e medievale. Il 'Romanzo di Alessandro' e altri scritti*, ed. R. B. Finazzi and A. Valvo (Alessandria: Edizioni dell'Orso, 1998), pp. 1–20.

22 Though, as Godefroid de Callataÿ has recently remarked, it is not clear whether this was merely a common opinion, or the Ikhwān's proper position. See *Les arts libéraux en Islam*, a paper presented at the Symposium 'Une lumière venue d'ailleurs. Héritages et ouvertures dans les encyclopédies d'Orient et d'Occident au Moyen Age' (Louvain-la-Neuve, 2005), forthcoming in the *Proceedings*.

increases damaging opposition among learned men.[23] Grammar is, of course, the basis of verbal communication, but the Ikhwān recognise this only implicitly,[24] instead confining themselves to the technique of the 'universal' form of language (i.e., logic) in siding with the Christian logician Abū Bishr Mattā ibn Yūnus rather than with the grammarian Abū Saʿīd al-Sīrāfī in the well-known tenth-century debate on the superiority of grammar or logic. Elsewhere, they appear interested in spoken language and even phonetics. Arabic in particular is held in high esteem as the language of Revelation, as recalled in Epistles 31 and 40.

It is noteworthy that although the first section should be — and is — the most uniform in terms of content, it nevertheless contains inconsistencies. After the treatment of the quadrivium sciences, Epistle 7 outlines the Ikhwān's epistemological vision in a sort of table of human sciences (*Rasāʾil*, vol. 1, pp. 266.14–274.21).[25] Here we find the sciences divided into three types: the 'propaedeutical' (*al-riyaḍiyya*), the 'legal' (*al-sharʿiyya al-waḍʿiyya*), and the 'philosophically true' (*al-falsafiyya al-ḥaqīqiyya*). It is on the basis of the sciences of this third group that the Ikhwān affirm that they composed their *Rasāʾil.*

This last group is further divided into propaedeutical, logical, natural, and metaphysical/theological sciences. The propaedeutical (the same word used to describe the sciences of the lowest level) are dealt with in Epistles 1, 2, 3, and 5. Only some of the logical sciences listed in Epistle 7 are delineated in the encyclopaedia;[26] the lowest of these such as writing, language, computation and business, magic and

23 This is discussed below in the present chapter.

24 See e.g., *Rasāʾil*, Epistle 12, vol. 1, pp. 414.14–415.8 and esp. p. 415.7–8.

25 See Carmela Baffioni, 'Oggetti e caratteristiche del *curriculum* delle scienze nell'*Enciclopedia* dei Fratelli della Purità', in *Studi arabo-islamici in memoria di Umberto Rizzitano*, ed. G. di Stefano (Trapani: Istituto di Studi Arabo-Islamici 'Michele Amari', 1991), pp. 25–31; Carmela Baffioni, 'Valutazione, utilizzazione e sviluppi delle scienze nei primi secoli dell'Islām: Il caso degli Ikhwān al-Ṣafāʾ", in *La civiltà islamica e le scienze*, ed. C. Sarnelli Cerqua, O. Marra, and P. G. Pelfer (Naples: CUEN, 1995), pp. 23–35; Carmela Baffioni, 'L'Islām e la legittimazione della filosofia. I "curricula scientiarum" del secolo X', in *La filosofia e l'Islām,* ed. G. Piaia (Padua: Gregoriana, 1996), pp. 13–34. The treatise is translated by Godefroid de Callataÿ, 'Ikhwān al-Ṣafāʾ: des arts scientifiques et de leur objectif (*Épître VII* des Frères de la Pureté)', *Le Muséon*, 116 (2003), pp. 231–258.

26 Alessandro Bausani argues that the text is corrupted, and emends it to read: (i) poetics, (ii) rhetoric, (iii) topics, (iv) analytics, and (v) sophistic; see Bausani,

enchantments, alchemy, crafts, and arts are dealt with in the corpus, or at least mentioned. Grammar, which is not considered by the Ikhwān as a science, is also placed in the lowest group. The 'natural' sciences indicated are those described in Epistles 15–19 and 21–22. The theological sciences are examined in great depth here, with more subjects actually covered than are indicated in the table: knowledge of the Creator, of spiritual beings, of angels and souls that inhabit heavenly and terrestrial bodies, the five kinds of politics, and resurrection. Among the 'legal' sciences, revelation, hermeneutics, tradition, asceticism, and the interpretation of dreams are also addressed in the *Rasā'il*; and there are numerous references to law throughout.

The table can be considered the best proof of the 'stratification' of the whole: it might have been retained from a 'first draft' before the corpus was further elaborated.

The content of Epistle 7 is broadly based on Aristotle: the introduction of the nine 'philosophical questions' directly recalls the Aristotelian categories.[27] But, ultimately, the scientific *methodos* is identified with the path towards the hereafter, and is therefore connected with asceticism; linking the Aristotelian *epistêmê* to Ismaili teachings (*ta'līm*) reveals that scientists and prophets proceed along the same way.[28]

The authors' Ismaili commitment might be proved by the fact that in the subsequent and complementary treatise, 'On the Practical Arts and their Goals', ancient *banausic* activities are given prominence, as the high esteem in which practical arts were held by the Ismailis is well known.[29]

It may be that the multiplicity of theoretical references affected the structure of the encyclopaedia, but whatever the reason, we can

---

L'enciclopedia dei Fratelli della Purità, p. 69. Cf. Godefroid de Callataÿ's chapter in the present volume.

27 See *Rasā'il*, Epistle 7, vol. 1, pp. 262.14–266.7; ibid., Epistle 29, vol. 3, p. 35.6–7; ibid., Epistle 40, vol. 3, pp. 345.6–346.6; ibid., Epistle 42, vol. 3, p. 513.14–16; ibid., pp. 514.13–516.5.

28 See ibid., Epistle 7, vol.1, pp. 274.20–275.4.

29 See Bernard Lewis, 'An Epistle on Manual Crafts', *Islamic Culture*, 17 (1943), pp. 141–151; Yves Marquet, 'La place du travail dans la hiérarchie ismā'īlienne d'après *L'encyclopédie des Frères de la Pureté*', *Arabica*, 8, 3 (1961), pp. 225–237; Farhad Daftary, *The Ismā'īlīs: Their History and Doctrines*, 2nd ed. (Cambridge: Cambridge University Press, 2007), p. 15.

hypothesise changes to the order of the treatises. For instance, the 'programmatic' aim of Epistle 7 makes it, potentially, the ideal opening for the whole corpus; similarly, Epistle 20 could have been the perfect introduction to the second section.

The variety of theoretical models also has a bearing on the religious dimension of the encyclopaedia. This is the case with Epistle 9, which breaks the series of the scientific treatises in a different spirit. A variety of moral behaviour is considered, supported by a long series of anecdotes about prophets and wise men; Sufi positions are especially emphasised. Several of the qualities and attitudes here described strongly resemble those that the Ikhwān set out in order to guide their community.[30]

We must postpone discussion as to precisely which kinds of 'guide' and 'community' the Pure Brethren might have had in mind, but that they held a degree of assent to ascetic tendencies cannot be dismissed. Consider, for instance, their enhancement of 'purification' (*tahdhīb* or *iṣlāḥ*) of the soul in the actual titles of some propaedeutical and natural treatises (such as Epistles 6, 16, and 24), which also indicates the moral aims of the encyclopaedia as a whole.

If the Ikhwān find asceticism the most appealing means to attaining purification,[31] there may be a case for reconciliation with the theological/doctrinal perspectives discussed above. The matter of the authors' religious and ideological allegiance seems to pose an urgent question from the very beginning of the encyclopaedia.

Use and Elaboration of Ancient Sources

Even though the Ikhwān's outlook can be said to be mostly Aristotelian,[32] it is clear that they have deeply and variously reworked Aristotle's

30 See Carmela Baffioni, 'The "Friends of God" in the *Rasā'il Ikhwān aṣ-Ṣafā*", in *Proceedings of the 20th Congress of the Union Européenne des Arabisants et Islamisants (UEAI), Part Two: Islam, Popular Culture in Islam, Islamic Art and Architecture*, ed. A. Fodor, *The Arabist, Budapest Studies in Arabic*, 27 (2003), pp. 17–24.

31 See e.g., *Rasā'il*, Epistle 27, vol. 3, p. 8.16–17. It is interesting to note that in certain manuscripts that I have consulted, the Ikhwān are included among the *Ṣūfiyyūn*.

32 Consider, for instance, their descriptions of the terrestrial world or the theory of knowledge as rooted in sensation.

thought. One example is their approach to logic: it is no longer an instrument of science but a science in itself — perhaps even the highest one, given that the Pure Brethren called it 'the scale of sciences' through which wise men are able to discern 'true from false in speeches, right from wrong in opinions, worth from futility in beliefs, and good from evil in acts'.[33] In fact, elsewhere also, moral and spiritual superiority is recognised in appropriate knowledge, dispositions, opinions, and works;[34] and physical pleasures are given by (appropriate) opinions, doctrines, habits, and works,[35] while (moral) illnesses designate 'accumulated ignorance, bad habits, corrupted opinions and evil deeds'.[36] This is perfectly consistent with the previously mentioned view of philosophy as *imitatio dei* — imitation of God's knowledge, work, deeds, morals, opinions, and actions.[37] This is stated plainly in one of the treatises devoted to logic, Epistle 13: 'On the Prior Analytics'.[38]

The selection of logical works in the *Rasā'il* shows that the Ikhwān followed the line of the 'old' tradition that limited research to the first books of the *Organon*. Although the explanation ascribing such delimitations to the religiously motivated Syriacs has since been recognised as a 'forgery' (i.e., falsely attributed to al-Fārābī), the fact remains that the Brethren were not in line with their great contemporary, who broadened inspection to include that entire corpus. The reason the Brethren used only the first books might be due to their religious aims with the encyclopaedia, which could only be grounded on a (theologically) sound (*ṣaḥīḥ*) method for discovering truth.

Other well known Aristotelian themes are reinterpreted by the Ikhwān: an example of this is the theory of the immovable mover, mentioned in Epistle 37,[39] and again in Epistle 40. In the former of

33 See *Rasā'il*, Epistle 7, vol. 1, p. 268.14–16.

34 See ibid., Epistle 27, vol. 3, p. 16.14–17; ibid., p. 30.6–8

35 See ibid., Epistle 30, vol. 3, p. 71.5–10.

36 See ibid., Epistle 27, vol. 3, p. 11.9–10.

37 See, e.g., ibid., Epistle 41, vol. 3, p. 371.14–17, where God's 'emanation' is added at the end.

38 See ibid., Epistle 13, vol. 1, pp. 427.14–428.3.

39 Translated by Albert R. Ricardo-Felipe, 'La "*Risāla fī māhiyyat al-ʿišq*" de las *Rasā'il Ikhwān al-Ṣafā*", *Anaquel de Estudios Árabes*, 6 (1995), pp. 185–207.

these, *orexis* is the basis of the movement imparted to the stars by the heavenly Universal Soul; in the latter, it takes on a religious aspect, given that its purpose is to confirm the love human beings feel for permanence (with God as the most beloved in that He is the everlasting cause of all beings) and their hatred for death.

Non-Aristotelian concepts are reworked in the *Rasā'il*, too. For instance, the Pythagorean and Ikhwānian terms respectively idea of 'the one' as the principle of numbers rather than a number itself is used to explain that God is the origin of beings but not a being Himself. Number is considered the best way to clarify the nature of God and His relation to created beings, and can be conceived of as a kind of 'divine writing', the Ikhwān say, which emphasises God's absolute distinction from other beings and presents Him as the sole cause of everything.[40]

Epistles 32 and 33 in the third section, which deal with rational beings as understood in Pythagorean and Ikhwānian terms respectively, demonstrate beyond doubt that the Ikhwān were true Muslim Pythagoreans. When they come to discuss human beings,[41] they adopt a numerological approach on the basis of the Pythagorean tenet that 'existing beings correspond to the nature of number'. This same idea can be observed on a larger scale in Epistle 40, where the whole of existence, including religion, is considered from a numerological perspective.

It might be noted at this juncture that we have not yet discussed Neoplatonism as such. This is because the Ikhwān's interpretation of Neoplatonism, merged with Neopythagoreanism, can be taken as their *own* philosophy rather than merely an outside source or influence. This seems to be borne out by the fact that the Ikhwān frequently used Neoplatonic emanationism to give a 'conceptual' reading of Creation. In so doing, the Ikhwān determine in an original fashion the link between ontology and gnosiology, often employed in Islamic

40 See *Rasā'il*, Epistle 40, vol. 3, p. 347.5. If God is like the unit 1, the Active Intellect is like the number 2, the Universal Soul like the number 3, and Nature (or Matter) like the number 4.

41 In Epistle 23, where humanity is considered as a *micropolis*; cf. Epistle 26, along the same lines.

philosophy to explain the 'mode' of the emanative process — that is, the 'mode' of creation.

Not only are the divine will and word (*kun*) substituted by knowledge, but the whole concept is stated once again in terms of a 'philosophy of light':

> Intellect is the light of God — may He be exalted — and His emanation that He first emanated; Soul is the light of Intellect and his emanation that the Creator emanated from him; Prime Matter is the shadow of Soul and abstract forms are the engravings, colours, and figures which Soul assembled in Matter through God's permission and help [*ta'yyīd*] to her through Intellect. All these objects are outside [*bilā*] time and space, but [they were given existence] by His saying: 'Be!' And it is, as He said: 'Our commandment is but one word, as the twinkling of an eye' [Qur'an, 54:50].[42] (*Rasā'il*, Epistle 40, vol. 3, p. 352.13–18)

There are recognisable echoes of the Ismaili approach of al-Sijistānī in the terminology and content in this passage. The Ikhwān's version of Neoplatonism is also likely to have been close to that of al-Nasafī.[43] The link with Ismailism is confirmed by the fact that the Brethren's formulations were to be accepted and reworked by a 'true' Ismaili philosopher, the *dā'ī* Ḥamīd al-Dīn al-Kirmānī.[44]

Religious Commitment

If, like the Ismailis, the Ikhwān adopted emanationism as the best explanation for the Creation of the world by God, then, like the Ismailis, they would have rejected the view that salvation is attainable through

42 Arthur J. Arberry's translation in *The Koran Interpreted* (Oxford: Oxford University Press, 1964), p. 556.

43 His work is not extant, but he is considered the first — and perhaps the most faithful — reviver of Neoplatonism in the Islamic world. See Paul E. Walker, *Early Philosophical Shiism: The Ismaili Neoplatonism of Abū Ya'qūb al-Sijistānī* (Cambridge: Cambridge University Press, 1993), pp. 55–60.

44 See Daniel De Smet, *La quiétude de l'intellect. Néoplatonisme et gnose ismaélienne dans l'œuvre de Ḥamīd al-Dīn al-Kirmānī (Xe/XIe S.)* (Leuven: Peeters, 1995).

philosophy alone. For, just as the *imitatio dei* concept of philosophy implies that the disciple must have not only an acute mind but also a pure heart through which God will be recognised as the sole and the supreme teacher in knowledge and deeds,[45] the Ikhwān also considered the prophetic messages, esoteric and exoteric, to be essential as the second means necessary for human salvation and happiness.

This brings us to the second aspect of the encyclopaedia to consider: not only were the Ikhwān philologists and historians of the ancient world, but they gave their work a religious orientation. This is proved mainly by their attitude towards the basic Islamic tenet of divine Creation. It is repeatedly and strongly confessed, appearing in Epistle 29 as a fascinating allegory destined for the masses.

But there are other interesting and perhaps less-explored elements through which the religious concerns of the Ikhwān can be discerned. Consider their system of epistemology, which the Ikhwān developed from ancient philosophy in pursuit of the specifically Muslim aim of explaining abstract knowledge and of reconciling the autonomy of that knowledge with Revelation and religious faith.[46]

Epistle 27 demonstrates another instance of the merging of Greek philosophy with Islam, whereby souls (which, according to the Ancients, are self-subsistent) unite themselves with human bodies in order to ensure salvation; hence wise men and prophets are said to be 'the physicians of souls'.[47] And in Epistle 28 there is a curious interpretation of the Aristotelian *medietas*,[48] also with reference to the acquisition of knowledge.

---

45 Cf. Qur'an 2:31, which is recalled in *Rasā'il*, Epistle 28, vol. 3, p. 18.14–16; ibid., Epistle 31, vol. 3, p. 112.19; ibid. p. 141.13–14. Cf. also ibid., Epistle 23, vol. 2, p. 381.17.

46 See Carmela Baffioni, 'From Sense Perception to the Vision of God: A Path Towards Knowledge According to the Ikhwān al-Ṣafā'', *Arabic Sciences and Philosophy*, 8 (1998), pp. 213–231; Carmela Baffioni, 'The Concept of Science and its Legitimation in the Ikhwān al-Ṣafā'', in *Religion versus Science in Islam: A Medieval and Modern Debate*, ed. C. Baffioni, *Oriente Moderno*, 19 (2000), pp. 427–441.

47 Cf. *Rasā'il*, Epistle 27, vol. 3, p. 13.7ff.

48 The thesis of Aristotle's 'doctrine of the mean' (from his *Nicomachean Ethics*) is that moral virtue (*aretê*) is a mean that is between the extreme excesses and defects of pleasure or action or temperament.

The central religious topic, however, is resurrection. The theme is introduced in Epistle 29 with the observation that there is a recognisable wisdom in death in that it permits the vision of God; hence the highest sorrow is damnation. Along the same lines, we read in Epistle 30 that earth is the real hell and paradise is heaven.[49] But the broadest approach to eschatology occurs in Epistle 38, where epistemology is also refined, in that knowledge of resurrection is said to be 'the heart of hearts [*lubb al-albāb*] and the secret of the Friends of God'.[50]

In Epistle 40, the Ikhwān frame Aristotle's statement that 'Nature does not make anything in vain' in a religious context, applying his teleological approach to the development of the world to the problem of theodicy. Although the Muslim faith does not allow God to be identified as the cause of evil, divine will and power are considered in respect of the existence of evil in the world. Ultimately, the Ikhwān assign responsibility for evil to human beings, or, rather, to creatures in general, and they do so from a Muʿtazilī perspective. Earlier in the corpus, in Epistle 19, a slightly different explanation for this problem is given: in the hierarchical taxonomy of the universe there are 'intermediate' beings to which everything that happens in the world may be attributed. God, a remote 'First Beloved', is beyond all this, like kings whose orders are behind everything built during their reigns (without causing evil deeds).[51]

From the foregoing, it is clear that revealed theology, or prophetism, constitutes a fundamental part of the Ikhwān's ideological system. Epistle 40, where knowledge is emphasised as the highest divine gift,

49 See *Rasāʾil*, Epistle 30, vol. 3, p. 68.4–6. It is interesting in this regard that the Ikhwān considered ignorance of an afterlife to be 'providential' (see ibid., Epistle 40, vol. 3, p. 370.13–19), but it is difficult to isolate a particular doctrine or time associated with this idea. Here the Ikhwān have diverged far from the Pythagorean idea of the body as the 'tomb' of the soul.

50 Ibid., Epistle 38, vol. 3, p. 288.3–5. Translated by Jean (Yahya) Michot, 'L'épître de la résurrection des Ikhwān al-Ṣafāʾ', *Bulletin de Philosophie Médiévale*, 16–17 (1974–1975), pp. 114–148.

51 See Carmela Baffioni, *Appunti per un'epistemologia profetica. L'Epistola degli Ikhwān al-Ṣafāʾ Sulle cause e gli effetti* (Naples: Università degli Studi di Napoli "L'Orientale", Dipartimento di Studi e ricerche su Africa e Paesi Arabi-Guida, 2006), pp. 125–128.

is especially meaningful in this regard.[52] Consequently, cosmogony is recognised as the highest attainment of knowledge; hence the Ancients are defended against the charge of eternalism.[53] Philosophy as *imitatio dei* remains one of the main guiding principles of the Ikhwān's system, given that guidance by divine revelation is equally emphasised. The letters placed at the beginning of twenty-nine Qur'anic suras are mentioned, whose meaning can be known only by 'the pure'.

This epistle is rich in Ismaili themes, both familiar ones, such as ontological and epistemological hierarchies, the 'metaphysics of light', astral cycles and the ending of the world, the links between earth and heaven, and less familiar ones, such as the myth of the soul's fall, salvation, and return to God. There are also influences from al-Sijistāni; for example, the Ikhwān's approach to divine law in terms of numerology, the reciprocal assistance of souls in terms of *ta'yyīd* ('divine help'), and the comparison of the divine establishment of reality (*ibdāʿ*, meaning literally 'extra-temporal innovation') with language.[54] Above all, Ismaili *taʿlīm* finds here a special development founded on the hadith: 'Wise men and the learned are the heirs of the prophets'.[55] So even if its main topic — causality — is Aristotelian, Epistle 40 is a 'programmatic' treatise: it provides the conceptual basis of the whole system of the Ikhwān, and might thus have made a suitable opening for the third section of the *Rasāʾil.*

Political Concerns and Ideological Orientation

We cannot consider here whether the *Rasāʾil* constituted an instance of Shiʿi support of the Muʿtazila. However, the fusion of Ismaili and

52 This results from the Ikhwān's reading of the Qur'an (3:7) which precedes the much better known interpretation of Ibn Rushd (Averroes) by two centuries; see *Rasāʾil*, Epistle 40, vol. 3, pp. 344.10–345.5.

53 Cf. ibid., p. 356.11–18.

54 See Carmela Baffioni, 'The "Language of the Prophet" in the Ikhwān al-Ṣafāʾ', in *Al-kitāb: La sacralité du texte dans le monde de l'Islam*, ed. Daniel De Smet, Godefroid de Callataÿ, and Jan van Reeth (Brussels, Louvain-la-Neuve, and Leuven: Brepols [Acta Orientalia Belgica Subsidia III], 2004), pp. 357–370.

55 See Kalābāḏī, *Il sufismo nelle parole degli antichi*, ed. and trans. P. Urizzi (Palermo: Officina di Studi Medievali, 2002), p. 21, note 58.

Muʿtazilī ideas therein leads us directly to the Ikhwān's political theories.

The authors were deeply involved in political debates of the Islamic world. According to them, religious creeds change as a result of transformations in language, a particular case of development of knowledge. The learned debate on the identity of the Messenger's deputy had been the main cause of divergence within the *umma* up to their time.[56]

The Ikhwān'a approach to the history of early Islam is made from a Shiʿi perspective. Certain important events of the Prophet Muhammad's life and mission are recalled, and attention soon shifts to two fatal events that led to the deposition of Shiʿism: the battle of Ṣiffīn and the slaughter of Karbalāʾ.[57]

Confirmation of the Ikhwān's Shiʿi orientation can be seen in their discussion of the Imamate. This brings us to the other 'programmatic' treatise of the encyclopaedia: Epistle 42, which opens the fourth section. Here, the Ikhwān do not confine themselves to epistemological expositions, but also develop their apologetic. First, they list the doctrines against which Islam must fight; then, they approach Islam itself.

The Ikhwān were in agreement with the Shiʿi tradition that after the Prophet's death an Imam had to be appointed to preserve the law, revive the tradition, and *encourage good and forbid evil*;[58] for such activities as levying land tax, collecting tithes and the *jizya*, maintaining harbours, fortifying countries, subduing enemies, protecting roads, and defending the weak, other people were appointed as his deputies in Islamic countries. Jurists and learned men were to address the Imam on religious matters as he would be responsible for establishing such undertakings as prayers, rituals and feasts, pilgrimage, wars, and the appointment of judges.[59]

To this day, everyone agrees on the tasks of the Imam, but there is disagreement as to his identity because *imāma* is of two kinds: prophetic and regal. Usually, the tasks of the king are clearly distinguished

56 See *Rasāʾil*, Epistle 31, vol. 3, p. 153.8–10; ibid., p. 165.8–12.

57 See ibid., Epistle 44, vol. 4, p. 17.13–16; ibid., p. 33.2–6.

58 This is one of the quotations of the fifth Muʿtazilī *aṣl* ('principle') in the encyclopaedia (see, e.g., *Rasāʾil*, Epistle 7, vol. 1, p. 273.23; ibid., Epistle 42, vol. 3, p. 493.13–14).

59 See ibid., Epistle 42, vol. 3, pp. 493.12–494.2.

from those of the prophet, because kingship is a mundane activity, whereas prophecy is related to the spiritual.[60] However, sometimes these qualities are combined in a single person, who is then the delegated prophet and also the king.[61] The fact that men in whom kingship and prophecy are united do not crave after worldly things is proof of God's tenderness towards His community.[62] The Prophet Muhammad was both prophet and king of the Muslim *umma*, thus ensuring its best defence, but his successors did not always match him in nobility: a prophet requires 'about forty qualities', whereas a king needs qualities different from these.[63]

Yet prophethood and caliphate (or kingship) are sometimes found in two mutually supporting people, one of whom is the prophet delegated to a community and the other, the person who has been given political power over it.[64] The real problem seems to be the dignity and piety of the person appointed caliph, which alone is the condition for legitimate succession. But the Ikhwān's own ideas regarding the identity of such a person are not entirely clear, and for them there is apparently no distinction between prophet, Imam, or king with regard to the qualities required in a ruler. The Ikhwān speak of a 'Lawgiver', who might be identified from time to time with all of these figures. Epistle 47 ('On the Essence of the Divine Law') does not clarify the matter either, in that it deals with the 'Lawgiver' in three different senses: prophet, Imam, and king.[65] The sole 'doctrinal' solution offered in the *Rasā'il* seems to be that a king with ascetic tendencies can legitimately take the place of the Imam.

60 Ibid., p. 497, 5–6.
61 Ibid., p. 495.18–19.
62 Ibid., p. 497.10–22 passim.
63 Ibid., lines 2–3.
64 Ibid., p. 495.19–21.
65 The Ikhwān usually use for 'Lawgiver' the name '*waḍi' al-sharī'a*' (which designates a 'prophet'), and sometimes also the name '*ṣāḥib al-sharī'a*' (*Rasā'il*, Epistle 47, vol. 4, p. 136.15; ibid. p. 137.20). Themes proper to the role of the Imam rather than to that of the prophet are addressed, for instance, in the *ta'wīl* or the *da'wa ilā Allāh*, though they are referred to the '*waḍi al-sharī'a*' (ibid., p. 130.18; ibid., p. 131.21–22). If '*ṣāḥib al-sharī'a*' could actually fit the Imam better, it is also true that such contexts refer to the *waḍi al-sharī'a*. A more complete examination of the issue has to be postponed for the time being.

In political terms, however, there are further elements that allow better definition of the Ikhwān's position. In a passage on the struggles for Imamate, the 'representatives of the Muḥammadī law and of the Hāshimī federation',[66] are opposed; if the latter were the 'Abbāsids or even the supporters of Abū Hāshim, son of Muḥammad ibn al-Ḥanafiyya, the former should be seen as the direct descendants of the Prophet through Fāṭima, which accounts for the Ikhwān's unwavering 'Alid devotion.[67]

A long passage from Epistle 31 leads me to suppose that the authors' target here was the caliph al-Ma'mūn, despite his pro-'Alid policy and political support for Mu'tazilism. According to the Ikhwān's report, he encouraged political and religious controversies between *'ulamā'* of different persuasions, claiming for himself the right to interpret the law on the basis of the doctrine of the created Qur'an. To him, the *ahl bayt al-nubuwwa* are opposed as the guarantors of the true tradition concerning prophetic succession. The Ikhwān could then still be identified with the supporters of the strict 'Alid conception of the *imāma*, according to which the science of (and the eligibility for) the caliphate are proper to the family of the Prophet only.

In line with this hypothesis, Epistle 31 contains the core not only of the Ikhwān's political thought but also of their Ismaili commitment. Developing this idea further, to the Ikhwān al-Ṣafā' themselves I would ascribe the function of being the Imam's helpers in ruling the Muslim *umma*.

In the *Rasā'il* a *madīna fāḍila* is featured, less known but no less worthy than that proposed by al-Fārābī.[68] On this occasion, the Ikhwān did not have recourse to the ancient Platonic heritage, but openly adapted a distinctly Greek idea to a Muslim situation.[69]

66 *Rasā'il*, Epistle 31, vol. 3, p. 165.11–12.

67 See Carmela Baffioni, 'Ideological Debate and Political Encounter in the *Ikhwān al-Ṣafā'*', in *Mağāz. Culture e contatti nell'area del Mediterraneo. Il ruolo dell'Islam*, ed. A. Pellitteri (Palermo: Università di Palermo, Facoltà di Lettere e Filosofia, 2003), pp. 33–41.

68 See Carmela Baffioni, '*Al-madīnah al-fāḍilah* in al-Fārābī and in the Ikhwān al-Ṣafā': A Comparison', in *Studies in Arabic and Islam*, ed. S. Leder et al. (Leuven: Peeters, 2002), pp. 3–12.

69 See Carmela Baffioni, 'The "General Policy" of the Ikhwān al-Ṣafā': Plato and Aristotle Restated', in *Words, Texts and Concepts Cruising the Mediterranean*

Most scholars seem to conceive of this 'perfect city' as being as *utopian* as al-Fārābī's.[70] In introducing mutual love as both the basis and the goal of this community, Epistle 45 helps to ascertain whether the Ikhwān had in mind a spiritual or true governmental rule such as that of the Fāṭimid caliph, whom they might have served as his *duʿāt*.[71] In describing their 'perfect city', the Ikhwān give details that could refer to an actual location; on the other hand, they often foreshadow the defeat of evil people.[72] In Epistle 4, they even present a cyclical conception of the alternation of ruling dynasties, and say that 'the dynasty of Evil reached its apex', with the result that only decline and diminution can be expected in the future.[73] The dynasty of the Good begins when the learned agree 'on a unique school and a sole religion', and also when they commit themselves to mutual help as well as desiring the vision of God, and so act according to His Will as their reward.[74] These are just the qualities combined in the king-Imam, and the Ikhwān introduce themselves as the wise men involved in his rule, endowed with qualities analogous to those mentioned in Epistle 9.[75]

Sea: Studies on the Sources, Contents and Influences of Islamic Civilization and Arabic Philosophy and Science, ed. A. Arnzen and J. Thielmann (Leuven, Paris, and Dudley, MA: Peeters, 2004), pp. 575–592.

70 See, e.g., Yves Marquet, *La philosophie des Iḫwân aṣ-Ṣafâ': L'imâm et la société* (Dakar: Université de Dakar, Faculté des Lettres et Sciences Humaines, Département d'arabe [Travaux et documents no. 1], 1973), p. 153; or, more recently, Abbas Hamdani, 'Brethren of Purity, a Secret Society for the Establishment of the Fāṭimid Caliphate: New Evidence for the Early Dating of their Encyclopaedia', in *L'Égypte fatimide: Son art et son histoire*, ed. M. Barrucand (Paris: Presses de l'Université de Paris-Sorbonne, 1999), pp. 73–82, esp. p. 81.

71 See Carmela Baffioni, 'Temporal and Religious Connotations of the "Regal Policy" in the Ikhwān al-Ṣafā'', in *The Greek Strand in Islamic Political Thought*, ed. E. Gannagé et al., *Mélanges de l'Université Saint-Joseph*, 57 (2004), pp. 337–365.

72 See, e.g., *Rasā'il*, Epistle 31, vol. 3, pp. 154.19–24, 156.2–3.

73 According to ʿAbd al-Laṭīf Ṭībāwī, the Ikhwān here foreshadow the fall of the ʿAbbāsids. See ʿAbd al-Laṭīf Ṭībāwī, 'Ikhwān aṣ-Ṣafā' and their *Rasā'il*: A Critical Review of a Century and a Half of Research', *Islamic Quarterly*, 2 (1955), p. 37, note 4. See also ʿAbd al-Laṭīf Ṭībāwī, 'Further Studies on Ikhwān aṣ-Ṣafā'', *Islamic Quarterly*, 20–22 (1978), p. 60.

74 *Rasā'il*, Epistle 4, vol. 1, pp. 181.14–182.2.

75 It is worth noting that in Epistle 42 the Ikhwān regard the question of the *imāma* as still unresolved (*Rasā'il*, Epistle 42, vol. 3, p. 498.20–22). The Brethren who co-operated for the final version of the corpus may have been, in broader terms, disguised adversaries of the Būyids. I have given a more detailed exposition on

From Epistle 43 onwards, they construct their own system in terms of religion. Well-known topics are focused on more closely — or, rather, are reconciled with the overall general theme of the *Rasāʾil*; this is the case with, for instance, asceticism, which is again emphasised in Epistle 43.[76] A 'history of religions' from the time of the *jāhiliyya* (the pre-Islamic era of 'ignorance') onwards is sketched out in Epistle 44, but the apex is reached in the description of ancient doctrines confirming the immortality of the soul, of which the strongest example is the memory of Shiʿi martyrdom. Epistle 46 reconfirms faith as the second condition for salvation; use of the term *daʿwa* in the title and, above all, the contents of Epistle 48 might further support the hypothesis of a project for militancy by the Ikhwān.[77] Epistle 49 is perhaps the most complete representation of a hierarchical structure of the universe, this time with references to 'angelic entities', as we have seen. The issue considered in Epistle 50: 'On the Kinds of Political Administration' is in line with the sketch of knowledge given in Epistle 7; but the 'ecumenism' displayed in reporting religious rites of different origins needs to be reconciled with the Fāṭimid experience, which, as we know, supported a very particular form of ecumenism. Epistle 51 duly gathers up the threads of the whole, and Epistle 52 deals with magic.

Consideration of such a religious system might, consequently, show that even if we are not yet fully certain of their ideological identity, the Ikhwān's spiritual legacy can hardly be charged with inconsistency. After all, their reworking of the 'foreign' heritage, which was also accurately reproduced in terms of philology, was anything but inconsistent. In the ninth and tenth centuries their opus represented the only way of supporting the notion of a dual source of knowledge, based on revelation *and* reason.

---

these issues in 'History, Language and Ideology in the Ikhwān al-Ṣafāʾ View of the Imāmate', in *Authority, Privacy and Public Order in Islam*, ed. B. Michalak-Pikulska and A. Pikulski (Leuven: Peeters, 2006), pp. 17–27. See also Muḥammad Jalūb Farhān, 'Philosophy of Mathematics of Ikhwan al-Safa [sic]', *Journal of Islamic Science* [Aligarh], 15, 1–2 (1999), pp. 25–53, esp. pp. 30–31.

76 Translated by D. H. Yūsufjī, 'The Forty-Third Treatise of the Ikhwān al-Ṣafāʾ', *Muslim World*, 33 (1943), pp. 39–49.

77 Addressing this matter must be postponed for another occasion, and may perhaps be elucidated by the forthcoming Arabic edition of this epistle, which is part of the series which the present volume initiates.

FIVE

The Rasā'il Ikhwān al-Ṣafā' *in the History of Ideas in Islam*

Ian Richard Netton

Introduction

The *Rasā'il* (*Epistles*) of the Ikhwān al-Ṣafā' inhabits a number of universes of discourse. In the first place, these fifty-two epistles constitute an encyclopaedia in the true sense of the word. Elsewhere I have noted: 'Their writings, presented in the form of epistles (*rasā'il*) are frequently complicated, repetitive and, at the same time, impressively encyclopaedic. Their subject matter is vast and ranges from mathematics, music and logic, through mineralogy, botany and embryology, to philosophical and theological topics which are concluded by a treatise on magic.'[1]

The division of knowledge is 'into four main sections, comprising fourteen *Rasā'il* on Mathematical Sciences, seventeen on Natural Sciences, ten on Psychological and Rational Sciences, and eleven on Theological Sciences'.[2]

While the structure of the Arabic text of the *Rasā'il* may not always

1 Ian Richard Netton, *Muslim Neoplatonists: An Introduction to the Thought of the Brethren of Purity (Ikhwān al-Ṣafā')* (1982; repr., London: Routledge Curzon, 2002), p. 1.

2 Netton, *Muslim Neoplatonists*, p. 2. The most accessible Arabic edition, which

resemble that of other great encyclopaedias, it is clear from a comparative glance that its intentions, at least, form part of a 'great tradition', to deploy a phrase beloved of the literary critic F. R. Leavis. It is true, furthermore, that a theological intent (such as is to be found in the *Rasāʾil*) will be lacking from a great many other encyclopaedic works. Nonetheless, while the latter may not share the Ikhwan's *daʿwa* ('mission') from a theological perspective, they will, universally, share an intent to inform. David Crystal, in his single-volume *Penguin Encyclopedia*, observed: '*The Penguin Encyclopedia* provides a succinct, systematic and readable guide to the facts, events, issues, beliefs and achievements which make up the sum of human knowledge.'[3] He adds: 'Internationalism is a major focus of *The Penguin Encyclopedia*.'[4] It will be readily apparent that many of these features are shared by the *Rasāʾil*. It is true that each 'entry', or 'epistle' (*risāla*), lacks succinctness, but all are informed by a desire to cover major and minor issues of both local and international significance. All have a fully epistemological dimension. And none of this is invalidated by the idea, first advanced by Abdul Latif Tibawi, and accepted by the present author as entirely plausible, that 'the *Rasāʾil* were the product of meetings (*majālis*) convened by the Brethren for the purpose of philosophical discussion. One author has aptly likened their content to the draft of deliberations by a learned society composed by a well-educated secretary, and this could be very close to the truth: the authors of the *Rasāʾil* insist that their Brethren hold special meetings at set times, to which none but they are to be admitted, and where their secrets and esoteric knowledge can be discussed in peace. Elsewhere it is suggested that such a meeting should take place every twelve days.'[5]

Even if the *telos*, or goal, was not a formal encyclopaedia, there

will be cited hereafter, is: Ikhwān al-Ṣafāʾ, *Rasāʾil Ikhwān al-Ṣafāʾ*, ed. Buṭrus Bustānī, 4 vols. (Beirut: Dār Ṣādir, 1957).

3 David Crystal, ed., *The Penguin Encyclopedia*, 2nd ed. (2002; repr., London: Penguin Books, 2004), p. vii.

4 Ibid.

5 Netton, *Muslim Neoplatonists*, p. 3; Abdul Latif Tibawi, 'Ikhwān aṣ-Ṣafāʾ and their *Rasāʾil*: A Critical Review of a Century and a Half of Research', *Islamic Quarterly*, 2 (1955), p. 37; *Rasāʾil*, vol. 4, pp. 41, 168; Ikhwān al-Ṣafāʾ, *al-Risāla al-jāmiʿa*, ed. Jamīl Ṣalībā, 2 vols. (Damascus: Maṭbaʿat al-Taraqqī, 1949–1951), vol. 2, p. 395.

can be few neater ways of thrashing out the content of such a volume than by means of their *majālis*, which sound much akin to modern university seminars. Such a series of meetings, of course, would also help to account for some of the contradictions which we find within the text.

As I indicated earlier, the *Rasā'il* forms part of a great tradition down the ages, which also includes the systematic works of Aristotle, particularly his biological and zoological works; *A Dictionary of the English Language* by Dr Samuel Johnson (1709–1784); the *Encyclopédie* of Denis Diderot (1713–1784), which appeared in France between 1751–1776; and the famous *Encyclopaedia Britannica*, the first edition of which came out between 1768–1771. It is, of course, true that works of lexicography differ from works of encyclopaedism but all are united by an epistemological passion for classification and information. In this respect, as in others, the *Rasā'il* is a worthy ancestor of Diderot, Johnson and the *Britannica*.[6]

Secondly, as well as belonging to the universe of encyclopaedism, the *Rasā'il* belongs to that of *adab* ('*belles lettres*'). The Ikhwān attemped to propagate their views and their theology, not just in a didactic, schoolmasterly fashion, but in an entertaining, anecdotal fashion through the medium of storytelling. Such stories are drawn from sources as diverse as Qur'anic prophetology to others of Greek, Persian, Indian and Buddhist provenance. It is an eclectic range whose each individual element aims to tell a story well in order to underline a point or sustain a moral argument.[7]

In the *Rasā'il* there is a passion then, not just for classification but for narrative as well, which many of the other great mediaeval Arab *belletrists* — like Abū 'Uthmān 'Amr ibn Baḥr al-Jāḥiẓ (ca. 776–868/869) and Yāqūt ibn 'Abd Allāh al-Ḥamawī (1179–1229), with his magisterial surveys of places and authors — would have recognised.[8]

---

6 See Crystal, *The Penguin Encyclopedia*, for relevant entries.

7 See Netton, *Muslim Neoplatonists*, pp. 78–94.

8 See Yāqūt ibn 'Abd Allāh al-Ḥamawī, *Mu'jam al-buldān*, 10 vols. (Cairo, 1906–1907), vol. 5; Yāqūt, *Mu'jam al-udabā'*, ed. Iḥsān 'Abbās, (Beirut: Dār al-Gharb al-Islāmī, 1993).

Plato

It is with Plato that we discover the concept of the *ideai* ('forms'), or *eidê*.[9] Although this was not an epistemological concept which held great appeal for the Ikhwān, it is clear that the idea of archetypes was much more attractive for certain other Islamic philosophers, most notably the Shaykh al-Ishrāq (meaning 'Master of Illumination'), Shihāb al-Dīn Abū al-Futūḥ Yaḥyā Ḥabash ibn Amīrak al-Suhrawardī (1153–1191).[10]

However, the Ikhwān did not make Plato's doctrine of *ideai* part of their intellectual apparatus. They were aware of it, as the following quotation amply demonstrates:

> Another said: the various kinds of animals in this world are only pictures and images of those forms and creatures in the world of the spheres and the compass of the heavens, just as the paintings and pictures which appear on the surfaces of walls and ceilings are pictures and images of the forms of these animals made of real flesh. The relationship of beings made of flesh to those creatures with pure essences is like the relationship of those painted, embellished pictures to those flesh and blood animals.[11]

But this very Platonic passage is left without any commentary or remark which might indicate, here at least, that the Ikhwān shared such sentiments. A similar statement is articulated by a *jinnī* philosopher during the 'Great Debate of the Animals and Man' in Epistle 22, but again the Ikhwān do not seize the opportunity to state their views.[12]

9 See Nicholas P. White, *A Companion to Plato's Republic* (Indianapolis, IN and Cambridge, MA: Hackett Publishing, 1979), pp. 31ff; J. C. B. Gosling, *Plato*, The Arguments of the Philosophers (London: Routledge and Kegan Paul, 1983), pp. 140ff. White cites Plato's *Republic*, (sections) 476–480, 523–525 for 'Plato's initial notion of a Form' (White, *Companion*, p. 31).

10 See Ian Richard Netton, *Allāh Transcendent: Studies in the Structure and Semiotics of Islamic Philosophy, Theology and Cosmology* (Richmond: Curzon Press, 1994), esp. pp. 256–268.

11 *Rasāʾil*, vol. 1, p. 238, trans. in Netton, *Muslim Neoplatonists*, p. 18.

12 *Rasāʾil*, vol. 2, p. 276, Netton, *Muslim Neoplatonists*, p. 18.

Plato, then, provides heroes like Socrates whom the Ikhwān admire;[13] yet the latter's epistemology and theology diverge. Plato's *ideai* are not the ideas which we will pursue in what follows, as we trace and survey the *Rasā'il* in the history of ideas in Islam. We will examine the *Rasā'il* according to five 'universes of knowledge'. These are by no means intended to be exclusive nor to signal that they are the only headings according to which the *Rasā'il* could be analysed or discussed. It is hoped, however, that the five chosen headings will illustrate the sheer diversity, eclecticism and encyclopaedism of the *Rasā'il*, whilst covering the most important areas of discourse within that extensive text.

Our five ideas are as follows: (1) The Idea of God; (2) The Ideas of Emanation and Creation; (3) The Idea of Man; (4) The Idea of Knowledge; and (5) The Idea of Salvation.

The Idea of God

The idea of deity has been variously conceived, imagined and articulated by mankind down the ages. It has ranged from Aristotle's Unmoved First Mover to the incarnate God of mainstream Christianity. It has embraced the pure monotheism of the three traditional Abrahamic faiths of Judaism, Christianity, and Islam, as well as the Hindu counterpoint of monotheism *and* polytheism, which should not be unduly simplified.[14] The religion of ancient Egypt was also rather less simple than it might appear at first sight, even if one discounts the theology of the monotheist Akhenaten (r. 1350–1334 BCE) as an aberration.[15] With the eighteenth dynasty, Amenhotep IV succeeded Amenhotep III (r. 1386–1349 BCE), and his change of name, focused on the Aten (the sun disc) — his new name meant 'servant of the Aten'— together with his

13 See *Rasā'il*, vol. 4, pp. 34, 58, 73, 271; refer also to Plato, *Phaedo*, (sections) 115-118.

14 See Karel Werner, *A Popular Dictionary of Hinduism* (Richmond: Curzon Press, 1994), especially 'God' and 'Gods', pp. 69–71. I am indebted to Professor Kim Knott of the Department of Theology and Religious Studies, University of Leeds, for warning me against undue simplification of Hindu theology and pointing out that Hinduism is both monotheistic and polytheistic.

15 See Erik Hornung, *Conceptions of God in Ancient Egypt: The One and the Many*, trans. John Baines (London, Melbourne, and Henley: Routledge and Kegan Paul, 1983), esp. pp. 244–250.

inauguration of a cult of sun worship, has led scholars to query whether he was really a heretic within a polytheistic milieu, a real monotheist, or just a 'religious maniac'.[16] Yet in the reign of Akhenaten, Hornung reminds us that 'syncretism too was very much alive'.[17]

In sum, the examples of both Hinduism and ancient Egypt, not to mention that of ancient Greece, are a reminder that our sometimes simplistic classifications of world religions into neat boxes labelled 'polytheistic' or 'monotheistic' are less than helpful when studying an individual nation's attempt to make sense of the origins of the universe and mankind.

And it is the term 'syncretism' which can be of particular help in evaluating the complex theology of the Ikhwān al-Ṣafā', and surveying it against the theologies of their contemporaries and those who came before and after them, like Abū Yūsuf Yaʿqūb ibn Isḥāq al-Kindī (d. after 866), Abū Naṣr Muḥammad ibn Muḥammad ibn Tarkhān ibn Awzalagh al-Fārābī (870–950), Abū ʿAlī al-Ḥusayn ibn Sīnā (979–1037), and the great 'Master of Illumination', al-Suhrawardī (1153–1191).[18]

Firstly, however, it is useful to survey and elaborate what I have termed elsewhere 'the Qur'anic Creator Paradigm'. For Islam, 'the Qur'anic Creator Paradigm embraces a God who (1) creates *ex nihilo*; (2) acts definitively in historical time; (3) guides his people in such time; and (4) can in some way be known indirectly by His creation'.[19]

Examples of all these aspects of this paradigm are not difficult to find: 'God's creative activity is His *leitmotiv* par excellence throughout the entire text of the Qur'ān.'[20] God operates upon and within human history, sending the last revelation via His 'Seal of the Prophets', Muhammad, over a period of years and, to give another example, sending angels to fight on the side of Muhammad at that first key battle

16 Peter A. Clayton, *Chronicle of the Pharaohs: The Reign-by-Reign Record of the Rulers and Dynasties of Ancient Egypt* (London: Thames and Hudson, 1994), pp. 99–121.
17 Hornung, *Conceptions of God*, p. 245.
18 See the relevant chapters in Netton, *Allāh Transcendent*.
19 Ibid., p. 22.
20 Ibid., p. 23.

between the Prophet and the Meccans, the Battle of Badr (624).[21] All this is recorded in the Qur'an. Elsewhere the sacred text lauds God's guidance again and again.[22] Finally, despite His sublime and unimaginable transcendence, God has revealed Himself to mankind in certain ways and can thereby be recognised: '*Sa-nurīhim āyātinā fī al-āfāq wa fī anfusihim* [We shall show them Our signs on the horizons and deep within themselves].'[23]

This, then, is the mainstream paradigm. How do the Ikhwān diverge from this and from the theological models of divinity proposed both by their contemporaries and successors? As I suggested earlier, syncretism is the key; this is neatly illustrated in the following:

> In one of the many didactic stories which appear in the *Rasā'il Ikhwān al-Ṣafa'*, two men from India, one blind and the other crippled, enter a garden: the owner has pity on them and allows them to take their fill. However, they become greedy and plunder the garden in the owner's absence, the cripple mounted on the shoulders of the blind man. They are discovered, forgiven once by the supervisor of the garden but, on doing the same thing again, they are expelled from the garden and cast into the desert, in the Ikhwān's words, 'as was done with Adam and Eve, peace be upon them, when they tasted the tree'. An elaborate exegesis of this heavily symbolic tale is provided by the Ikhwān: the body is the blind man and the soul is the cripple. The body is led where the soul wishes. The garden is the world whose owner is God while the garden's fruits are the good things of this world. The supervisor or warden [*al-nāṭūr*] of the garden is the Intellect [*al-ʿAql*, i.e., the Universal Intellect].
>
> This parable is a typical example of the syncretic nature of the thought of the Ikhwān al-Ṣafā' and, indeed, it is no exaggeration to say that syncretism is a keynote of the *Rasā'il* [as well as being behind much of the history of Ideas in that multifaceted text]. The parable is set in India but the symbolism of the body and the soul in the *tafsīr* could be labelled Platonic in inspiration. The ethics of the parable are also

21 Ibid., pp. 23-24; see Qur'an 8:9, 8:17, 8:45–46.

22 Qur'an 92:12, 6:70.

23 Qur'an 41:53.

> Judaeo-Christian in their emphasis on charity, forgiveness and final damnation. The whole is given a further Neoplatonic dimension by the introduction of the Intellect as the warden in that garden. Finally, the parable is not unique to the *Rasā'il*: it is to be found in the Talmud where two watchmen, one blind and one lame, combine to raid the figs of the king's orchard. They are therefore jointly judged as one by the owner.[24]

Syncretism is the key to unlocking the complexities of the Ikhwān's Idea of God which is, by turns, Qur'anic and Neoplatonic. We shall explore this view more deeply in the next section which follows.

The Ideas of Emanation and Creation

It is to Plotinus (ca. 204/205–270) and the complex series of writings which he produced under the title of the *Enneads*, that we must turn for our primary understanding of the theory of emanation.[25] In Qur'anic Islam, God creates *ex nihilo*:

> It is He who hath created for you
> All things that are on earth;
> Moreover His design comprehended the heavens,
> For He gave order and perfection
> To the seven firmaments;
> And of all things
> He hath perfect knowledge.[26]

24 Ian Richard Netton, 'Foreign Influences and Recurring Ismā'īlī Motifs in the *Rasā'il* of the Brethren of Purity', *Convegno sugli Ikhwān aṣ-Ṣafā', Roma, 1979* (Rome: Accademia Nazionale dei Lincei, 1981), pp. 49–50. For the original texts, see *Rasā'il*, vol. 3, pp. 156–160; *Hebrew-English Edition of the Babylonian Talmud*, ed. Isidore Epstein (London: Soncino Press, 1969), Sanhedrin 91a–b.

25 Plotinus, *Enneads*, ed. and trans. A. H. Armstrong, 7 vols. The Loeb Classical Library (Cambridge, MA: Harvard University Press, 1966–1988). See also Plotinus, *The Enneads*, trans. Stephen MacKenna, 2nd rev. ed. (London: Faber and Faber, 1956); Plotinus, *Plotini Opera*, ed. Paul Henry and Hans-Rudolph Schwyzer (Paris: Desclée de Brouwer and Brussels: Edition Universelle, 1951–1959); Lloyd P. Gerson, *Plotinus*, The Arguments of the Philosophers (London and New York: Routledge, 1994); Dominic O'Meara, *Plotinus: An Introduction to the Enneads* (Oxford: Clarendon Press, 1993).

26 Qur'an 11:29, *The Holy Qur'an: Text, Translation and Commentary*, trans.

To Him is due
The primal origin
Of the heavens and the earth:
When He decreeth a matter,
He saith to it: 'Be',
And it is.[27]

In Plotinus' *Enneads*, all things *emanate* from the One:

> Resembling the One thus, Intellect produces in the same way, pouring forth a multiple power – this is a likeness of it – just as that which was before it poured it forth. This activity springing from the substance of Intellect is Soul, which comes to be this while Intellect abides unchanged: for Intellect too comes into being while that which is before it abides unchanged.[28]

And in later classical Neoplatonism, the hypostases multiply beyond the 'simple' Plotinian triad of the One, Intellect, and Soul. For example, if we turn to Proclus Diadochus (410–485) and his influential *Elements of Theology* (*Stoikheiôsis Theologikê*), we find that Plotinian metaphysics, by comparison with those of Proclus, 'have an endearing simplicity!'[29]

As I have elaborated elsewhere: 'Proclus introduced or, at least, employed the terms "*henad*" and "*monad*".' The first indicated 'a class of participated forms of the One which proceed from it and are present primarily in Intellect but also in each hypostasis below the One and all the processions of each hypostasis';[30] the second term 'was normally reserved for the defining term or "leader" and so normally the "imparticipable" of any order or series'.[31] 'Thus four major types of

Abdullah Yusuf Ali (Kuwait: Dhāt al-Salāsil, 1984), p. 29.

27 Qur'an 11:117, trans. Yusuf Ali, pp. 49–50.

28 Plotinus, *Ennead V*, trans. A. H. Armstrong (1984; repr., Cambridge, MA and London: Harvard University Press, 1994), 2.1, pp. 58–61.

29 Netton, *Allāh Transcendent*, p. 10.

30 A. C. Lloyd, 'The Later Neoplatonists', in *The Cambridge History of Later Greek and Early Medieval Philosophy*, ed. A. H. Armstrong (Cambridge: Cambridge University Press, 1970), p. 307.

31 Ibid.

henad derive from the One: Intelligible, Intellectual, Supercosmic, and Intracosmic. These correspond to various types of Being and Life.'[32]

When we turn to the Ikhwān's own 'Hierarchy of Being', and set it against what we have outlined above, the role of syncretism becomes abundantly clear. No longer are we confronted by the classically simple triad of the One, Intellect, and Soul within which the divine mechanism of emanation operates according to the parameters outlined, albeit very laboriously, in the *Enneads*. It is true that the Ikhwān do not confront us with a universe or hierarchy of being composed of *monads* and *henads*. But their participation in the classical Qur'anic Creator Paradigm,[33] outlined above, is paralleled, without any attempt at harmonisation, by a Neoplatonic universe of discourse in which nine members or levels of Being have prominence and in which emanation is the primary motor.

In my volume *Muslim Neoplatonists*, I describe this process as follows: '[This universe] comprised nine members or levels of being: the Creator [*al-Bārī*], the Intellect [*al-'Aql*], the Soul [*al-Nafs*], Prime Matter [*al-Hayūlā'l-Ūlā*], Nature [*al-Tabī'a*], the Absolute Body [*al-Jism al-Muṭlaq*], the Sphere [*al-Falak*], the Four Elements [*al-Arkān*], and the Beings which live in this world [*al-Muwalladāt*], divided among the mineral, plant and animal kingdoms.'[34]

This syncretic view of divinity fits neatly into the history of ideas in Islam if we assess the way in which Neoplatonic thought developed generally in the Islamic Middle Ages. Whereas al-Kindī works mainly with a Qur'anic Creator Deity, with a few tinges of Neoplatonism, al-Fārābī and Ibn Sīnā espouse a fully-fledged emanationist hierarchy of Ten Intellects, while the Shaykh al-Ishrāq, Shihāb al-Dīn al-Suhrawardī, presents us with an emanationist hierarchy of multiple lights headed by a divinity characterised as *Nūr al-anwār* ('the Light of Lights').[35] There

32 See R. T. Wallis, *Neoplatonism* (London: Duckworth, 1972), pp. 151–152; Netton, *Allāh Transcendent*, pp. 10–11; Proclus, *The Elements of Theology*, (bilingual Greek–English text) ed. and trans. E. R. Dodds (Oxford: Clarendon Press, 1933; 2nd ed. 1963; repr. 1992); Lucas Siorvanes, *Proclus: Neoplatonic Philosophy and Science* (Edinburgh: Edinburgh University Press, 1996).

33 See Netton, *Muslim Neoplatonists*, pp. 78–83.

34 Ibid., p. 35; see *Rasā'il*, vol. 3, pp. 56, 181–182, 184, 285.

35 See Netton, *Allāh Transcendent* for a development of the theme of emanation

is a gradual development from the unitary model such as is to be found classically in the Qur'an, to a Neoplatonic model with multiple facets in which the Universal Intellect, or diverse Intellects, and the Universal Soul play a prominent role. Frequently, what sets the thought of the Ikhwān al-Ṣafā' aside from some of this, and gives it individuality, is its syncretism, particularly evident in the way that fundamental Aristotelian aspects are refashioned in a Neoplatonic mould.[36]

The Idea of Man

The distinguished scholar of the Ikhwān al-Ṣafā', Professor Seyyed Hossein Nasr, has shown clearly how the Ikhwān's world view, particularly as it relates to a hierarchy of being, is integrated with 'the analogy of the microcosm and macrocosm'.[37] He reminds us that the macrocosm-microcosm analogy is universal and was by no means limited to the civilisations and cosmologies of Greece, Islam, or the Christian West, but was to be found in India and China as well.[38] This is extremely important from the perspective of the history of ideas. Nasr points out how the entire text of the *Rasā'il* is saturated 'with reference to the analogy between man and the universe',[39] and draws attention to an important passage in which the sages characterise the universe as 'the great man' (*al-insān al-kabīr*), because the world has a single body which embraces 'all its spheres, gradations of heavens, its generating elements (*arkān*) and their productions'. The Ikhwān, Nasr notes, also hold that the Universe has a single Soul (*Nafs*) analogous to the human soul which animates the human body. And the death of the universe is brought much closer psychologically, and made to appear more real, if one likens it to human death.[40]

---

in various chapters.

36 See Netton, *Muslim Neoplatonists*, esp. pp. 44–45.

37 Seyyed Hossein Nasr, *An Introduction to Islamic Cosmological Doctrines: Conceptions of Nature and Methods Used for Its Study by the Ikhwān al-Ṣafā', al-Bīrūnī and Ibn Sīnā*, rev. ed. (1964; London: Thames and Hudson, 1978), p. 66.

38 Ibid.

39 Ibid., p. 67.

40 Ibid., pp. 67–68 (citing *Rasā'il*, vol. 2, p. 20 in the Cairo edition of 1928).

The Idea of Knowledge

The search for knowledge, and its stored accumulation, is a universal construct that has probably existed since the dawn of thinking man. Plato expounded a precisely articulated epistemology which focused on the archetypal forms and was profoundly suspicious of the senses.[41] For the Brethren of Purity, however, real knowledge could be derived from a wide variety of sources, and those sources included the senses.[42]

Their quest for, and squirrel-like accumulation of, knowledge stands within the general framework of the Islamic tradition, and conformed to a paradigm which, from the earliest days, might be characterised as 'a thirst for global knowledge'. The Qur'an inaugurates the quest with its famous invocation: *Ya rabbī, zidnī ʿilm*[an] ('O my Lord! Increase me in knowledge').[43]

It is continued in the well-known hadith according to which the Prophet Muhammad is said to have advised that knowledge should be sought even as far as China. Even if this hadith is ultimately 'weak' (*ḍaʿīf*) or even 'invented' (*mawḍūʿ*), its very existence still conforms to the basic paradigm. The famous collectors of the tradition (*ḥadīth*) corpus, like Muḥammad ibn Ismāʿīl al-Bukhārī (810–870) and Abū al-Ḥusayn Muslim ibn al-Hajjāj (ca. 817–875), are known to have travelled many miles in search of authentic or 'sound' traditions. The university claimed by many Arabs to be the oldest in the world, al-Azhar, was founded by the victorious Fāṭimid dynasty in Cairo, following their conquest of Egypt in 969. Originally, it was designed to be a beacon of Ismaili doctrine and scholarship, but later it became the principal bastion and custodian of Sunni knowledge and learning.

Finally, the *Riḥla*, or *Travelogue*, immortalised in the two famous *Riḥlas* of the Spanish Muslim Ibn Jubayr (1145–1217) and the Moroccan Ibn Baṭṭūṭa (ca. 1304–1368 or d. ca. 1377),[44] became a vehicle, not

41 See Plato, *Phaedo*, 64b–66e. For a comparative Plato–Ikhwān epistemology, see Ian Richard Netton, 'Private Caves and Public Islands: Islam, Plato and the Ikhwān al-Ṣafā'', *Sacred Web*, 15 (2005).

42 See *Rasāʾil*, vol. 3, p. 424; Netton, *Muslim Neoplatonists*, p. 17.

43 Qur'an 20:114.

44 See Netton, *Seek Knowledge: Thought and Travel in the House of Islam* (Richmond: Curzon, 1996), pp. 95 ff; Ibn Jubayr, *Riḥla* (Beirut: Dār Ṣādir, 1964); Ibn

only for recounting of pilgrimage to Mecca, but also for the transmission of much learning about, and knowledge of, the House of Islam (*Dār al-Islām*). In the light of all this, and our remarks earlier about the encyclopaedism of the Ikhwān, it is clear that the *Rasāʾil*, in an entirely neat fashion, fits into the generally accepted epistemological paradigm of mediaeval Islam.[45]

The Idea of Salvation

In their magisterial volume *Heaven: A History*, Colleen McDannell and Bernhard Lang begin as follows:

> In the ancient world, belief in life after death was widespread, considered normal, and not generally weakened by scepticism. Death ended the visible form of our life on earth, but did not extinguish existence altogether. While the images provided by tradition and learned speculation may not have promised an idealized or 'better' life, the complete denial of an afterlife remained the exception rather than the rule. The majority of ancient authors, in spite of periods of doubt, assumed that some form of life existed after death.[46]

That 'form of life' might be literally 'heavenly' or 'hellish'; on the one hand, certain deeds might need to be performed to merit Heaven, together with the acceptance of certain creeds. On the other, justification and consequent salvation might be dependent on faith alone. In the different faith traditions, soteriology was multifaceted and the route to salvation, usually offered to all, was manifold. Islam's way, as we shall see, combined both faith and works.

In antiquity and the mediaeval period, there were many paradigms of the afterlife and the routes thereto. These ranged from the gloom

---

Baṭṭūṭa, *Riḥla* (Beirut: Dār Ṣādir, 1964); Ian Richard Netton, 'Riḥla', *EI2*, vol. 8, p. 528.

45 See also Franz Rosenthal, *Knowledge Triumphant: The Concept of Knowledge in Medieval Islam* (Leiden: E. J. Brill, 1970).

46 Colleen McDannell and Bernhard Lang, *Heaven: A History* (New Haven, CT and London: Yale University Press, 1988), p. l.

of the Greek Hades and the Hebrew Sheol,[47] to the jousting halls of the Scandinavian Valhalla and the sublime barque of (Amon-) Re.[48] Within the Abrahamic paradigm, the Judaic 'apocalyptic perspective, [according to which] God would release the dead from Sheol so that they could appreciate a renewed earth'[49] was replaced, after the birth of Christianity, by at least three other Jewish perspectives with the Pharisees (*contra* the Sadducees) adhering to a belief in life after death.[50]

Christianity itself, of course, held that redemption via Jesus Christ was required, in consequence of the Original Sin of Adam and Eve, and that the merit Christ gained through His Passion would open the gates of Paradise to all who sought that vision with pious beliefs and works.

By contrast, the last of the three great Abrahamic religions, Islam, had no doctrine of Original Sin, and therefore no precept of redemption as necessary for all mankind. Salvation history, then, in the Christian sense, is absent from the specifically Islamic aspect of the Abrahamic paradigm — a paradigm which is, as we can already see, multifaceted and by no means coherent between its three members. This is not to say that Islam lacks a soteriology; far from it. That soteriology, however, depends on man's salvation of himself: he is his own saviour, though guided, of course, to right intentions and right actions by the Holy Qurʾan. There is no direct concept of saving grace such as might be bestowed by an incarnate God, as in Christianity, for example.

That persistent union in Islam of intention and action is neatly expressed in the following hadith:

> Actions are but by intention and every man shall have but that which he intended. Thus he whose migration was for Allāh and His Messenger, his migration was for Allāh and His Messenger, and he whose migration was to achieve some worldly benefit or to take some woman in marriage, his migration was for that which he migrated.[51]

47 See Job 10:21–22, 14:13; McDannell and Lang, *Heaven: A History*, pp. 1–18.

48 See 'Re' in George Hart, *A Dictionary of Egyptian Gods and Goddesses* (1986; repr., London and New York: Routledge and Kegan Paul, 1988), esp. p. 182.

49 McDannell and Lang, *Heaven: A History*, p. 13.

50 Ibid., pp. 19–20.

51 Al-Nawawī, *Matn al-arbaʿīn al-nawawiyya*, (Arabic–English text) trans. Ezze-

In one of the later *sūwar* ('chapters') of the Qur'an (Sura 95), whose title is *Sura of the Fig* (*Sūrat al-Tīn*), this is powerfully reinforced:

> We have indeed created man
> In the best of moulds,
> Then do We abase him
> [To be] the lowest
> Of the low —
> Except such as believe
> And do righteous deeds:
> For they shall have
> A reward unfailing.[52]

It is clear, then that mainstream Islam espouses a soteriology of success,[53] in right faith, and combined with good works: this is the fundamental Islamic paradigm. Nonetheless, there is always an element of 'gift' as well. As Martin Borrmans neatly puts it: 'Salvation [*najāt*] is always God's gift granted to faithful people in the present time and in the hereafter.'[54] The fundamental analogy of Islamic 'gift' and Christian 'grace' will not be missed here, although, it must be emphasised that Islam is very far from the Protestant Reformation paradigm of justification by faith alone.

So, to which paradigm did the Ikhwān al-Ṣafā' adhere? Their primary image is the 'Ship of Salvation' (*Safīnat al-najāt*).[55] In a brilliant and skilfully deployed image, the Ikhwān maintain that 'the Brotherhood of Purity which they established was their "Ship of Salvation" from the sea of matter which included the world, its material aspects and a large number of its inhabitants'.[56]

For the Ikhwān, faith and deeds count in equal measure. Their Islamic faith, albeit 'Neoplatonised' in its metaphysics at times, is on

---

din Ibrahim and Denys Johnson-Davies, 3rd ed. (Damascus: The Holy Koran Publishing House, 1977), Hadith 1: p. 26 (Arabic text), p. 27 (English trans.).

52 Qur'an 95:4–6, trans. Yusuf Ali.

53 See Maurice Borrmans, 'Salvation', in *Encyclopaedia of the Qur'ān*, ed. Jane Dammen McAuliffe, vol. 4 (Leiden and Boston: E. J. Brill, 2004), p. 522–533.

54 Ibid., p. 524.

55 See Netton, *Muslim Neoplatonists*, pp. 105–108.

56 Ibid., p. 108; *Rasā'il*, vol. 4, p. 18.

full display throughout the *Rasāʾil*; their works are articulated in a philosophy of Brotherhood and *taʿāwun*, co-operation. Salvation is ultimately the gift of Allah, it is true, but it is earned by fraternity and the co-operative impulse.

SIX

Misled and Misleading . . . Yet Central in their Influence: Ibn Taymiyya's Views on the Ikhwān al-Ṣafāʾ

Yahya Jean Michot

New editions and studies confirm that the famous Mamlūk *mufti* and theologian Taqī al-Dīn Aḥmad ibn Taymiyya (d. in Damascus, 728/1328) had a more complex relationship to philosophy than one might conclude from his often-quoted *Refutation of the Logicians* (*al-Radd ʿalā al-manṭiqiyyīn*).[1] He knew of the early and late *falāsifa*, as well as of several other Muslim intellectuals, for example, the Ismaili Abū Yaʿqūb al-Sijistānī (d. ca. 390/1000).[2] Moreover, he explored, commented on, or expressed views about texts as diverse as Ibn Sīnā's (also known as Avicenna) *Risāla aḍḥawiyya*,[3] the abridgement (*talkhīṣ*) of Aristotle's *Metaphysics* by Thābit ibn Qurra, or the *Commentary*

1 See, for example, Yahya J. Michot, 'Vanités intellectuelles . . . L'impasse des rationalismes selon le *Rejet de la contradiction* d'Ibn Taymiyya', *Oriente Moderno*, 19, 80 (2000), pp. 597–617 (hereafter cited as 'Vanités').

2 See Yahya J. Michot, 'A Mamlūk Theologian's Commentary on Avicenna's *Risāla aḍḥawiyya*: Being a Translation of a Part of the *Darʾ al-taʿāruḍ* of Ibn Taymiyya', *Journal of Islamic Studies*, 14, 2–3 (2003), pp. 149–203, 309–363; see esp. pp. 199–203 (hereafter cited as 'Mamlūk').

3 See Michot, 'Mamlūk'.

[*sharḥ*] *on the Ishārāt* written by al-Naṣīr al-Ṭūsī, in 644/1246, for the Ismaili Muḥtasham Shihāb al-Dīn.[4]

The two earliest bibliographers of Ibn Taymiyya, Ibn ʿAbd al-Hādī (d. 744/1343)[5] and Abū ʿAbd Allāh Muḥammad Ibn Rushayyiq (d. 749/1348),[6] do not mention any title that he would have devoted to the Ikhwān al-Ṣafāʾ. He nevertheless refers to them and their *Epistles* in several passages of his *fatwa*s and of his main works. As a precise chronology of these writings is impossible to establish, no effort will be made here to trace an eventual evolution of Ibn Taymiyya's thought. Moreover, the survey of these writings proposed here cannot, of course, claim to be exhaustive. Based on more than forty texts in which the name of the Ikhwān explicitly appears, it should, however, make possible a better understanding of the theologian's opinion on the illustrious Brethren of Purity.[7]

In Taymiyyan texts, the place mentioned most frequently in relation to the Ikhwān is Cairo; the names of personalities are Jaʿfar al-Ṣādiq, Ibn Sīnā, and Abū Ḥāmid al-Ghazālī; and the ideological trends or movements within Islam, are the philosophers (*faylasūf*), or philosophizers (*mutafalsif*), then the Qarmaṭīs and 'esotericists' (*bāṭinī*), then the Ismailis and 'heretics' (*mulḥid*).

Ibn Taymiyya refers at least ten times to the construction of Cairo in connection with the dating of the composition of the *Rasāʾil*:[8] 'The

4 See Yahya J. Michot, 'Vizir "hérétique" mais philosophe d'entre les plus éminents: al-Ṭūsī vu par Ibn Taymiyya', *Farhang*, 15–16, notes 44–45 (2003), pp. 195–227 (hereafter cited as 'Hérétique').

5 See Abū ʿAbd Allāh ibn ʿAbd al-Hādī (d. 744/1343), *al-ʿUqūd al-durriyya min manāqib Shaykh al-Islām Aḥmad bin Taymiyya*, ed. Muḥammad Ḥamīd al-Fiqī (Cairo: Maṭbaʿat Ḥijāzī, 1357/1938), pp. 26–67 (hereafter cited as *ʿUqūd*).

6 Ibn Rushayyiq's bibliography of Ibn Taymiyya is usually wrongly attributed to Ibn Qayyim al-Jawziyya (d. 751/1350). On this problem, see M. ʿU. Shams and ʿA. ibn M. al-ʿImrān, *al-Jāmiʿ li-sīrat shaykh al-Islām Ibn Taymiyya khilāl sabʿat qurūn, Āthār shaykh al-Islām Ibn Taymiyya wa-mā laḥiqa-hā min aʿmāl*, 8 (Mecca: Dār ʿĀlam al-Fawāʾid li-al-Nashr wa'l-Tawzīʿ, AH 1422), pp. 56–61 (hereafter cited as *Jāmiʿ*). See also Ibn Rushayyiq's bibliography, which is edited in ibid., pp. 282–311.

7 More texts concerning the Ikhwān are thus taken into consideration here than are indexed in R. Y. al-Shāmī, '*Ibn Taymiyya: Maṣādiru-hu wa-manhaju-hu fī-taḥlīli-hā*', *Journal of the Institute of Arabic Manuscripts*, 38, 1 (1994), p. 211 (hereafter cited as 'Ibn Taymiyya').

8 See the following texts by Ibn Taymiyya: *Bughyat al-murtād fī al-radd ʿalāʾl-*

scholars know that they were only composed after the third century, at the time of the construction of Cairo',[9] which 'was built around 360[/970], as [reported] in the *Taʾrīkh al-jāmiʿ al-Azhar* [History of the al-Azhar Mosque]'.[10] To justify his affirmation, Ibn Taymiyya does not offer a clue linking the *Rasāʾil* specifically to the foundation of Cairo but, rather, to certain events contemporaneous with the latter, which, he says, the Ikhwān allude to.

> The person who composed them indeed mentions in them an event that happened in Islam: the conquest of the coasts of Syria by the Nazarenes, and similar events that happened after the third century.[11]
>
> They also mention, in them, something that happened to the Muslims: the conquest of the coasts of Syria by the Nazarenes. Now, this only happened after the third century.[12]

This 'entry of the Nazarenes into the countries of Islam'[13] 'at the beginning of the fourth century'[14] does of course not refer to the Crusades, which did not start until 488/1095, but to the military successes of the

mutafalsifa wa'l-Qarāmiṭa wa'l-Bāṭiniyya, ahl al-ilḥād min al-qāʾilīn bi'l-ḥulūl wa'l-ittiḥād, ed. Musa ibn Sulayman al-Duwaysh ([Medina?]: Maktabat al-ʿUlūm wa'l-Ḥikam, 1408/1988), p. 329 (hereafter cited as *Bughya*); *Darʾ taʿāruḍ al-ʿaql wa'l-naql aw muwāfaqat ṣaḥīḥ al-manqūl li-ṣarīḥ al-maʿqūl*, ed. Muḥammad Rashad Sālim, 11 vols. (Riyadh: Dār al-Kunūz al-Adabiyya, [1399/1979]), vol. 5, pp. 10, 26–27 (hereafter cited as *Darʾ*) — this reference is translated by Michot in 'Mamlūk', part 1, p. 189; *Majmūʿ al-fatāwā*, ed. ʿAbd al-Raḥmān ibn Muḥammad ibn Qāsim, 37 vols. (Rabat: Maktabat al-Maʿārif, 1401/1981), vol. 11, p. 581 (hereafter cited as *MF*); ibid., vol. 27, p. 174; ibid., vol. 35, p. 134; ibid., vol. 35, p. 183; *Majmūʿat fatāwā*, 5 vols. (Beirut: Dār al-Fikr, 1403/1983), vol. 1, p. 333 (hereafter cited as *MaF*); ibid., vol. 4, p. 234; *Minhāj al-sunna al-nabawiyya fī naqḍ kalām al-shīʿa al-qadariyya*, ed. Muḥammad Rashad Sālim, 9 vols. (Cairo: Maktabat Ibn Taymiyya, 1409/1989), vol. 2, p. 466 (hereafter cited as *Minhāj*); ibid., vol. 4, pp. 54–55. See also Yahya J. Michot, 'Ibn Taymiyya on Astrology: Annotated Translation of Three Fatwas', *Journal of Islamic Studies*, 11, 2 (2000), pp. 176–177 (hereafter cited as 'Astrology').

9 Ibn Taymiyya, *MF*, vol. 35, p. 134; also in Ibn Taymiyya, *MaF*, vol. 4, p. 234.

10 Ibn Taymiyya, *MF*, vol. 35, p. 134; also in Ibn Taymiyya, *MaF*, vol. 4, p. 234.

11 Ibn Taymiyya, *MF*, vol. 35, p. 134; also in Ibn Taymiyya, *MaF*, vol. 4, p. 234.

12 Ibn Taymiyya, *Minhāj*, vol. 2, p. 466.

13 Ibn Taymiyya, *Darʾ*, vol. 5, p. 27, trans. in Michot, 'Mamlūk', part 1, p. 189.

14 Ibn Taymiyya, *Minhāj*, vol. 4, p. 55.

Byzantine Nicephorus Phocas and John Tzimisces over the Ḥamdānids of Aleppo after 350/961, just a few years before Cairo was built by the Fāṭimids in 358/969.

The theologian's insistence on this dating of the *Rasā'il* is motivated by his will to provide a final refutation of their attribution to the Shiʿi Imam, Jaʿfar al-Ṣādiq who died 'more than two hundred years before the construction of Cairo',[15] or 'some two hundred years before the composition of these epistles'.[16] As he puts it: 'Every individual who wanted to sell his lies well attributed them to Jaʿfar.'[17]

> Lies were told about Jaʿfar such as were not told about anybody [else] because, in regard to knowledge and the religion, there was something in him by which God had distinguished him. He, his father — Abū Jaʿfar — and his grandfather — ʿAlī ibn al-Ḥusayn — were among the most prominent of the Imams, in regard to knowledge and religion. And, after Jaʿfar, among the People of the [Prophetic] House, there was nobody [who could be esteemed] equal to him. Many of the adepts of heresy and innovations thus started to attribute to him what they [themselves] were saying. The authors of the epistles of the Ikhwān al-Ṣafā' even attribute them to him.[18]

Ibn Taymiyya elaborates this argument, claiming that to him were also attributed, for example, 'words on the stars and on the quivering of the limbs, falsified commentaries [on the Qur'an], and various vain things from which God exculpates him'.[19]

> A group of people even hold the opinion that the epistles of the Ikhwān al-Ṣafā' come from him. This is a well-known lie. Jaʿfar passed away in the year 148[/765] whereas these epistles were composed some two hundred years afterwards.[20]
>
> They attribute that to him in order to present that as a legacy

15 Ibn Taymiyya, *MF*, vol. 35, p. 134; also in Ibn Taymiyya, *MaF*, vol. 4, p. 234.
16 Ibn Taymiyya, *Bughya*, p. 330.
17 Ibn Taymiyya, *Minhāj*, vol. 4, p. 54.
18 Ibn Taymiyya, *MF*, vol. 11, p. 581.
19 Ibn Taymiyya, *Dar'*, vol. 5, p. 26, trans. in Michot, 'Mamlūk', part 1, p. 189.
20 Ibn Taymiyya, *Minhāj*, vol. 6, p. 54.

> coming from the People of the [Prophetic] House. This is among the ugliest and most blatant of lies.[21]

Ibn Taymiyya further states: 'Every intelligent person who understands the [*Rasāʾil*] and knows Islam, knows that they contradict the religion of Islam.'[22] According to him, a simple examination of the content of the *Rasāʾil* would therefore suffice to refute its attribution to such an eminent religious scholar as Jaʿfar al-Ṣādiq. Ibn Taymiyya is, however, keen to present more 'factual', not just doctrinal, evidence; hence the historical criticism that characterises his approach to the matter. As for the true identity of the Ikhwān, the Ḥanbalī theologian is aware of Abū Ḥayyān al-Tawḥīdī's famous testimony, and mentions it twice. One passage is quite general, but the second is more explicit:

> In the book *al-Imtāʿ waʾl-muʾānasa*, Abū Ḥayyān al-Tawḥīdī makes mention of the conversations of Abū al-Faraj ibn Ṭirāz with some of the authors of the [*Rasāʾil*] and his discussion with them, as well as of the words of Abū Sulaymān the logician about them, and matters thanks to which the situation somehow becomes clear.[23]

> Those who composed the [*Rasāʾil*] are known, like Zayd ibn Rifāʿa, Abū Sulaymān ibn Maʿshar al-Bistī (known as al-Maqdisī), Abū al-Ḥasan ʿAlī ibn Hārūn al-Zanjānī, Abū Aḥmad al-Nahrajūrī,[24] and al-ʿAwfī. Abū al-Futūḥ al-Muʿāfī ibn Zakariyāʾ al-Jarīrī, the author of the book *al-Jalīs waʾl-anīs*, had a discussion with them. Abū Ḥayyān al-Tawḥīdī referred to that in the book *al-Imtāʿ waʾl-muʾānasa*.[25]

Ibn Taymiyya's linking of the composition of the *Rasāʾil* to the construction of Cairo must also be understood in the context of his views on the ideological allegiance of the authors. Incidentally, and quite

21 Ibn Taymiyya, *Bughya*, p. 329.

22 Ibn Taymiyya, *Minhāj*, vol. 2, p. 465.

23 Ibn Taymiyya, *Bughya*, pp. 329–330.

24 Cf. 'al-Mihrajānī', in Abū Ḥayyān al-Tawḥīdī, *Kitāb al-Imtāʿ waʾl-muʾānasa*, 2nd edition, ed. Aḥmad Amīn and Aḥmad al-Zayn, 2 vols. (Beirut: Manshūrāt Dār Maktabat al- Ḥayāt, 1965), vol. 2, p. 5 (hereafter cited as *Imtāʿ*).

25 Ibn Taymiyya, *Minhāj*, vol. 2, p. 466. See Tawḥīdī, *Imtāʿ*, vol. 2, pp. 3–5.

inconsequently, he writes twice that the *Rasāʾil* was composed 'under the dynasty of the Būyids'.[26] More generally and, in fact, even in one of the two passages mentioning the Būyids, it is to the 'Ubaydids, 'the descendants of 'Ubayd Allāh ibn Maymūn al-Qaddāḥ',[27] that he connects them.

> These epistles were composed under the dynasty of the Būyids, during the fourth [/tenth] century, at the beginning of the dynasty of the 'Ubaydids who built Cairo.[28]

The theologian speaks of 'Ubaydids rather than of Fāṭimids, because he would not accept that the latter belonged to the Family of the Prophet or, even, that they were faithful to him.

> The Sons of 'Ubayd — whom they call 'al-Qaddāḥ' — who used to say that they were Fāṭimids, built Cairo, and remained kings [there], claimed that they were 'Alids [*'Alawī*]. [They reigned for] about two hundred years, and achieved supremacy over half of the empire [*mamlaka*] of Islam. They even achieved supremacy, at certain points, over Baghdād . . . The people of knowledge all know that the ['Ubaydids] were not of the children of Fāṭima. Rather, they were from among the descendants of the Magi — it has also been said that they were the descendents of a Jew. They were among the people who are the most distant from the Messenger of God, God bless him and grant him peace, as far as his Sunna and his religion are concerned.[29]

Ibn Taymiyya nevertheless speaks explicitly of the 'Ubaydids as 'Ismailis' also. He notes that the *Rasāʾil* was 'composed at the time of the appearance of the doctrine of the Ismaili 'Ubaydids who built Cairo',[30] 'when the dynasty of the Ismaili esotericists who built the Cairo of

26 Ibn Taymiyya, *Bughya*, p. 329; Ibn Taymiyya, *MF*, vol. 35, p. 183, trans. in Michot, 'Astrology', p. 177; and also in Ibn Taymiyya, *MaF*, vol. 1, p. 333.

27 Ibn Taymiyya, *MF*, vol. 35, p. 134; also in Ibn Taymiyya, *MaF*, vol. 4, p. 234.

28 Ibn Taymiyya, *MF*, vol. 35, p. 183, trans. in Michot, 'Astrology', p. 177; also in Ibn Taymiyya, *MaF*, vol. 1, p. 333.

29 Ibn Taymiyya, *MF*, vol. 27, 174.

30 Ibid., vol. 11, p. 581.

al-Muʿizz, in the year 350-something, appeared'.[31] Moreover, the content of the *Rasā'il* was itself understood by him as being of an Ismaili nature:

> These [epistles] were composed during the fourth century, when the ʿUbaydid dynasty appeared in Egypt and [when] they built Cairo. They were composed according to the doctrine of those Ismailis, as is demonstrated by their content.[32]

By connecting the epistles to Egypt and labelling them 'Ismaili', did Ibn Taymiyya mean to say that they are not Qarmaṭī works? Not at all, since he also writes that 'this book is the foundation [*aṣl*] of the doctrine of the Qarmaṭī philosophers'.[33] Elsewhere, he also argues that:

> It is for these Qarmaṭīs that the *Rasā'il* of the Ikhwān al-Ṣafā' was composed — i.e., those who are called the Ismailis due to the fact that they relate themselves to Ismāʿīl ibn Jaʿfar.[34]

The theologian does not in fact see any real doctrinal difference between the Ismailis and the Qarmaṭīs.

> [These people] have nicknames that are well known among the Muslims. Sometimes they are called the 'heretics' [*mulḥid*], sometimes they are called the 'Qarmaṭīs', and sometimes they are called the 'esotericists' [*bāṭinī*]; sometimes they are called the 'Ismailis', and sometimes they are called the 'Nuṣayrīs';[35]

31 Ibn Taymiyya, *Minhāj*, vol. 4, pp. 54–55.

32 Ibid., *Minhāj*, vol. 2, p. 466.

33 Ibn Taymiyya, *Bughya*, p. 329.

34 Ibn Taymiyya, *Sharḥ al-ʿaqīda al-Iṣfahāniyya*, ed. Ḥ. M. Makhlūf (Cairo: Dār al-Kutub al-Islāmiyya, 1386/1966), p. 170 (hereafter cited as *Iṣfahāniyya*).

35 It is worth noting that the Nuṣayrīs in fact represent a Shiʿi tradition other than that of the Ismailis, deriving their name from Muḥammad ibn Nuṣayr al-Fihrī al-Numayrī, a disciple of the tenth or eleventh Twelver Imam, and which still exists today (the ʿAlawīs of Syria); see Abū al-Fatḥ al-Shahrastānī, *Livre des religions et des sects*, ed. and trans. Daniel Gimaret and Guy Monnot, vol. 1 (Leuven and Paris: Peeters and UNESCO, 1986), p. 542, note 225 (hereafter cited as *Religions*). Ibn Taymiyya expounds and refutes their doctrines in the *fatwa* from which this text is taken; see Stanislas Guyard, 'Le *fetwa* d' Ibn Taymiyyah

> sometimes they are called the 'Khurramiyya'[36] and sometimes they are called the 'Muḥammira'.[37] There are some of these names that are common to them whereas others are peculiar to some of their kinds; just as 'Islam' and 'faith' are common to the Muslims although some of them have a name which is peculiar to them by reason of their lineage, or their rite [*madhhab*], or their country, etc.[38]

For the Ḥanbalī theologian, the Ismailis are thus 'a type of Qarmaṭīs' among others — the Khurramiyya, for example. As for the Qarmaṭīs, he maintains that they are themselves part of the wider ideological ensemble of the esotericists. Esotericism is indeed present within Shiʿism as within Sufism or *kalām* ('theology'). The likes of the Shiʿi esotericists, i.e., the Qarmaṭīs, are, in Sufism, 'the unionists — the adepts of the oneness of existence — such as Ibn Sabʿīn, Ibn ʿArabī, and their like'[39] and, 'among the straying exponents of *kalām*',[40] 'the deniers of the [divine] attributes . . . like the Muʿtazilīs and others'.[41] Speaking once of 'groups of esotericists', Ibn Taymiyya thus adds, more explicitly, 'Shiʿi esotericists like the authors of the *Rasāʾil Ikhwān al-Ṣafāʾ* and Sufi esotericists like Ibn Sabʿīn, Ibn ʿArabī, and others'.[42] And because he considers this shared esotericism 'a shared heresy', he can also liken 'the heretics . . . among the followers of the Sons of

---

sur les Nosairis', *Journal Asiatique*, 6, 18 (1871), pp. 158–198 (hereafter cited as '*Fetwa*').

36 *Khurramiyya*, or *Khurramdīniyya* (from the Persian *khurram-dīn*, 'joyous', or 'pleasant', 'religion'), originally denoted the religious movement of Mazdak in general and, later on, of various Iranian, anti-Arab, and frequently rebellious sects, that were influenced by certain Mazdak and Manichaean beliefs as well as by some Shiʿi 'extremist' doctrines. See Wilferd Madelung, 'Al-Khurramiyya', *EI2*, vol. 5, p. 63–65.

37 *Muḥammira* seems originally to have been another name for the religious movement of Mazdak in general. Later on, the word came to be used for various rebellious factions in Iran. See ibid.

38 *Nuṣayriyya*, in Ibn Taymiyya, *MF*, vol. 35, p. 152 (also in Ibn Taymiyya, *MaF*, vol. 4, p. 212). French trans. in Guyard, '*Fetwa*', p. 189; also in Michot, 'Hérétique', p. 205.

39 Ibn Taymiyya, *Iṣfahāniyya*, p. 52.

40 Ibn Taymiyya, *MF*, vol. 4, p. 346.

41 Ibn Taymiyya, *Iṣfahāniyya*, p. 52.

42 *Kitāb al-Radd ʿalā al-mantiqiyyīn*, ed. M. Ḥ. M. Ḥ. Ismāʿīl (Beirut: Dār al-Kutub al-ʿIlmiyya, 1423/2003), p. 433 (hereafter cited as *Radd*).

ʿUbayd, like the authors of the *Rasāʾil Ikhwān al-Ṣafāʾ* and others',[43] to 'the Sufi heretics walking out of the way of the earlier shaykhs who belonged to the people of the Book and the Tradition, like Ibn ʿArabī, Ibn Sabʿīn, Ibn Ṭufayl (the author of *Ḥayy ibn Yaqẓān*), and many other creatures'.[44]

It is with Ibn Sīnā (Avicenna), rather than these late unionist Sufis with whom he sometimes puts them, that Ibn Taymiyya most often likes to connect the Ikhwān. His main reason for doing so is the famous passage of his autobiography, in which the Shaykh al-Raʾīs recalls:

> My father was one of those who responded to the propaganda of the Egyptians and was reckoned among the Ismāʿīliyya. From them, he, as well as my brother, heard the account of the soul and the intellect in the special manner in which they speak about it and know it. Sometimes they used to discuss this among themselves.[45]

Just like others before him, the Ḥanbalī theologian forgets about the rest of this testimony — 'I was listening to them and understanding what they were saying, but my soul would not accept it'[46] — and considers that these conversations led Ibn Sīnā to become, not only an Ismaili philosopher, but a faithful of the Egyptian 'ʿUbaydids'.

> Ibn Sīnā said: 'My father and my brother were among the adepts of their missionary propaganda [*daʿwa*], and this is why I occupied myself with philosophy.'[47]

> Ibn Sīnā mentioned that his father was among the adepts of their missionary propaganda [*daʿwa*], among the adepts of the missionary calling of the Egyptians among them, who at

43 Ibn Taymiyya, *MF*, vol. 17, p. 333.

44 Ibn Taymiyya, *Darʾ*, vol. 1, p. 11. French trans. in *Ibn Taymiyya: Lettre à Abū al-Fidā*, ed. and trans. Yahya J. Michot (Louvain-la-Neuve: Université Catholique de Louvain, 1994), pp. 24–25 (hereafter cited as *Lettre*).

45 *The Life of Ibn Sina: A Critical Edition and Annotated Translation*, ed. and trans. William E. Gohlman (Albany, NY: SUNY Press, 1974), p. 19 (hereafter cited as *Life*).

46 Gohlman, *Life*, p. 19.

47 Ibn Taymiyya, *Iṣfahāniyya*, p. 170.

> that time had gained possession of Egypt and conquered it. 'It is for this reason', Ibn Sīnā says, 'that I occupied myself with philosophy.'[48]
>
> Ibn Sīnā and the members of his house were indeed among the followers of al-Ḥākim the Qarmaṭī, the 'Ubaydid who was [ruling] in Egypt.[49]
>
> The members of Ibn Sīnā's house were among the followers of these Qarmaṭīs, among those who had answered to al-Ḥākim, who was [then reigning] in Egypt. 'It is for that reason', Ibn Sīnā says, 'I embarked on philosophy.'[50]

Ibn Taymiyya does not affirm that Ibn Sīnā read the *Rasā'il*, but works out a narrative which, by linking them, the Persian thinker, Cairo, al-Ḥākim, and the Ismaili ''Ubaydids', leads to a picture that makes sense within the structure of his ideology if not in historical terms: that of a time during which 'the situation of the Muslim . . . was seriously troubled',[51] not just because of external enemies but because of the growing 'threat against Islam' represented, among Muslims themselves, by that idea whose interests esotericism served so well: philosophy. Modern historians of Islamic classical thought are sometimes reluctant to consider the Ikhwān as *falāsifa*. As for Ibn Taymiyya, there is no doubt at all in his mind:

> Intrinsically, their ideology [*amr*] is the doctrine of the philosophers, and it is according to this ideology that these epistles

48 Ibn Taymiyya, *Bayān talbīs al-jahmiyya fī ta'sīs bida'i-him al-kalāmiyya aw naqd ta'sīs al-jahmiyya*, ed. Muḥammad ibn 'Abd al-Raḥmān ibn Qāsim, 2 vols. ([Cairo?]: Mu'assasat Qurṭuba, 1392[/1972]), vol. 1, p. 374 (hereafter cited as *Bayān*).

49 Ibn Taymiyya, *MF*, vol. 13, p. 249.

50 Ibn Taymiyya, *Dar'*, vol. 5, p. 10; Ibn Taymiyya, *MF*, vol. 11, p. 571; ibid., vol. 27, p. 175; ibid., vol. 35, p. 135; Ibn Taymiyya, *MaF*, vol. 4, p. 235. See also Ibn Taymiyya, *Majmū'at al-rasā'il al-Kubrā*, 2 vols. (Cairo: al-Maṭba'at al-'Āmirat al-Sharqiyya, 1323[/1905]), vol. 2, pp. 288–289 (hereafter cited as *MRK*). See also *Musique et danse selon Ibn Taymiyya: Le Livre du* samā' *et de la danse (*Kitāb al-Samā' wa'l-raqṣ*) compilé par le Shaykh Muḥammad al-Manbijī*, Études musulmanes, 33, ed. and trans. Yahya J. Michot (Paris: J. Vrin, 1991), pp. 78–79 (hereafter cited as *Musique*).

51 Ibn Taymiyya, *Dar'*, vol. 5, p. 10.

> were composed. A group of well-known philosophizers composed them.[52]

He therefore has no difficulty in speaking of 'the authors of the *Rasā'il Ikhwān al-Ṣafā'* and *similar philosophizers*' [italics added].[53] Now:

> The people who are the closest [to the 'Ubaydids] are the philosophers, albeit they did not also adopt the rule [*qā'ida*] of a particular philosopher. This is why groups of the philosophizers related themselves to them. Ibn Sīnā and the members of his house were indeed among their followers. Ibn al-Haytham and similar people were among their followers. Mubashshir ibn Fātik and his like were among their followers.[54]

So, also, 'the authors of the *Rasā'il Ikhwān al-Ṣafā'* composed their epistles in a way similar to theirs'.[55]

But what type of philosophy did the Ikhwān, whose epistles, as already mentioned, Ibn Taymiyya says 'were composed according to the doctrine of those Ismailis',[56] have in common with Ibn Sīnā, Mubashshir ibn Fātik, Abū 'Alī ibn al-Haytham and other 'followers of al-Ḥākim'?[57] Quoting al-Ghazālī's *al-Munqidh*, the theologian at one point speaks of:

> Insipid crumbs of Pythagoras' philosophy. The latter was one of the earliest of the Ancients and his doctrine was the first of the doctrines of the philosophers. Aristotle refuted him or, more [precisely], corrected what he was saying and despised it. It is what is talked about in the book of the *Rasā'il Ikhwān al-Ṣafā'* and it is really the refuse of philosophy.[58]

52 Ibn Taymiyya, *Minhāj*, vol. 4, p. 55.
53 Ibn Taymiyya, *MF*, vol. 18, p. 336.
54 Ibid., vol. 27, p. 175.
55 Ibid.
56 Ibn Taymiyya, *Minhāj*, vol. 2, p. 466; Ibn Taymiyya, *MF*, vol. 27, p. xxx.
57 Ibn Taymiyya, *MF*, vol. 13, p. 249.
58 Ibn Taymiyya, *Iṣfahāniyya*, p. 111. See al-Ghazālī, *Al-Munqidh min al-ḍalāl (Erreur et délivrance)*, ed. and [French] trans. Farid Jabre (Beirut: Librairie Orientale, 1969), p. 33 (hereafter cited as *Munqidh*). The version quoted by Ibn Taymiyya often differs from the text prepared by Jabre.

More regularly, however, it is to Aristotle and 'Greek Peripatetic philosophy'[59] that Ibn Taymiyya traces the Ikhwān's philosophical views, although not exclusively, as he also notes that the epistles 'were composed according to their doctrine which they assembled from what was said by the Greek philosophers, the Persian Magi, and, among the people of the Qibla, the Shiʿis'.[60] Or, that they were partly influenced by 'the things said by the later Ṣābiʾans, that is, the innovated philosophy'.[61]

> Sometimes, they rest what they say on the doctrines of the naturalist or divinalist philosophers, and sometimes they rest it on what is said by the Magi who worship the light. And to this they join [Shiʿi] 'rejectionism' [*rafḍ*].[62]

Ibn Taymiyya is somehow clearer about what he thinks concerning these Magi, Ṣābiʾans or 'Ṣābiʾan philosophizers relating themselves to Islam,[63] in a passage concerning the famous translation movement under the ʿAbbāsids:

> At the end of the second century [200/815], before it and after it, the books of the Greeks and other *Rhomaioi* were imported from the countries of the Nazarenes, translated into Arabic, and so the doctrine of the Ṣābiʾan substitutors [*mubaddil*], such as Aristotle and his kin, spread.[64] The Khurramiyya appeared at that time, that is, the first esotericist Qarmaṭīs who, inwardly, were adopting some elements of the religion of the Ṣābiʾan substitutors and some of the religion of the Magi. For example, they took from the former what they said about the intellect and the soul and they took from the latter what they said about light and darkness; they dressed that up with

59 Ibn Taymiyya, *Bughya*, p. 179.

60 Ibn Taymiyya, *MF*, vol. 11, p. 581.

61 Ibid., *MF*, vol. 12, p. 23.

62 *Nuṣayriyya*, in Ibn Taymiyya, *MF*, vol. 35, p. 153 (also in Ibn Taymiyya, *MaF*, vol. 4, p. 212); French trans. in Guyard, '*Fetwa*', p. 190.

63 Ibn Taymiyya, *MF*, vol. 4, p. 314.

64 A few lines earlier, Ibn Taymiyya speaks of those who 'substitute [*baddala*] and change the primordial state of the creatures [*fiṭra*] of God and His way [*sharīʿa*]'.

> [various Islamic] idioms, tampered with it and presented it to the Muslims. Consequently, there were in Islam [all] the Qarmaṭīs that there were, and they leaned much towards the way of the Ṣābi'an substitutors! It is also in their time that the *Rasā'il Ikhwān al-Ṣafā'* was composed.[65]

As briefly alluded to in this excerpt, what Ibn Taymiyya found particularly 'dangerous for the faith' in the *Rasā'il* and similar philosophies was defined by him as their 'deceitful and cunning ways':

> Those who revere the [philosophers] want an agreement between what they say and what the prophets came with.[66]
>
> Someone who travels along the [philosophers'] way wants to have a coherence between what they say and what the Messengers came with. By sophistry and 'Qarmaṭising' he thus embarks on [various] kinds of absurdities which no intelligent being can be satisfied with, as was done by the authors of the *Rasā'il Ikhwān al-Ṣafā'* and their like. It is from here that the Qarmaṭīs, the esotericists, and those who became their associates in some of those [ideas] went astray.[67]

Supposedly, 'the [esotericists] want to make a synthesis between what the Messengers have told [us] about and what those [philosophers] say, as the authors of the *Rasā'il Ikhwān al-Ṣafā'* did'.[68] Or, as he elaborates:

> [The philosophers] claimed that they make a synthesis between the divine Law and Greek philosophy. This is for example claimed by the authors of the *Rasā'il Ikhwān al-Ṣafā'* and their like among these heretics.[69]

Ibn Sīnā is classed by Ibn Taymiyya as one of these, as is Ibn Rushd (Averroes):

65 Ibn Taymiyya, *Bayān*, vol. 1, p. 374.
66 Ibn Taymiyya, *Radd*, p. 366.
67 Ibn Taymiyya, *MF*, vol. 9, pp. 36–37.
68 Ibid., *MF*, vol. 35, pp. 133–134; also in Ibn Taymiyya, *MaF*, vol. 4, p. 234.
69 Ibn Taymiyya, *Dar'*, vol. 6, p. 242.

> [Ibn Sīnā made] a synthesis between the Law and philosophy — and likewise did the Ismaili esotericists proceed in their book called the *Rasāʾil Ikhwān al-Ṣafāʾ*. After him, Ibn Rushd also did so.[70]

> Ibn Sīnā, Ibn Rushd ('the grandson'),[71] and their like started to bring the fundamentals of those [Aristotelian philosophers] and the way of the prophets closer together, and to state publicly that their fundamentals did not conflict with the prophetic Laws.[72]

> In the [*Rasāʾil*, the Ikhwān] claim, they made a synthesis between the things said by the later Ṣābiʾans — that is, the innovated philosophy — and what the Messengers brought from God. They thus come up with things which they claim to be intelligible, although there is no proof for many of them. Sometimes, they also mention that it is something traditionally transmitted [*manqūl*], although it contains grave lies and alterations.[73]

Consequently, Ibn Taymiyya wondered how could one speak of synthesis 'when Aristotle is in fact preferred to the Qurʾan', and the result of it all was, he felt, rather than an agreement between religion and reason, 'a mixture of distorted tradition and erring rationalism'? As he wrote:

> This Book [the Qurʾan] which [is such that], under the surface of the sky, there is no book nobler than it, and if the knowledge of this is not learned from the Messenger who is the most eminent creature of God, Exalted is He, in everything — knowing, teaching, etc. — will this then be mentioned in what is said by Aristotle and his kin, by the authors of the *Rasāʾil Ikhwān al-Ṣafāʾ* and their like, who establish things by means of syllogisms containing mere claims, [based on] no soundly

70 Ibn Taymiyya, *Bughya*, p. 199.

71 Perhaps Ibn Taymiyya is hinting in this context at an intellectual lineage connecting Aristotle, Ibn Sīnā, and Ibn Rushd.

72 Ibn Taymiyya, *Kitāb al-Ṣafadiyya*, ed. M. R. Sālim, 2 vols. (Mansoura and Riyadh: Dār al-Hady al-Nabawī and Dār al-Faḍīla, 1421/2000), vol. 1, p. 237 (hereafter cited as *Ṣafadiyya*).

73 Ibn Taymiyya, *MF*, vol. 12, p. 23.

> transmitted tradition nor clear rationality, and which rather resemble the empty syllogisms which are devoid of effect and which, when submitted to verification, reduce to phantasms that have no reality in the outside [world]?[74]

Ibn Taymiyya was not, of course, the first to attack the Ikhwān's syncretism. Three centuries before him, Abū Sulaymān al-Sijistānī had already criticised their attempt to synthetise the truths of philosophy and the ways of faith.[75] As he refers to *al-Imtāʿ wa'l-mu'ānasa*,[76] the theologian most probably knew of al-Tawḥīdī's report on this, but, in the texts here analysed, it is not referred to. He quotes another condemnation of the *Rasā'il*, that of Abū ʿAbd Allāh al-Māzarī (d. 536/1141),[77] which he must have particularly appreciated as it implicitly indicates a continuity between the Ikhwān and Ibn Sīnā.

> The [one] who composed the *Epistles* was a man, a philosopher, who plunged into the sciences of the Law [*sharʿ*], formed some mixture between the two sciences [i.e., Law and philosophy] and embellished philosophy in the hearts of the adepts of the Law by means of [Qur'anic] verses and *ḥadīth* that he quoted to them. Then, in this later period, there was a philosopher known as Ibn Sīnā who filled the world with writings concerning the sciences of philosophy, quoted the Law as his authority, and adorned himself with the ornaments of the Muslims. His strength in the science of philosophy led him subtly to make every effort in order to reduce the foundations of the creeds to the science of philosophy and he achieved, regarding this, things that were not achieved by the other philosophers.[78]

The theologian's point against the Ikhwān somehow differs from that of

74 Ibn Taymiyya, *Bughya*, p. 222.

75 Tawḥīdī, *Imtāʿ*, vol. 2, pp. 6–14; ʿAbd al-Raḥmān Badawī, *Quelques figures et thèmes de la philosophie islamique* (Paris: G. P. Maisonneuve et Larose, 1979), pp. 104–111 (hereafter cited as *Figures*).

76 Tawḥīdī, *Figures*, p. xxx.

77 On the Sicilian al-Māzarī, see Asin Palacios, M., 'Un faqīh siciliano, contradictor de Al Ghazzālī (Abū ʿAbd Allāh de Māzara),' in *Centenario della Nascita di Michele Amari*, vol. 2 (Palermo: 1910), pp. 216–244 (hereafter as 'Faqīh').

78 Ibn Taymiyya, *Iṣfahāniyya*, p. 133.

al-Sijistānī and al-Māzarī. In Ibn Taymiyya's opinion, first, the *Rasā'il*'s so-called 'synthesis between the Law and philosophy'[79] does not just 'conflict with religion', it is 'irrational'; second, their philosophy does not 'contradict' Islam exclusively, but also Judaism and Christianity.

> Will anyone, among those who know the religion of the Muslims, or [that] of the Jews, or [that] of the Nazarenes, deny that what the authors of the *Rasā'il Ikhwān al-Ṣafā'* say conflicts with the three religions [*milla*]? . . . As for opposing the Messengers concerning what they informed [us] about and commanded, accusing much of what they came with of being lies, and substituting [new things] for the Laws of all the Messengers, there are in these [epistles] things that do not remain hidden to whoever knows [even] one of the religions! Those people are in fact walking out of the three religions.[80]

> The matter with them is really that they believe neither in any of the prophets and the Messengers (neither Noah, nor Abraham, nor Moses, nor Jesus, nor Muḥammad, the blessings of God and His peace be upon all of them), nor in any of the Books of God sent down [to us] (neither the Torah, nor the Gospel, nor the Qur'an).[81]

> Someone holding the opinion that what the Messengers say corresponds to [what is said by] these Greeks, thus proves his ignorance of what the Messengers came with and of what these [Greeks] say. Such an [opinion] is only found in what is said by the heretics among the adepts of the [various] religions [*milla*] — the heretics of the Jews, the Nazarenes, the Muslims and others — like the authors of the *Rasā'il Ikhwān al-Ṣafā'* and their like among the heretics relating themselves to Shiʿism or to Sufism, like Ibn 'Arabī, Ibn Sabʿīn, and those similar to them.[82]

79 Ibn Taymiyya, *Bughya*, p. 199.

80 Ibn Taymiyya, *MF*, vol. 35, p. 134; also in Ibn Taymiyya, *MaF*, vol. 4, p. 234.

81 *Nuṣayriyya*, in Ibn Taymiyya, *MF*, vol. 35, p. 152; also in *MaF*, vol. 4, p. 212. French trans. in Guyard, '*Fetwa*', p. 190.

82 Ibn Taymiyya, *al-Jawāb al-ṣaḥīḥ li-man baddala dīn al-Masīḥ*, ed. 'Alī ibn Ḥasan ibn Nāṣir, 'Abd al-ʿAzīz ibn Ibrāhīm al-ʿAskar and Ḥamdan ibn Muḥammad al-Ḥamdān, 7 vols. (Riyadh : Dār al-ʿĀṣima li'l-Nashr wa'l-Tawzīʿ, 1419/1999), vol. 5, p. 37 (hereafter cited as *Jawāb*).

In sum, for Ibn Taymiyya, the Ikhwān 'strayed and led astray'.[83] More specifically:

> They confess neither that the world has a Creator who created it, nor that He has a religion which He commanded [to be adopted], nor that He has an abode, other than this abode, in which He retributes people for their actions.[84]

Without saying it explicitly, the theologian extends to the Ikhwān a judgement taken from al-Ghazālī, which he uses repeatedly against the Ismailis ("Ubaydids') of Cairo.

> The Sons of 'Ubayd . . . were just as Abū Ḥāmid al-Ghazālī said they were: outwardly, their doctrine was [Shi'i] rejectionism [*rafḍ*] whereas, inwardly, it was sheer unbelief [*al-kufr al-maḥḍ*]. . . . Inwardly, their religion is assembled from the religion of the Magi and the Ṣābi'ans. And what they practise outwardly of the religion of the Muslims is the religion of the [Shi'i] rejectionists [*rāfiḍ*]. The best among them are those who adhere to the religion of the [Shi'i] rejectionists [*rāfiḍ*] — yet they are the ignorant and the commoners among them, and every one of them thinks that he is a Muslim and believes that his religion is truly Islam. As for their elite — their kings and their scholars — they know that they are walking out of the religion of all the communities [*milla*]: the religion of the Muslims, the Jews, and the Nazarenes.[85]
>
> It is in those times that these epistles were composed, because of the appearance of this doctrine which, outwardly, is [Shi'i] rejectionism [*rafḍ*] whereas, inwardly, it is sheer unbelief. They were making out as if they were following the Law and [said] that it has an inward [meaning] which conflicts with its outward one.[86]

83 Ibn Taymiyya, *Kitāb al-Istighātha fī al-radd 'alā al-Bakrī*, ed. 'Abd Allāh ibn Dujayn al-Suhaylī, 2 vols. (Riyadh: Dār al-Waṭan, 1417/1997), vol. 2, p. 479 (hereafter cited as *Istighātha*).

84 *Nuṣayriyya*, in Ibn Taymiyya, *MF*, vol. 35, p. 152; also in Ibn Taymiyya, *MaF*, vol. 4, p. 212). French trans. in Guyard, '*Fetwa*', p. 190.

85 Ibn Taymiyya, *MF*, vol. 27, p. 174.

86 Ibn Taymiyya, *Minhāj*, vol. 4, p. 55. See also Ibn Taymiyya, *MF*, vol. 11, p. 581; *Nuṣayriyya*, in Ibn Taymiyya, *MF*, vol. 35, p. 152; also in Ibn Taymiyya, *MaF*,

The whole approach revealed by these texts is fairly general, and one wonders what Ibn Taymiyya effectively knew of the structure and actual content of the *Rasāʾil*. Quoting al-Māzarī, he writes: 'There are fifty-one of these epistles, each of them being an independent treatise.'[87] Acknowledging elsewhere that not everything in the *Rasāʾil* is 'to be condemned', he alludes to some of the disciplines they relate to.

> Concerning mathematics, physics, some [parts] of logic, *divinalia*, and the sciences of ethics, politics, and domestic economy, there are in these [epistles] things that are not to be rejected [*lā yunkaru*].[88]

He nevertheless suspects a 'devious intention' in the Ikhwān's interest for these disciplines. 'They lead many astray' by their philosophy:

> [It] contains subjects of physics and mathematics that have no connection to the subject of prophethood and messengership, neither negatively, nor positively, but from which one benefits in terms of living well in this world — for example, the arts of agriculture and weaving, construction, tailoring, and others.[89]

Nowhere does Ibn Taymiyya comment on specific pages from the *Rasāʾil* as he does with works by Ibn Sīnā, al-Ghazālī, al-Ṭūsī, and several other authors. Generally, he does not even consider them in themselves but refers to them as an illustration of an ideological perspective, philosophical doctrines, and particular views, shared by a wider group of thinkers, which he is busy exposing and refuting. Syncretism, as already been shown, was the main ideological approach for which the theologian attacks the Ikhwān. As for philosophical doctrines, there seem to be two main ones he has problems with: (a) the

vol. 4, p. 212. French trans. in Guyard, '*Fetwa*', p. 189. See also Ibn Taymiyya, *Ṣafadiyya*, vol. 1, p. 2.

87 Ibn Taymiyya, *Iṣfahāniyya*, p. 133.

88 Ibn Taymiyya, *MF*, vol. 35, p. 134; also in Ibn Taymiyya, *MaF*, vol. 4, p. 234.

89 Ibn Taymiyya, *MF*, vol. 12, p. 23.

Ikhwān's prophetology and hermeneutics; (b) their philosophy of the intellect and the soul.

For Ibn Taymiyya, the Ikhwān's prophetology and hermeneutics are directly influenced by their syncretism. As philosophical truth is supposed to correspond to revealed truth, they say, the different style of language apparent in prophetic messages must be explained by the obligation that the Messengers have to address the commonalty, not just the intellectual elite. Hence, their symbolist approach to scriptures, which other thinkers shared also, and which the theologian analyses mainly through the prism of Ibn Sīnā's thought: the utilitarian hermeneutics set forth in the *Risāla aḍḥawiyya* of the Shaykh al-Raʾīs, which Ibn Taymiyya knew well, having written a commentary on it.[90] From this perspective, philosophers are able to recognise in the revealed books images, similitudes, and parables of what reason teaches them; prophets are also aware of this philosophical truth, but do not teach it as such because the vulgar would not be able to grasp it if they were given a transparent version of it. Moreover, a transparent teaching of the truth, i.e., philosophy, to the vulgar would stir up sectarian discussions and threaten public order. The Qurʾan's abstention from doing so is thus in the greatest interest of human society, according to the Ikhwān.

> There is no doubt about it, the Qarmaṭīs and their like among the philosophers say that the [Prophet] said publicly the opposite of what he knew inwardly, and that he addressed the commoners [*ʿāmma*] with things by which he meant the opposite of what he was making them understand in their interest [*maṣlaḥa*], as it was not possible for him to reform them but in this way. This is what was claimed by Ibn Sīnā, the authors of the *Rasāʾil Ikhwān al-Ṣafāʾ*, and their like among the philosophers and the esotericist [*bāṭinī*] Qarmaṭīs.[91]

Ibn Taymiyya grasps this interpretation of the Prophet's pedagogic method as a process of 'substitution', encouraging estimation and imagining.

90 See the text translated in Michot, 'Mamlūk', parts 1 and 2.

91 Ibn Taymiyya, *MF*, vol. 13, p. 249.

The people [speaking] of estimation [*wahm*][92] and imagining are those who say that the prophets gave [us] information about God and the Last Day, the Garden and the Fire, and even about the angels, that does not conform to the matter in itself. They addressed people with things thanks to which they could imagine and estimate that God is an enormous body, that the bodies will be returned [to life], and that there will be for them a bliss and a chastisement perceptible through the senses, even if the matter, in itself, is not like that. It is indeed in the interest [*maṣlaḥa*] of the vulgar that they be addressed in such a way that they estimate and imagine that that is the case. Even if this is a lie, it is a lie in the interest of the vulgar, since inviting them [to follow the path of God] and [securing] their interest was not possible except in this way. On this principle Ibn Sīnā and his like based the rule [*qānūn*] they [follow when reading the revealed texts], as [is the case with] the rule he mentions in his *Epistle for the Feast of the Sacrifice* [*al-Risāla al-aḍḥawiyya*]. By these terms, those people say, the prophets meant their outward meanings. They wanted the vulgar to understand, from them, these outward meanings, even if these outward meanings, as far as the matter itself is concerned, are a lie, something vain, opposed to the truth. Their purpose was to make the vulgar understand by means of lies and vain things, in their [own better] interest [*maṣlaḥa*].[93]

Intrinsically [*fī al-bāṭin*], they [the philosophizers] say that what the Messengers informed [us] about concerning God and the Last Day has no truth in itself and only constitutes images [*takhyīl*], similitudes [*tamthīl*], and parables [*amthāl*] that are given in order to make the commonalty understand things from which, they claim, they profit, even if it is contrary to reality as far as the thing itself is concerned. They may also consider that the characteristic of prophethood is to make up images [*takhyīl*].[94]

This is, in sum, what the philosophizers and the esotericists say,

92 On the translation of '*wahm*' by 'estimation' and its role, close to 'imagination', in this internal sense, see Deborah L. Black, 'Estimation (*wahm*) in Avicenna: The Logical and Psychological Dimensions', *Dialogue*, 32 (1993), pp. 219–258 (hereafter cited as 'Estimation').

93 Ibn Taymiyya, *Dar'*, vol. 1, pp. 8–9; French trans. in Michot, *Lettre*, pp. 21–22.

94 Ibn Taymiyya, *Ṣafadiyya*, vol. 1, p. 237.

> those such as the Ismaili heretics, the authors of the *Rasā'il* of the Shi'i, al-Fārābī and Ibn Sīnā, al-Suhrawardī (the one who was killed), Ibn Rushd (the grandson), and the Sufi heretics walking out of the way of the earlier shaykhs who belonged to the People of the Book and the Tradition, like Ibn 'Arabī, Ibn Sab'īn, Ibn Ṭufayl (the author of *Ḥayy ibn Yaqẓān*), and many other creatures.[95]

For Ibn Taymiyya, three important distinctions must then be made at this stage of reflection. The first one deals with 'respect', or 'disrespect', to the intelligence of the Prophet:

> [Some philosophers] say that the Prophet knew the truth but publicly said the opposite of it in the [people's best] interest, whereas some others say that he did not know the truth as the speculative philosophers and their like know it.[96]

Thus, he concludes that the latter type 'give pre-eminence to the perfect philosopher over the Prophet' — this was the case with 'al-Fārābī, Mubashshir ibn Fātik, and others' — or 'to the perfect Friend [of God; *walī*], to whom this [spiritual] spectacle offers itself'[97] — this was the case with, for example, Ibn 'Arabī and Ibn Sab'īn. As for the philosophers saying that the Prophet secretly knew the truth:

> They say that the Prophet is more eminent than the philosopher because he knows what the philosopher knows, and something more. It is also possible for him to address the vulgar in a way that the philosopher is unable to adopt. Ibn Sīnā and his like are among these.[98]

What about the Ikhwān? In a text already quoted, Ibn Taymiyya clearly indicated that he put them in the latter group: 'The Qarmaṭīs and their like among the philosophers say that the [Prophet] publicly said the

95 Ibn Taymiyya, *Dar'*, vol. 1, pp. 10–11; French trans. in Michot, *Lettre*, pp. 24–25.
96 Ibn Taymiyya, *Dar'*, vol. 1, p. 9; French trans. in Michot, *Lettre*, pp. 22–23.
97 Ibn Taymiyya, *Dar'*, vol. 1, pp. 9–10; French trans. in Michot, *Lettre*, p. 23.
98 Ibn Taymiyya, *Dar'*, vol. 1, p. 10; French trans. in Michot, *Lettre*, p. 23.

opposite of *what he knew inwardly*' [italics added].[99] Elsewhere, Ibn Taymiyya also writes:

> The prophet, as seen by these philosophizers, is like a scholar practising *ijtihād*, whom one follows [*al-mujtahid al-matbūʿ*], as seen by the exponents of *kalām*. This is why those who associate the [philosophers] with the prophets, like the authors of the *Rasāʾil* of the Shiʿi and their like, say 'The prophets and the wise are agreed,' or they speak of 'the prophets and the philosophers'. Just as those studying the principles [of the religion] [*uṣūlī*] say 'The prophets, the jurists, and the exponents of *kalām* are agreed . . . ' and 'This is what the jurists and the *kalām* theologians say.'[100]

A second distinction concerns doctrinal and practical scriptural teachings. Is it only the literal meaning of the scriptural statements related to creed and beliefs, specially tailored to the vulgar, that the elite can dispense with, or can such an approach to the Prophet's message be extended to encompass ritual and practical prescriptions? According to Ibn Taymiyya, the 'esotericists' and Qarmaṭīs took the latter position:

> Anyone hearing the Qur'an, the abundantly certified [*mutawātir*] *ḥadīth*s and the commentary of the Companions and the Followers on these, is bound to know that the Messenger, God bless him and grant him peace, informed [us] of the return of the bodies [in the hereafter] and [also] that to belittle that is like belittling the fact that he came with the five prayers, the fasting of the month of Ramaḍān, the pilgrimage to the Ancient House, etc. The esotericist Qarmaṭīs, who are among the philosophers, denied this and that, and claimed that these [things] are all symbols [*ramz*] of, and allusions [*ishāra*] to, esoteric sciences. So they say that 'praying is knowing our secrets', 'fasting is concealing our secrets', 'going on pilgrimage is visiting our sanctified shaykhs', and other similar things that are mentioned in the books composed in order to uncover

99 Ibn Taymiyya, *MF*, vol. 13, p. 249; Ibn Taymiyya, *Dar'*, vol. 1, p. xxx.
100 Ibn Taymiyya, *Radd*, p. 366.

> their secrets and rend their veils.[101] It is for these Qarmaṭīs that the *Rasā'il Ikhwān al-Ṣafā'* was composed.[102]

In another text, the theologian seems to imply that 'these [esotericists and Qarmaṭīs] reject the outward meanings of [scriptural] matters relating to practice and knowledge altogether',[103] even for the masses. Nevertheless, he does not count the Ikhwān among these 'extreme rejectionists'.

> As for others than these, they do not reject the outward meaning of the abundantly certified prescriptions [*mutawātir*] relating to practice but consider them aimed at the commonalty of humans, not at their elite — just as they say something similar concerning [scriptural] matters relating to information. The axis of what they say is that, as far as knowledge and practice are concerned, the [prophetic] message assures the [best] interest [*maṣlaḥa*] of the commonalty; as for the elite, no. This is also the axis of what is said by the authors of the *Rasā'il Ikhwān al-Ṣafā'* and the rest of the most eminent of the philosophizers.[104]

Ibn Taymiyya almost certainly considers Ibn Sīnā one of these 'most eminent of the philosophizers', and this is not fair towards him since the Shaykh al-Ra'īs does speak explicitly of the usefulness of religious practices for the philosophical elite as for the commonalty. Defending the veracity of the prophets, Ibn Sīnā also explains how, in the hereafter, people will live in the reality of their paradisial promises and infernal threats by way of imagination.[105] Ibn Taymiyya does not allude to

101 'The *qāḍī* Abū Bakr ibn al-Ṭayyib composed his book, which he titled *The Uncovering of the Secrets and the Tearing of the Veils* [*Kashf al-asrār wa hatk al-astār*], in order to uncover their condition; it was similarly the case for an amazing number of Muslim scholars, intended by God, like the *qāḍīs* Abū Ya'lā and Abū 'Abd Allāh Muḥammad ibn 'Abd al-Karīm al-Shahrastānī.' (Ibn Taymiyya, *MF*, vol. 27, p. 174)

102 Ibn Taymiyya, *Iṣfahāniyya*, pp. 169–170.

103 Ibn Taymiyya, *MF*, vol. 4, pp. 99–100.

104 Ibid., *MF*, vol. 4, p. 100.

105 See Yahya J. Michot, *La destinée de l'homme selon Avicenne. Le retour à Dieu* (ma'ād) *et l'imagination* (Leuven: Peeters, 1986) (hereafter cited as *Destinée*).

this imaginal eschatology when he attributes the same hermeneutics of 'socially beneficial "prophetic lies"' to both the Ikhwān and the author of the *Risāla aḍḥawiyya*'. In a particularly interesting passage devoted to it, however, he provides a 'clue' as to why Ibn Sīnā had to develop such an eschatology; if 'there was no truth' in the Prophet's teachings:

> Those among people who possess intelligence and acumen inevitably would have known about it. And if they had known about it, it would have been generally impossible for them to continue to connive to conceal it, just as it is impossible for them to continue to connive in order to deceive. Indeed, just as it is impossible for most people to continue to connive in order to deceive, it would also be impossible for them to continue to connive in order to conceal something, the disclosure and mention of which there would be numerous interests and motives for; especially in such a case as knowledge of these grave matters.[106]

A third and final distinction relates to the exact purpose of the Prophet's education of the common people. An ambiguity indeed subsists regarding what the latter are effectively supposed to do with these symbols [*ramz*], allusions [*ishārāt*], images, and other metaphors of the philosophical truth which the philosophers have discovered in the revealed messages. Are the crowds expected to remain beneath the veil by sticking as much as possible to the apparent meaning of these messages or are they invited to pass through the veil, beyond this outward meaning, and thereby to access the realm of spiritual wisdom? Are prophets just sent to reform societies or, as far as possible, also to introduce them to the truth? In the texts examined, Ibn Taymiyya does not seem to be fully aware of the problem — possibly because he does not distinguish sufficiently between Ibn Sīnā's realistic and pragmatic hermeneutics and al-Fārābī's idealistic one.[107] In the various passages quoted above, he appears to have somewhat hastily equated

106 Ibn Taymiyya, *MF*, vol. 13, p. 249.

107 See Michot, *Destinée*, pp. 33–43; and Miriam M. Galston, 'Realism and Idealism in Avicenna's Political Philosophy', *The Review of Politics*, 41 (1979), pp. 561–577 (hereafter cited as 'Realism').

the Ikhwān's hermeneutics with that of Ibn Sīnā: 'By these terms, those people say, the prophets meant their outward meanings.'[108] There is, however, another text in which, in a way more akin to Fārābī, he considers that, for the Ikhwān, the function of the revealed parables is not to be an opaque screen but a translucent film.

> As for the hypocrites in this community, who do not confess the well-known terms of the Qur'an and the Sunna, they detach the words from their contexts and say that these [descriptions of paradise and hell] are parables given so that we may understand the spiritual return [of the soul in the hereafter]. They are similar to the esotericist Qarmaṭīs whose sayings are composed from things said by the Magi and the Ṣābi'ans, and similar to the Ṣābi'an philosophizers relating themselves to Islam, as well as to a group of people who resemble them: secretaries [*kātib*], or physicians [*mutaṭabbib*], or *kalām* theologians, or Sufis, like the authors of the *Rasā'il Ikhwān al-Ṣafā'* and others, or hypocrites.[109]

Particular aspects of the Ikhwān's philosophically concordant prophetology are also criticised by Ibn Taymiyya. This is the case with their conception of the epistemic and practical properties of prophethood, which the theologian says they share with various other thinkers, including al-Ghazālī, and in which Ibn Sīnā's ideas also clearly play some part. The revealing of the Qur'an is thus reduced, in his eyes, to an intellectual overflowing upon the soul of the Prophet, followed by specific operations of the imagination and common sense. The prophetic miracles are explained as psychic powers, the angels become intelligible forms and Mount Sinai, where Moses was spoken to, the active intellect. Moreover, such powers are of course accessible to humans other than God's Messengers, and effectively actualised to various degrees among philosophers and others.

> [The philosophers] considered the properties of prophethood to be of two species: (1) The faculty of knowledge by which

108 Ibn Taymiyya, *Dar'*, vol. 1, p. 9; French trans. in Michot, *Lettre*, p. 22.
109 Ibn Taymiyya, *MF*, vol. 4, p. 314.

knowledge is obtained — either by means of the logical syllogism or by means of the detachment [*tajarrud*] that is like the detachment of the sleeper, so that it is joined to the soul of the celestial sphere. (2) The practical faculty, which consists in the soul having the power to dispose freely of the world's *hylê* in such a manner that prodigies arise from it. The first species comprises two things: the first is the knowledge of universal sciences by the logical syllogism, and the second is the knowledge of the particulars by this junction [*ittiṣāl*]. The imagination thereafter represents the intelligibles as forms corresponding to them and engraves them into the common sense. Man then sees forms and hears sounds inside of himself. These forms, according to [these philosophers], are the angels of God, and these sounds are the speech of God. This is why the heretics among the Sufis adopted their way, like Ibn ʿArabī, Ibn Sabʿīn, and others. They travelled along the road of the Shiʿi heretics — like the authors of the *Rasāʾil Ikhwān al-Ṣafāʾ* — and followed what they found of the words of the author of The Books Kept Away From Those Who Are Not Worthy of Them [*al-Kutub al-maḍnūn bi-hā ʿalā ghayr ahli-hā*] and of other [writings] corresponding to that. Some of them, therefore, got to the point where they thought that the door of prophethood was [still] open and could not possibly be shut, saying as Ibn Sabʿīn used to say, 'The son of Āmina was presumptuous when he said: "No prophet after me".' Or he thinks, because he gives a greater importance to the Law, that the door of prophethood is shut, and pretends that friendship [*walāya*] is greater than prophethood and that the Seal of the Friends is more knowledgeable of God than the Seal of the Prophets.[110]

[The philosophizers] say that this Qurʾan is the Speech of God and that what the Messengers came with is the Speech of God. What they mean, however, is that it flowed upon the soul of the Prophet, God bless him and grant him peace, from the active intellect. Sometimes they also said that the intellect is Gabriel, who is not lowly, that is, not mean with what is hidden [*al-ghayb*] because he is overflowing. They say that God spoke to Moses from the heaven of his intellect, and that the adepts of exercise and purity get to a point where they hear what Moses heard, just as Moses heard it. These sayings led astray many

110 Ibn Taymiyya, *Radd*, p. 410.

> of the well-known [thinkers] such as Abū Ḥāmid al-Ghazālī, [who] discussed this idea in some of his books. They also composed the *Rasāʾil Ikhwān al-Ṣafāʾ*, and other texts.[111]

> To say so, that is, to say that the miracles of the prophets, God bless them and grant them peace, are psychic powers, is vain. Or, even, it is unbelief. Anyone saying it will be called to repent and the truth will be expounded to him. If he persists in believing it after the legal proof has been demonstrated to him, he is an unbeliever. If he persists in holding this [opinion] publicly after having been called to repent, he will be killed. This is among the things said by a group of the philosophizers, the esotericist [*bāṭinī*] Qarmaṭīs, the Ismailis, and those similar to them, such as Ibn Sīnā and his like, the authors of the *Rasāʾil Ikhwān al-Ṣafāʾ*, and the ʿUbaydids who were in Egypt — the Ḥākimīs — and those resembling them.[112]

> [The philosophizers] consider the suggestion [*īḥāʾ*] and inspiration [*ilhām*] that take place in the waking state and in sleep similar to the hearing of God's speech by Moses: they are equal, without any difference between them, apart from the fact that the words addressed to Moses were meant for him, whereas there are those who hear things that are addressed also to those other than them. Afterwards, when they truly express themselves, [however,] they go back to pure philosophy, [saying] that there is no difference at all between Moses and others. These philosophizers, indulging in interpretations, similarly consider 'taking off the two sandals' as an allusion to [one's] abandonment of the two worlds, the 'Mount' as an expression meaning the active intellect, and similar things [that are] among the interpretations of the Ṣābiʾan philosophers and of those who take after them — the Qarmaṭīs, the esotericists, the authors of the *Rasāʾil Ikhwān al-Ṣafāʾ*, and their like.[113]

These texts do not relate to prophethood exclusively but also involve specific opinions on the intellect and the soul. They thus invite one to move on to another central doctrine about which Ibn Taymiyya takes the Ikhwān to task: their philosophical ontology and psychology, with

111 Ibn Taymiyya, *MF*, vol. 12, p. 23.
112 Ibn Taymiyya, *Ṣafadiyya*, vol. 1, pp. 1–2.
113 Ibn Taymiyya, *MF*, vol. 6, pp. 180–181.

regard to the Creator, the angels, the cosmos, our present life, and the hereafter.

There is a tradition attributed to the Prophet which the Damascene theologian repeatedly comes back to in relation to the Ikhwān: 'The first [thing] that God created was the intellect.' His reason for doing so is the importance that he believes they give to this tradition in their syncretist philosophy, in this case in order to have Muslims accepting 'the saying of the philosophizers — the followers of Aristotle — that the first of the [things] emanating from the Necessary Existent is the intelligence'.[114] Other words attributed to the Prophet, remarks Ibn Taymiyya, are also cited by the Ikhwān, or by other thinkers, in order to legitimise some of their ideas from an Islamic viewpoint. For example, he recalls:

> This [man] also asked me about the *ḥadīth*s that the [unionists] put forward as arguments, for example the *ḥadīth* concerning the intellect that was referred to and [which says] that 'the first [thing] that God, Exalted is He, created was the intellect'. [Another] example was the *ḥadīth*: 'I was a treasury and was not known; so I loved to become known', etc. I therefore wrote him an elaborate answer and said that these *ḥadīth*s are invented.[115]

There indeed lies 'the problem':

> As arguments in favour of that they make use of prophetic words. These may be untrue sayings that they relate, as [for example] they relate that the Prophet, God bless him and grant him peace, said: 'The first [thing] that God created was the intellect'. The *ḥadīth* is inauthentic [*mawḍūʿ*]; those who possess the knowledge of the Tradition [*ḥadīth*] are agreed on that. Moreover, its wording is: 'When God created the intellect, He said to it: "Turn forward!" and it turned forward. He then said to it: "Turn backward!" and it turned backward.' They

114 *Nuṣayriyya*, in Ibn Taymiyya, *MF*, vol. 35, p. 153; also in Ibn Taymiyya, *MaF*, vol. 4, pp. 212–213); French trans. in Guyard, '*Fetwa*', p. 191.

115 Ibn Taymiyya, *Kitāb al-Nubuwwāt* (Beirut: Dār al-Fikr, 1409/1989), p. 83 (hereafter cited as *Nubuwwāt*). Ibn Taymiyya explains that this exchange took place during his stay in Alexandria (709/1309–1310).

> alter its wording and say: 'The first [thing] that God created was the intellect.'[116]

> As an argument, they may put forward the *ḥadīth* reported [as follows]: 'The first [thing] that God created was the intellect. He said to it: "Turn forward" and it turned forward. He then said to it: "Turn backward" and it turned backward.' He said: 'By My might! I have created no creature more precious to Me than you. By you I take and by you I give. By you I reward and you have to punish! ' This *ḥadīth* is a lie, invented.[117]

> The *ḥadīth* ['The first thing that God created was the intellect'] is, for the people possessing [knowledge of] the science of *ḥadīth*, a lie, invented. It is not [found] in any of the reliable books of Islam. It is only reported by people such as Dāʾūd ibn al-Muḥabbar,[118] and those, similar to him, who write about the intellect. It is mentioned by the authors of the *Rasāʾil Ikhwān al-Ṣafāʾ* and similar philosophizers. Abū Ḥāmid [al-Ghazālī] also mentioned it in some of his books; also Ibn ʿArabī, Ibn Sabʿīn, and their like. For the people possessing [knowledge of] the science of *ḥadīth*, it is a lie against the Prophet, God bless him and grant him peace, as mentioned by Abū Ḥātim al-Rāzī, Abū al-Faraj ibn al-Jawzī, and other people who write about the science of *ḥadīth*.[119]

He also states:

> The [prophetic words which they use as arguments] may also be [reported by them] from the Prophet, God bless him and grant him peace, with a confirmed wording. [However, then] they detach them from their context as the authors of the *Rasāʾil Ikhwān al-Ṣafāʾ* and their like do. They are indeed among their imams.[120]

116 *Nuṣayriyya*, in Ibn Taymiyya, *MF*, vol. 35, p. 153; also in Ibn Taymiyya, *MaF*, vol. 4, p. 212); French trans. in Guyard, '*Fetwa*', p. 190.

117 Ibn Taymiyya, *Jawāb*, vol. 5, p. 37.

118 Dāʾūd ibn al-Muḥabbar ibn Qaḥdham ibn Sulaymān (d. Baghdad, 206/821), was an unreliable *ḥadīth* transmitter. On this, see Khayr al-Dīn al-Ziriklī, *Al-Aʿlām: Qāmūs tarājim li-ashharr al-rijāl wa'l-nisāʾ min al-ʿarab wa'l-mustaʿribīn wa'l-mustashriqīn* (Sidon: al-Maṭbaʿa al-ʿAṣriyya, 1956), vol. 2, p. 334.

119 Ibn Taymiyya, *MF*, vol. 18, pp. 336–337.

120 *Nuṣayriyya*, in Ibn Taymiyya, *MF*, vol. 35, p. 153; also in *MaF*, vol. 4, p. 213;

As for the 'inauthentic' tradition of the intellect, for Ibn Taymiyya:

> It is amazing that those who desire to make a synthesis between the divine Law and Greek Peripatetic philosophy took this *ḥadīth* for their main reference [*ʿumda*] concerning the fundamentals of the religion, knowledge, and realisation [of the truth] [*taḥqīq*]. This [*ḥadīth*] is inauthentic [*mawḍūʿ*] and, yet, they all changed it, and reported [it as] 'The first [thing] that God created was the intellect. He said to it: "Turn forward"', took this for an argument, and considered it to correspond to what the Peripatetic philosophers — the followers of Aristotle — say when stating that the first of the [things] emanating from the Necessary Existent is the first intelligence. This spread out in the words of many of the later [thinkers], after they saw it in the *Rasāʾil Ikhwān al-Ṣafāʾ*.[121]

The theologian's insistence on the inauthenticity of this *ḥadīth* of the intellect comes from his conviction that, once accepted by Muslims, as urged by the Ikhwān and others after them, then nothing less than the entire *Weltanschauung* of the Hellenizing philosophizers will have been welcomed into Islam, along with its concomitant metaphysical and cosmological doctrines — the eternity of the world, the reduction of creation to a process of emanation, the divinisation of heavenly beings — and the ensuing 'threats', as he sees them, to the scriptural Islamic angelology, psychology, and eschatology.

> These [philosophizers] claim that the intelligences are pre-eternal and sempiternal, that the active intellect is the lord of everything which is under this sphere, and that the first intelligence is the lord of the heavens, of the earth, and of that which is between them.[122]

> These [thinkers] claim that the intellect proves the truth of their fundamentals and most people do not combine the knowledge of what the Messengers really came with [the knowledge of] what these [thinkers] really say. They do not

French trans. in Guyard, '*Fetwa*', p. 191.

121 Ibn Taymiyya, *Bughya*, pp. 179–180.

122 Ibn Taymiyya, *MF*, vol. 17, pp. 332–333.

> perceive the necessary consequences of what these [thinkers] say, due to which the corrupt nature of what they say becomes obvious to a clear intellect. Moreover, many people have adopted the doctrines of these [thinkers] but changed the way they are expressed. They sometimes express them by means of Islamic expressions so that the listener thinks that what these [thinkers] are saying is the truth with which the Messengers were sent and which is proven by the intellects. The authors of the *Rasā'il Ikhwān al-Ṣafā'* and the propagandists of the missionary calling [*da'wa*] of the esotericist [*bāṭinī*] Qarmaṭīs acted in that manner when they used the expressions 'the preceding' [*sābiq*] and 'the following' [*tālī*] for [respectively] the intellect and the soul. As an argument for this, they put forward the *ḥadīth* which they report on the authority of the Prophet, God bless him and grant him peace, and according to which he said: 'The first [thing] that God created was the intellect.' Afterwards, these things spread among many of the Sufis and the *kalām* theologians, the devout [*ahl al-ta'alluh*] and the intellectuals [*ahl al-naẓar*].[123]

Ibn Taymiyya is aware of the evolution that the philosophical doctrine of 'God' underwent from Aristotle to Ibn Sīnā, and he far prefers the latter to the Greek. Nevertheless, he fundamentally rejects the reduction of God to an abstraction, which he sees prevalent in the *Rasā'il* as in that of many other Muslim thinkers, from the Mu'tazilīs negating the divine attributes to Ibn Sīnā's doctrine of the Necessary Existent to the unionism of the adepts of *waḥdat al-wujūd*.

> What this man [Aristotle] and his followers generally speak about only concerns physics. It was the science with which people occupied themselves in their time. As for *divinalia* [*ilāhīyāt*], the man and his followers speak extremely little about them. Ibn Sīnā and his like nevertheless mixed their words about *divinalia* with the words of many of the theologians [*mutakallim*] of the [various] religious communities [*milla*], and people started to talk about *divinalia*.[124]
>
> Moreover, [these philosophizers] claim that the [Necessary

123 Ibn Taymiyya, *Ṣafadiyya*, vol. 1, p. 237.
124 Ibid.

> Existent] is existence absolute on the condition of absoluteness. It is neither given an identity [*muta'ayyin*] nor particularised by a reality by which it would be distinct from the rest of the existents. Rather, its reality is sheer existence, absolute on the condition of denial of all bonds, identifying and particularising. Now, they are learned in logic, and every intelligent [person capable of] conceiving what is thereby said [knows], that such an [existent] has no reality and no existence but in the mind, not in the outside. The Necessary Existent, to which outside existence bears testimony, thus comes to not exist except in the mind, and this participates in the most blatant contradiction, confusion, and conjoining of two contraries. As a necessary result of a real demonstration, they indeed make it existing externally, and as a necessary result of the negation of the [divine] attributes — that is, the [kind of] monotheism [*tawḥīd*] which they have imagined — they make it non-existing in the outside; what they say thus comes necessarily to imply its existence and [at the same time] its non-existence! It is similar for what is said by those who travel along their path: the esotericist Qarmaṭīs like the authors of the *Rasā'il Ikhwān al-Ṣafā'* and their like among the unionists — the adepts of the oneness of existence — such as Ibn Sab'īn, Ibn 'Arabī, and their like; or even along the path of the deniers of the attributes among the adepts of *kalām* theology like the Mu'tazilīs and others; or even along the path of the rest of those who deny any of the attributes. What necessarily follows from what they say is the denial and reduction of the [Necessary Existent] to nothing [*ta'ṭīl*] while, [simultaneously,] affirming its being there [*thubūt*]. It is thus the conjoining of two contraries![125]

Although he does not explicitly say so, the theologian might also have accused the Ikhwān 'and their likes' of self-contradiction in what they say about the angels. Sometimes, they make angels divine, immortal intelligences and souls ruling the world; and, in other contexts, they reduce them to the virtuous or wicked faculties of the soul.

> [Ibn Taymiyya] was asked: 'Do all the creatures, even the angels, die?' He answered that most people believe that the

125 Ibn Taymiyya, *Iṣfahāniyya*, p. 52.

> entirety of creatures die, even the angels, including Azrael, the Angel of Death. A *ḥadīth* going back to the Prophet, God bless him and grant him peace, is reported about that; and the Muslims, the Jews, and the Nazarenes are agreed about the possibility of that and about God's power to [enact] it. It is only opposed by groups of the philosophizers — the followers of Aristotle and their like — and of those who embark with them [on such views] among the people belonging to Islam, or the Jews, or the Nazarenes, like the authors of the *Rasā'il Ikhwān al-Ṣafā'* and those who, similar to them, claim that the angels are the intelligences and the souls and that their death is in no case possible. On the contrary, for them, the [angels] are gods, and lords of this world.[126]

> Whoever says that not all the angels prostrated in front of [Adam], but only the angels of the earth, rebuts the Qur'an through lying and slander. This and similar sayings are not among the things said by the Muslims, the Jews, and the Nazarenes. It is among the things said only by the philosophizing heretics who deem 'the angels' to mean the virtuous faculties of the soul, and 'the demons' to mean the wicked faculties of the soul. They deem 'the prostration of the angels' to mean the obedience of the [soul's] faculties to the intellect, and 'the refusal of the demons' to mean the disobedience of the wicked faculties to the intellect, and similar things that are said by the authors of the *Rasā'il Ikhwān al-Ṣafā'*, their like among the esotericist Qarmaṭīs, and whoever travels along their path of the straying *kalām* theologians and the pious worshippers. Similar things can also be found in the [Qur'an] commentators' sayings for which there is no reliable chain of authority [*isnād*].[127]

For Ibn Taymiyya, the doctrine of the human soul and its perfection, which is shared by the Ikhwān and others, obviously bears the mark of their metaphysics of angelic intelligences, as well as of their prophetology, and is judged by him as 'fundamentally wrong'. He expands on the point:

126 Ibn Taymiyya, *MF*, vol. 4, p. 259.
127 Ibid., pp. 345–346.

> Groups of esotericists — Shiʿi esotericists like the authors of the *Rasāʾil Ikhwān al-Ṣafāʾ* and Sufi esotericists like Ibn Sabʿīn, Ibn ʿArabī, and others — claim, and one also finds this in the words of Abū Ḥāmid al-Ghazālī and others, that the adepts of the [spiritual] exercises, of the cleansing of the heart, and of the purification of the soul by means of the laudable moral characters, may know, without the mediation of the information [given by] the prophets, the real essences [*ḥaqīqa*] of what the prophets have told regarding the faith in God, the angels, the Book, the prophets, the Last Day, and the knowledge of *jinn* and demons. [Such an idea] is based on this corrupt foundation, namely, that when they cleanse their souls, that [knowledge] comes down upon their hearts, either from the active intellect or from somewhere else.[128]

In fact, the theologian remarks that the claims sometimes made by philosophizers about what he terms the 'quasi-revealed nature' and, hence, 'irrefutable character' of their doctrines could well result from nothing more than autosuggestion:

> There is also somebody who pretends that he learned that [i.e. the Throne of God being the ninth celestial sphere] by the way of [mystical] uncovering [*kashf*] and contemplation [*mushāhada*]. He is lying in what he pretends. He only took that from these philosophizers, by imitating [*taqlīd*] them or agreeing with them on their corrupt road, just as the authors of the *Rasāʾil Ikhwān al-Ṣafāʾ* and their like did. He imagines [*takhayyala*] in his soul what he gets from another by imitating him and thinks that it is [some mystical] uncovering, just as a Nazarene imagines the Trinity in which he believes. He can also see that in his sleep and think it is [some mystical] uncovering, although it is only imagining what he believes in. For many of those who have corrupt beliefs, when they exert themselves, this exertion polishes their souls, and similitudes of their beliefs present themselves [*tamaththala*] to them, which they think is [some mystical] uncovering![129]

128 Ibn Taymiyya, *Radd*, p. 433.

129 Ibn Taymiyya, *MRK*, vol. 1, p. 258; also in Ibn Taymiyya, *MF*, vol. 6, p. 547 (with variants).

Moreover, he believes that the Ikhwān and other philosophizers are also 'misled and misleading' in what might be seen as an intellectualist understanding of the soul's bliss in the hereafter, of its government and morality in this life, and, more generally, in the 'theoretical-practical wisdom', which they adopted from the Greeks and then spread, often 'wrapped in an Islamic garb'.

> [For] those Ṣābi'ans among the philosophers, the utmost happiness of the souls consists in reaching the active intellect. The authors of the *Rasā'il Ikhwān al-Ṣafā'* composed their epistles according to the fundamentals of these [thinkers], mixed with things that they drew from the religion of the original monotheists [*ḥanīf*] ... As for the original monotheists, they believe that there is no servant whose Lord will not talk to him [in the hereafter], without any chamberlain or interpreter between them.[130]

> The insight of many people falls short of knowing what God and His Messenger love, regarding the things that benefit the hearts and the souls and those that corrupt them, what is useful to them — the truths of the faith — and what harms them — negligence and lust. As God, exalted is He, said: 'Do not obey any whose heart We have made negligent of Our remembrance, who follows his own lust and whose affair is excess.'[131] The Exalted also said: 'So, shun those who turn away from Our remembrance and want nothing but the life of this world. That is their attainment of knowledge.'[132] You will find that many of these [people], in many of the rulings [*ḥukm*] [of the religion], as far as benefits and causes of corruption are concerned, see nothing but things that go back to a financial or bodily benefit. And the furthest that many of them go, when they go beyond that, is to pay attention to the government of the soul and the refinement of morals [*siyāsat al-nafs wa-tahdhīb al-akhlāq*], in correlation to their attainment of knowledge. Such things are notably mentioned by the philosophizers and the Qarmaṭīs, such as the authors of the *Rasā'il Ikhwān al-Ṣafā'* and their like. They talk of the government of the soul and the refinement of morals in correlation to their

130 Ibn Taymiyya, *Istighātha*, vol. 2, pp. 478–479.
131 Qur'an (*al-Kahf*) 18:28.
132 Qur'an (*al-Najm*) 53:29–30.

attainment of philosophical knowledge and, added to this, their opinions about the Law. Now, at the furthest point which they eventually reach, they are still far beneath the Jews and the Nazarenes![133]

The edifice of the [philosophers'] practical wisdom is based wholly on the fact that they know that the soul has the faculty of concupiscence and anger — concupiscence for attracting what is suitable, and anger for repelling what is incompatible. They have then based [their] edifice of ethical wisdom on that. One ought, they said, to refine concupiscence and anger in order for each of the two to be between neglect and excess. This is called 'continence', which is also called 'courage', and the balance between the two, 'justice'. These three [qualities] are sought for in order to perfect the soul by means of theoretical-practical wisdom. Perfection, for them, then becomes these matters: continence, courage, justice, and science [*ʿilm*]. [Various] groups of those who entered Islam spoke of this and used as testimonies to that, what they found in the Qur'an, the *ḥadīth*, and the sayings of the Ancients [*salaf*] in praise of these affairs. Those who wrote about morals and actions according to their way base what they say on this principle. It is, for example, the case with *The Balance of Action* [*Mawāzīn al-aʿmāl*] of Abū Ḥāmid [al-Ghazālī], the authors of the *Rasāʾil Ikhwān al-Ṣafāʾ*, the books of Muḥammad ibn Yūsuf al-ʿĀmirī, and others. However, they were wrong. What God and His Messenger mean by the science that He praises is indeed not the theoretical science which the Greek philosophers had in mind.[134]

There is no doubt that Ibn Taymiyya's harsh judgement on the Ikhwān al-Ṣafāʾ (which also extends to the philosophers and the Ismailis) is extremely unfair and excessively negative. This is particularly true as regards their main philosophical doctrines and the syncretist perspective from which they developed them; Ibn Taymiyya's objections rest as much on the grounds of rational consistency as on his own religious motives. Although, his aggressive condemnation is not aimed at them exclusively but at all 'the philosophizers' whose ideas they share and

133 Ibn Taymiyya, *MF*, vol. 32, p. 233; also in Ibn Taymiyya, *MaF*, vol. 2, p. 15.
134 Ibn Taymiyya, *Radd*, pp. 368–369.

illustrate particularly well; namely, the condemnation is of philosophically *radical* ideas.

There was also no doubt in Ibn Taymiyya's hypercritical mind that by his time, the *Rasā'il* had exerted a palpable influence on various Muslim thinkers. Some of the texts already quoted speak clearly of this influence, as well as of a convergence between their views and those of later philosophers such as Ibn Rushd or those he judged as 'heretics' connected to Shi'ism or to Sufism, like Ibn 'Arabī, Ibn Sab'īn, and 'those similar to them'.[135] It is Ibn Sīnā, however, and, even more so, Abū Ḥāmid al-Ghazālī, whom the theologian regards as 'disciples of the Ikhwān'.

The 'privileged link' which Ibn Taymiyya maintains existed between the Ikhwān and the Shaykh al-Ra'īs is confirmed by a clear passage that deserves to be quoted here:

> Ibn Sīnā invented a philosophy which he assembled from the things said by his Greek predecessors and from what he drew from the innovating Jahmī *kalām* theologians and their like. He travelled along the road of the Ismaili heretics in many of the matters — related to knowledge [*'ilmī*] and practical — about which they [spoke] and mixed it with some of the words of the Sufis. His [ideas] really go back to the things said by his brothers, the Ismaili Qarmaṭī esotericists. Indeed, the members of his house were Ismailis:[136] followers of al-Ḥākim, who was reigning in Egypt and in whose time they were living. Their religion was the religion of the authors of the *Rasā'il Ikhwān al-Ṣafā'* and similar imams of the hypocrites of the [various] communities who are neither Muslims, nor Jews, nor Nazarenes.[137]

As for al-Ghazālī, Ibn Taymiyya is aware of two important things: (1) The diversity of the views sometimes displayed in his many writings. (2) The multiplicity of the influences by which he was affected.

---

135 Ibn Taymiyya, *Jawāb*, vol. 5, p. 37.

136 Ibn Taymiyya, noted as *'Al-Ismā'īliyya'* in *MF*, and as *'were followers'* in *MRK*.

137 Ibn Taymiyya, *MF*, vol. 11, p. 571; also in Ibn Taymiyya, *MRK*, vol. 2, pp. 288–289; French trans. in Michot, *Musique*, pp. 77–79.

It is because of this Ghazālian 'lack of consistency' that Ibn Taymiyya, as seen earlier on,[138] can quote a passage of the *Munqidh* attacking the Ikhwān, or some other Ghazālian statement directed against the 'Ismailis of Cairo', and, on another occasion, quote this statement of al-Māzarī:

> Some of the Companions of [al-Ghazālī], al-Māzarī said, informed me that he was addicted [*'ukūf 'alā*] to reading the *Rasā'il Ikhwān al-Ṣafā'*.[139]

No wonder, then, that the 'inauthentic' *ḥadīth* of the intellect embodying the Ikhwān's syncretism is also found in certain works of al-Ghazālī, or attributed to him, especially his famous *Maḍnūns*.

> In *al-Kutub al-maḍnūn bi-hā 'alā ghayr ahli-hā* [The Books Kept Away From Those Who Are Not Worthy of Them] and other discourses of that sort that are attributed to Abū Ḥāmid al-Ghazālī, there is also a bit of that [syncretism].[140]
>
> 'The first [thing] that God created was the intellect.' In what is said by Abū Ḥāmid al-Ghazālī in *al-Kutub al-maḍnūn bi-hā 'alā ghayr ahli-hā* [The Books Kept Away From Those Who Are Not Worthy of Them] and others, there is also a great deal of the ideas of those.[141]

As if he wanted to clear al-Ghazālī from any 'wrongdoing' relating to the pseudo-*ḥadīth* of the intellect, Ibn Taymiyya reports once that 'it is said that he abjured [*raja'a 'an*],[142] such ideas', and, elsewhere, he writes that:

> [He] does not use this [material] deliberately but quotes it either from the *Rasā'il Ikhwān al-Ṣafā'* or from what is said by Abū Ḥayyān al-Tawḥīdī, or from this sort of [author].

138 Ibn Taymiyya, *MF*, vol. 11, p. xxx.
139 Ibn Taymiyya, *Iṣfahāniyya*, p. 133.
140 Ibn Taymiyya, *Jawāb*, vol. 5, p. 37.
141 Ibn Taymiyya, *MF*, vol. 17, p. 333; Ibn Taymiyya, *Radd*, p. 410.
142 Ibn Taymiyya, *Bughya*, p. 181.

> In reality, these people belong to the genus of the Ismaili esotericists.[143]

This last passage already confirms that, for Ibn Taymiyya, the Ikhwān are not the only thinkers to have influenced al-Ghazālī. Partly following al-Māzarī, he underlines in various texts the importance of their specific impact on the Ḥujjat al-Islam's philosophy, and judges it comparable to that of al-Tawḥīdī or, even more, of Ibn Sīnā, notably in the *Mishkāt al-Anwār*.

> 'I found,' [al-Māzarī] said, 'that this al-Ghazālī relies on [Ibn Sīnā] in most of what he alludes to concerning the sciences of philosophy; so much so that, at some moments, he textually quotes his words, without change, whereas, at other moments, he changes them and relates them more to questions pertaining to the Law than Ibn Sīnā had done, as he is more knowledgeable of the secrets of the Law than him. . . . Al-Ghazālī relied on Ibn Sīnā and the authors of the *Rasā'il Ikhwān al-Ṣafā'* concerning the science of philosophy. As for the doctrine of the Sufis,' [al-Māzarī] said, 'I don't know who he relies on concerning them, nor who he relates himself to in order to know it. . . . I think,' [al-Māzarī] said, 'that it is Abū Ḥayyān al-Tawḥīdī the Sufi that he relies on for the doctrines of the Sufis.'[144]

> The material [used by] Abū Ḥāmid in philosophy comes from the words of Ibn Sīnā — this is why it is said that Abū Ḥāmid was made sick by *al-Shifā'* [The Healing] — from the words of the authors of the *Rasā'il Ikhwān al-Ṣafā'*, from the epistles of Abū Ḥayyān al-Tawḥīdī, and others. As for Sufism, which is 'the most sublime' of the sciences that he [possessed] and by which he is ennobled, most of his material in it comes from the words of the Shaykh Abū Ṭālib al-Makkī, whom he mentions in [the part of the *Iḥyā'* called] *al-Munjiyyāt* [The Things That Save] — [what he writes] about patience, gratitude, hope, fear, love, and sincerity is generally drawn from the words of Abū Ṭālib al-Makkī, but Abū Ṭālib [al-Makkī] was more intense and superior. Most of what he mentions in the quarter [of the

143 Ibn Taymiyya, *Nubuwwāt*, p. 83.
144 Ibn Taymiyya, *Iṣfahāniyya*, pp. 133–134.

> *Iḥyā'* called] *al-Muhlikāt* [The Things That Make One Perish], he drew from the words of al-Ḥārith al-Muḥāsibī in *al-Ri'āya* [The Observance], as is the case with what he reports about blaming envy, amazement, pride, ostentation, haughtiness, and the rest.[145]
>
> There is much philosophical material [*mādda*] in what al-Ghazālī says, because of what Ibn Sīnā says in *al-Shifā'* [The Healing] and elsewhere, [because of] the *Rasā'il Ikhwān al-Ṣafā'* and what Abū Ḥayyān al-Tawḥīdī says.[146]
>
> [Al-Ghazālī] divided the book [entitled *The Niche of Lights — Mishkāt al-Anwār*] into three chapters. The first chapter expounds that the real light is God, exalted is He, and that, for others than Him, the word "light" is purely metaphorical, without reality. His words go back to [the idea] that 'light' has the meaning of 'existence'. Before him, Ibn Sīnā proceeded in a similar fashion to that, by making a synthesis between the Law and philosophy — and likewise did the Ismaili esotericists proceed in their book called the *Rasā'il Ikhwān al-Ṣafā'*. After him, Ibn Rushd also did so. And likewise for the unionists [*ittiḥādī*]: they make His appearance and His epiphany in forms that have the meaning of His existing in [these forms].[147]

To the names (i.e., al-Ghazālī, Ibn Sīnā, Ibn Rushd, and the rest) mentioned by Ibn Taymiyya in this last passage could be added numerous ones which he associates with the Ikhwān. In some of his other texts he mentions al-Fārābī or Mubashshir ibn Fātik, al-Suhrawardī, Ibn 'Arabī, and Ibn Sab'īn as scholars associated with the corpus of the Ikhwān. Unfavourable though he was to them, the theologian's opinion of the Ikhwān indirectly demonstrates that he acknowledged the centrality of the *Rasā'il* in the development of philosophy and spirituality in the classical period of Islam, to the point that he maintained that they influenced two of its major representatives: Ibn Sīnā and al-Ghazālī. Even today, some analysts would still be reluctant to give so prominent an intellectual role to the Ikhwān and to Ismaili 'esotericists' or to the Qarmaṭī 'heretics' in the history of mainstream Islamic thought.

145 Ibn Taymiyya, *Bughya*, p. 449.
146 Ibn Taymiyya, *MF*, vol. 6, p. 54.
147 Ibn Taymiyya, *Bughya*, p. 199.

Did Ibn Taymiyya in some way overstate the Ikhwān's intellectual importance? This possibility cannot be dismissed, as it is clear that he had his own agenda when talking about the Ikhwān, and perhaps he made them into 'bugbears' for the purposes of his wider and strictly anti-philosophical undertakings. This being so, two questions, which could not be addressed here, still deserve attention at some point: (1) What is the real relevance of the convergences detected, or of the connections established, by Ibn Taymiyya, between the Ikhwān and other Muslim thinkers or movements of thought? And (2) How did the Damascene theologian's views on the Ikhwān influence the reception of their *Rasā'il* in later Islamic thought?

SEVEN

Epistolary Prolegomena: On Arithmetic and Geometry

Nader El-Bizri

Exordium

The *Rasā'il Ikhwān al-Ṣafā'* (*Epistles of the Brethren of Purity*)[1] commences with two initiatory mathematical tracts: the first on arithmetic (*Risāla fī al-'adad*; *al-arithmāṭīqī*), and the second on geometry (*Risāla fī al-jūmaṭrīyā*). Based on the quadrivium classification system (which is 'conceptually' attributed to Pythagoras of Samos [d. ca. 480 BCE], and was 'institutionally' elaborated by Anicius Manlius Severinus Boethius [d. ca. 524]),[2] the Ikhwān's initiatory epistles served as propaedeutic

1 All references to the epistles on arithmetic and geometry in this chapter are taken from the first volume of the *Rasā'il Ikhwān al-Ṣafā' wa-Khillān al-Wafā'*, ed. Buṭrus Bustānī, 4 vols. (Beirut: Dār Ṣādir, 1957) (hereafter cited as *Rasā'il*). I also refer the reader to an annotated English translation with commentaries of the epistle on arithmetic in Bernard R. Goldstein, 'A Treatise on Number Theory from a Tenth-Century Arabic Source', *Centaurus*, 10 (1964), pp. 129–160, with text on pp. 135–159. See also, Sonja Brentjes, 'Die erste *Risāla* der *Rasā'il Ikhwān al-Ṣafā'* über elementare Zahlentheorie', *Janus*, 71 (1984), pp. 181–274.

2 Supposedly coined by Boethius, the 'quadrivium' was integrated as part of the curriculum of mediaeval monasteries, and grouped together four topics of initiatory study, namely: arithmetic, geometry, astronomy, and music. In

enquiries in mathematics (*riyāḍiyyāt*; or, in Greek, '*mathêmatikê*'). The pedagogic rationale behind their epistles in arithmetic and geometry aimed at supporting further training in natural philosophy (*ṭabīʿiyyāt*; *phusikê*) and 'psychology' (*nafsāniyyāt*; *peri psukhês*; and the Latin being '*de anima*'), with a view to becoming properly primed to explore topics in theology (*ilāhiyyāt*; *thelogikê*).[3]

The *Rasā'il Ikhwān al-Ṣafā'* is customarily judged as being akin in its synoptic content to an encyclopaedia, although the investigations in arithmetic and geometry therein were less advanced in their epistemic consequences than other contemporary accomplishments within mediaeval Islamic civilisation in the ninth and tenth centuries.[4]

---

reference to Boethius' reflections on arithmetic, geometry, and music, including his adaptation of the mathematical tradition of Nicomachus of Gerasa, see Boethius, *De Institutione Arithmetica, De Institutione Musica, Geometria*, ed. G. Friedlein (Leipzig: Teubner, 1867).

3 See *Rasā'il*, p. 49. After all, arithmetic and geometry were described by the Ikhwān as being initiatory disciplines within the philosophical sciences ('*al-ʿulūm al-falsafiyya*'), which provided the optimal access to natural philosophy and psychology in laying down the pedagogic and epistemic foundations needed for the study of theology. Moreover, as Yves Marquet has noted, the Pythagorean teachings did not only shape the Ikhwān's mystical prolongations of the results of the mathematical disciplines, but, moreover, had an impact on the composition of many of their epistles, including those grouped under the heading of theology: Yves Marquet, *Les 'Frères de la Pureté', pythagoriciens de l'Islam: La marque du pythagorisme dans la rédaction des Épîtres des Ikhwān aṣ-Ṣafā'* (Paris: EDIDIT, 2006).

4 The polymaths of ninth- and tenth-century Islamic civilisation introduced motion into geometry, which was unprecedented in mathematics, as well as developing the field of geometrical transformations. They also devised the rudiments of dioptrics (the science of refraction in optics) based on geometric models, as well as applying algebra to geometry, and they found significant applications of mathematics in astronomy and mechanics, along with the design and perfection of scientific instruments, like astrolabes, sundials, and compasses. I have examined related aspects elsewhere, particularly with reference to developments in optics, and as exemplified by the oeuvre of the polymath al-Ḥasan Ibn al-Haytham (also known as Alhazen; d. after 1041), including the following studies: Nader El-Bizri, 'In Defence of the Sovereignty of Philosophy: al-Baghdādī's Critique of Ibn al-Haytham's Geometrisation of Place', *Arabic Sciences and Philosophy*, 17 (2007), pp. 57–80; Nader El-Bizri, 'A Philosophical Perspective on Alhazen's *Optics*', *Arabic Sciences and Philosophy*, 15 (2005), pp. 189–218; Nader El-Bizri, 'La perception de la profondeur: Alhazen, Berkeley et Merleau-Ponty', *Oriens-Occidens: Sciences, mathématiques et philosophie de l'antiquité à l'âge classique (Cahiers du Centre d'Histoire des Sciences et*

However, one ought to add, in fairness, that the Ikhwān explicitly affirmed that their compendium represented 'the most concise of summaries' (*awradnāhā bi-awjaz mā yumkin min al-ikhtiṣār*);[5] and one should concede that it was supplemented by oral teachings in seminars (*majālis al-ʿilm*). Nonetheless, the Ikhwān's epistles on arithmetic and geometry gave basic highlights of the mathematics of their era, even though their mathematical knowledge was less complex and hardly groundbreaking when compared with the foundational efforts of their predecessors, such as polymaths of the calibre of Muḥammad ibn Mūsā al-Khwārizmī (d. 840), the Banū Mūsā ibn Shākir (fl. mid-ninth century), Thābit ibn Qurra (d. 910), and his grandson Ibrāhīm ibn Sinān (d. 946).[6]

Arithmetic[7]

Arithmetic I

The early development of arithmetic in mediaeval Islamic civilisation resulted partly from the systematisation of *ḥurūf al-hind* (namely, the Hindi numerals associated with the *ghubār* 'dust-board' method

des Philosophies Arabes et Médiévales, CNRS), 5 (2004), pp. 171–184; Nader El-Bizri, 'Ibn al-Haytham', in *Medieval Science, Technology, and Medicine: An Encyclopedia*, ed. Thomas F. Glick, Steven J. Livesey, and Faith Wallis (New York and London: Routledge, 2005), pp. 237–240.

5 *Rasāʾil*, p. 77.

6 One should also note the great contributions in mathematics that were made in the tenth century and beginnings of the eleventh century by the likes of al-Uqlīdisī (d. 980), al-Būzjānī (d. 998), Ibn Sahl (d. 1000), al-Qūhī (d. 1000), Ibn Yūnus (d. 1009), al-Sijzī (d. 1020), al-Karājī (d. 1029), Ibn al-Haytham (d. ca. 1040), Ibn Sīnā (also known as Avicenna; d. 1037), and al-Bīrūnī (d. 1048).

7 The epistle on arithmetic is divided into fourteen chapters (*fuṣūl*), which are as follows: 'Untitled' preamble (pp. 48–56); 'Properties of Numbers' (pp. 56–60); 'Whole [Numbers] and Fractions' (pp. 60–64); 'Perfect, Deficient, and Abundant Numbers' (pp. 64–65); 'Amicable [Numbers]' (p. 65); 'Multiplication [*duplation*; *taḍʿīf*] of Numbers' (p. 66); 'Subdivisions' (pp. 66–68); 'Whole Numbers' (pp. 68–69); 'Multiplication, Roots, and Cubes' (pp. 69–70); 'Square Numbers' (pp. 70–71); 'Root Numbers' (p. 72); 'Propositions from Euclid's Elements, Book II' (pp. 72–75); 'Arithmetic and the Soul' (p. 75); and 'The Purpose of the Sciences' (pp. 75–77). In this part of the chapter on arithmetic, we shall follow an approximately parallel subdivision suitable to the scope of this study, to be noted under the sub-title '*Arithmetic*' followed by a roman numeral.

of numeration). Moreover, following the Greek mathematical traditions, 'the science of number' (*arithmêtikê tekhnê*; *'ilm al-'adad*) was distinguished from 'the science of reckoning' (*logistikê tekhnê*; *'ilm al-ḥisāb*) in view of highlighting the pragmatic and quotidian utilities of calculation. For instance, Abū al-Wafā' al-Būzjānī (d. 998) composed his treatise *Kitāb Fī mā yaḥtāj ilayh al-kuttāb wa'l-'ummāl min 'ilm al-ḥisāb* (What Scribes and Labourers Require of the Science of Calculation) to serve mercantile needs; and this computational art was further elucidated by Abū al-Ḥasan Aḥmad ibn Ibrāhīm al-Uqlīdisī (d. ca. 980) in his *Fī al-ḥisāb al-hindī* (On Indian Calculation).[8]

The Ikhwān's 'number theory' was inspired by the received Pythagorean tradition, and it was principally derived from the *Arithmêtikê eisagôgê* (*Introductio Arithmetica* in Latin; *al-Madkhal ilā 'ilm al-'adad*)[9]

8 The science of calculation (*'ilm al-ḥisāb*) consisted of three main types, namely: hand reckoning (*ḥisāb al-yad*), finger reckoning (*ḥisāb al-'uqūd*), and mental reckoning (*al-ḥisāb al-hawā'ī*). For references, see Ahmad S. Saidan, 'The Earliest Extant Arabic Arithmetic: *Kitāb al-Fuṣūl fī al-Ḥisāb al-Hindī* of Abū al-Ḥasan Aḥmad ibn Ibrāhīm al-Uqlīdisī', *Isis*, 57 (1966), pp. 475–490; Ahmad S. Saidan, *The Arithmetic of al-Uqlīdisī* (Dordrecht: D. Reidel, 1978); Ahmad S. Saidan, 'Numeration and Arithmetic', in *Encyclopedia of the History of Arabic Science*, ed. Roshdi Rashed with Régis Morélon, vol. 2 (London: Routledge, 1996), pp. 331–348. I also refer the reader to Rida A. K. Irani, 'Arabic Numeral Forms', *Centaurus*, 4 (1955), pp. 1–12; and the synoptic survey in A. I. Sabra, ''Ilm al-ḥisāb', *EI2*, vol. 3, pp. 1138–1141.

9 Nicomachus' *Introduction to Arithmetic* was translated into Arabic by Thābit ibn Qurra, under the title *Kitāb al-Madkhal ilā 'ilm al-'adad*, in *Arabische Übersetzung der Arithmêtikê Eisagôgê des Nikomachus von Gerasa*, ed. Wilhelm Kutsch (Beirut: Imprimerie Catholique, St. Joseph, 1959), with a manuscript preserved at the British Museum, as per the particulars: MS 426.15, (add. 74731), ff. 198, dated 1242. Nicomachus' tract had a marked Pythagorean mystical penchant, a tendency that was more emphatically articulated in his *Theologumena arithmeticae* (The Theology of Numbers), ed. Fridericus Astius (Leipzig: Libraria Weidmannia, 1817). Nicomachus' *Introduction to Arithmetic* itself is published under the title *Nicomachi Geraseni Pythagorei Introductionis arithmeticae*, ed. Richard Hoche (Leipzig: Teubner, 1886). See also Nicomachus of Gerasa, *Introduction to Arithmetic*, trans. M. L. D'Ooge, with studies by F. E. Robbins and L. C. Karpinski (New York, 1926); [Nicomaque de Gérase,] *Introduction arithmétique*, trans. Janine Bertier (Paris: J. Vrin, 1978); Leonardo Tarán, 'Nicomachus of Gerasa', in *Dictionary of Scientific Biography*, ed. Charles Coulston Gillispie et al., vol. 10 (New York: Charles Scribner's Sons, 1974), pp. 112–114.

of Nicomachus of Gerasa (ca. 60–120, Roman Syria),[10] who emphasised the anteriority of arithmetic with respect to geometry; given that geometry presupposes arithmetic and already implies it within its own disciplinary bounds. It is moreover reported that Nicomachus alluded to the composition of an *Introduction to Geometry* that followed his *Arithmetica*, though that tract is lost.[11]

In emulating Nicomachus' tradition, the Ikhwān assigned a semantic surplus to the entailments of arithmetic as embodied in mystical symbolism. They also relied on induction rather than proceeding by way of proofs (*barāhīn*) in this particular mathematical discipline.[12] In addition, they believed that numbers corresponded foundationally with natural phenomena and worldly configurations in reference to signifiers associated with time, place, the elements, classes of existents (*al-mawjūdāt*), and the differential derivations of human language.[13] Their analogism, which was saturated with picture-based language, resulted in mistaking resemblances for explanations; consequently, their *meta-mathematical* speculation around arithmetic borders on 'empty verbalism', which produced pseudo-explanations that are hardly translatable into meaningful epistemic terms.[14]

---

10 *Rasā'il*, p. 49. See also Carmela Baffioni, 'Citazioni di autori antichi nelle *Rasā'il degli Ikhwān al-Ṣafā'*: il caso di Nicomaco di Gerasa', in *The Ancient Tradition in Christian and Islamic Hellenism: Studies on the Transmission of Greek Philosophy and Sciences Dedicated to H. J. Drossaart Lulofs on his Ninetieth Birthday*, ed. Gerhard Endress and Remke Kruk (Leiden: Leiden Research School, 1997), pp. 3–27; Muḥammed Jalūb Farhān, 'Philosophy of Mathematics of Ikhwān al-Ṣafā', *Journal of Islamic Science*, 15 [Aligarh] (1999), pp. 25–53.

11 See Bertier, *Introduction arithmétique*, pp. 8, 58.

12 In this context, one would evoke the structuring function assigned by the Ikhwān to the microcosm/macrocosm analogy and to magic. See *Risāla* 26 (vol. 2, p. 12) 'On the Microcosm' (*al-Insān 'ālam ṣaghīr*; the human being is a 'micro-*kosmos*'); *Risāla* 34 (vol. 3, p. 3) 'On the Macrocosm' (*al-'Ālam insān kabīr*; the world is a '*macro-anthropos*'); and *Risāla* 52 (vol. 4, p. 10) 'On Magic' (*fī al-siḥr*). See also Geo Widengren, 'Macrocosmos–Microcosmos Speculation in the *Rasā'il Ikhwān al-Ṣafā'* and Some Ḥurūfī Texts', *Archivio di Filosofia*, 1 (1980), pp. 297–312; Paul Casanova, 'Alphabets magiques arabes', *Journal Asiatique*, 18 (1921), pp. 37–55; ibid., 19 (1922), pp. 250–262.

13 For instance, the fourfold schema (i.e., the four *murabba'āt*) described the correspondence of the elements, humours, temperaments, and seasons, following in this a tradition associated with Galenism. *Rasā'il*, pp. 51–52.

14 For an illumination of Pythagorean mathematics, I refer the reader to Jules Vuillemin, *Mathématiques pythagoriciennes et platoniciennes* (Paris: Albert

The Ikhwān were explicit about their indebtedness to Pythagoras and Nicomachus in arithmetic,[15] to Euclid's *Elements* (*Stoikheia*; *Kitāb Uqlīdis fī al-uṣūl*) in geometry, and to the *Almagest* of Claudius Ptolemy (d. 165) in astronomy. And yet, although the influence of Nicomachus is evident throughout the Ikhwān's epistle on arithmetic, the subdivisions and their contents do not entirely overlap with those of Nicomachus' tract.[16] One of the determining distinctions between both texts may be attributed to the stress on monism in the Ikhwān's *risāla*, versus a more pronounced emphasis by Nicomachus on dualism and on the associated interplay between *identity qua sameness* (the unit 1 *qua* the same) and *difference qua otherness* (the number 2 *qua* the other). Furthermore, the Ikhwān did not even mention Nicomachus' focal distinction between what he designated as a *noêtos arithmos* ('intelligible number'; what in Arabic could be coined as *'adad 'aqlī*) and the *epistêmonikos arithmos* (namely, the 'mathematical/scientific number'; what could be coined in Arabic as *'adad 'ilmī*). The *noêtos arithmos* relates to numbers as they pertain to the workings of the Divine Artisan, the Greek Demiurge, and are ontologically distinguished from the realm of sensible beings; as, for instance, is also highlighted in reference to the intelligible in Plato's *Timaeus* (27d–28a) and Aristotle's *Topics* (100b19). As for the *epistêmonikos arithmos*, it designates numbers in their abstract form (*monadikos arithmos*; 'abstract number') as mathematical objects or entities that regulate all beings, and sustain the acquisition of rational knowledge as opposed to opinion.

Blanchard, 2001). See also Dominic J. O'Meara, *Pythagoras Revived: Mathematics and Philosophy in Late Antiquity*, 2nd ed. (Oxford: Clarendon Press, 1990).

15 The Ikhwān also acknowledged their indebtedness to Pythagoras in the mathematical application of harmonics in music, as noted in Epistle 5 of their *Rasā'il*.

16 There are no explicit indications that the Ikhwān directly benefited from the *Arithmetica* of Diophantus of Alexandria (fl. third century), which was translated into Arabic by Qusṭā ibn Lūqā. It is, moreover, unclear whether the Ikhwān fully integrated investigations on numbers as they figured in Plato's *Timaeus*, or in terms of the exaggerated version of 'realism' that was presented in the *Phaedo*. There is also no evidence that they were influenced by the objections raised by Aristotle against the Platonic thesis on the 'self-subsistence of numbers'. See Thomas L. Heath, *Diophantus of Alexandria: A Study in the History of Greek Algebra* (repr., New York, 1964); Aristotle, *Metaphysics*, ed. David Ross (Oxford: Clarendon Press, 1997), 1083b23–9, 1085b34.

Despite the existential-cum-metaphysical applications of numbers in modulating all beings as claimed by the Ikhwān, we do not find in their arithmetic any indication of Nicomachus' distinction between *noêtos arithmos* and *epistêmonikos arithmos*. This textual state of affairs may have resulted from the Ikhwān's theological inclination for a prudent eschewal of references to the 'Creator' of monotheism that may be suggestive of similarities with the pagan 'Artisan Deity' who generated the ordering of the universe and its beings from pre-existing and co-eternal worldly constituents.

Based on 'theological' (meta-mathematical) assessments of the 'merits' of arithmetic, the Ikhwān asserted that 'the one' (namely, the 'unit' 1, that preceded the 'first number', 2) was changeless and indivisible, and that it was the beginning and the end of numbers, 'similar to the way the Immutable Divine related to all beings'. Moreover, the Ikhwān believed that the propositions of arithmetic supported the unfolding of 'metaphysical' interrogations regarding 'how an infinite multitude begins with the One'; in response to this question, they explicitly affirmed that the generation of numbers *ad infinitum* from the unit 1 (*al-wāḥid*) offers lucid evidence (*dalīl*) of God's Unity (*waḥdāniyya*).[17]

Arithmetic II

The Ikhwān linked their mystical 'number theory' with an investigation of the nature of existents, and followed in this a pattern in thinking akin to the Platonic tradition in emphasising the importance of mathematics as a foundation for the study of philosophy and the teleological pursuit of happiness.

In commenting on the foundational unit 1, the Ikhwān described this arithmetic 'entity' in general terms as a *shay'* (namely, a 'thing').[18] Moreover, they noted in the introduction to their first epistle that arithmetic involves a study of the attributes of numbers (*khawāṣṣ al-ʿadad*) in view of elucidating the properties of (correlative) existents (*maʿānī al-mawjūdāt*). Ultimately, as the first of the abstract sciences,

17 *Rasā'il*, p. 54.
18 Ibid., p. 49.

arithmetic involves the investigation of numbers in view of grasping the reality of worldly beings (*al-mawjūdāt al-latī fī al-ʿālam*). Moreover, the general appellation 'the thing' (*al-shay'*) refers to the utterance 'one' (*wāḥid*), which is used either in concrete linguistic applications (*bi'l-ḥaqīqa*) or by way of metaphor (*bi'l-majāz*). In its proper-cum-actual application (or quotidian use), the 'one' designates a *thing* that cannot be partitioned or divided, whilst metaphorically, the 'one' is a multitude (*jumla*) that acts as a unity. Numerically, the 'one' encompasses in its *oneness* the idea of a unit (*waḥda*, or in Greek, *monas/monad*), which grounds the applications of the art of *ḥisāb* ('calculation' or 'reckoning') in the addition and subtraction of numbers (*jamʿ al-ʿadad wa tafrīquh*).

In this context, the Ikhwān's reflections on arithmetic are reminiscent of what is stated in Book VII ('On the Fundamentals of Number Theory') of Euclid's *Elements*; wherein, Definition 1 notes that: 'a unit [monad] is that by virtue of which each of the things that exist is called one'; and Definition 2, adds that: 'a number is a multitude of compound units [monads]'. Thus, the idea of number (*arithmos*; *ʿadad*) was associated with the notion of a unit (*monad*), and it was moreover connected to the concept of a *thing* (*ôn*; *shay'*) with the gathering of units into a multitude (*plêthos*; *jumla*); namely, a whole that is made up of units (*monads*), which refers to the quantitative 'iteration-number' that is needed in order to produce a given number *A* (*arithmos*; *al-ʿadad*).[19]

Arithmetic III

Following Nicomachus' precedent, the Ikhwān divided numbers (*al-aʿdād*) into two principal kinds: (positive) integers (*ʿadad ṣaḥīḥ*) and

19 The quantitative 'iteration-number' [$a \geq 2$], designates the cumulation of unit-*monads* [*E*] 'as often as' (*hosakis*) it is needed to result in the number-*arithmos* [*A*]. This may be notated as follows: $A \Rightarrow [aE]$, $a \geq 2$; which means 'as many units as there are in it' (*hosai eisin en autô monades*). For further particulars, see Ioannis M. Vandoulakis, 'Was Euclid's Approach to Arithmetic Axiomatic?', *Oriens-Occidens*, 2 (1998), pp. 143–144. See also, in the same volume of the journal, Jean-Louis Gardies, 'Sur l'axiomatique de l'arithmétique Euclidienne', pp. 125–140.

fractions (*kusūr*). They also reaffirmed that 'the one' (*al-wāḥid*; i.e., the unit 1) is 'the origin of numbers' (*mabda' al-ʿadad*), and that integers start with the smallest quantity (that is, 2) and increase *ad infinitum* by way of addition (*yadhhab bi'l-tazāyud bilā nihāya*).[20] An integer is thus generated out of the 'one' and is led back by way of analysis to that unit, following a 'natural progression' (*yansha' al-ʿadad al-ṣaḥīḥ min al-wāḥid wa-yanḥall ilayh ʿalā al-naẓm al-ṭabīʿī*).

Integers themselves are also classed according to four ranks (*marātib*), namely: the scale of ones (unities), of tens, of hundreds, and the scale of thousands (respectively: *aḥād, ʿasharāt, mi'āt, ulūf*). These were designated by twelve simple utterances (*lafẓa basīṭa*) from which all numbers are named (i.e., the ten numbers from *wāḥid* to *ʿashara*, plus *mi'a*, plus *alf*).[21] The Ikhwān also observed that each integer (whole number; *ʿadad ṣaḥīḥ*) has the property of being equal to half the quantity that results from the sum of its two immediately adjacent numbers (*ḥāshiyatāh*), e.g., 5 = ½ (4 + 6); 6 = ½ (5 + 7); 7 = ½ (6 + 8); 8 = ½ (7 + 9); 9 = ½ (8 = 10); 10 = ½ (9 + 11). This can be expressed generally in the algebraic form: $n = \frac{1}{2}\,[(n + m) + (n - m)]$.

The Ikhwān's system of correspondence between numbers and letters of the alphabet was based on the common mathematical practices of their age, which emulated in part the classical Phoenician and Greek models, and in part the system, more familiar to us, of Roman numerals. Deploying Arabic numerals, the Ikhwān represented numbers through the following 'abjad' sequence (*ḥurūf al-jumal*):[22]

20 *Rasā'il*, pp. 50, 68–69. The Ikhwān echoed a Greek conception of number, wherein '2' (*duas*) is taken to be the smallest number without qualification, as was noted in, for instance, Aristotle, *Physics*, ed. W. David Ross (Oxford: Clarendon Press, 1998), Delta, 220a27–28.

21 In addition to these ranks, we also highlight the tradition inherited from the sexagesimal system that is believed to be of Babylonian provenance (in other words, the scale of 60, as opposed to our decimal system). This numerical order denoted numbers by letters, and was used in finger reckoning in reference to the divisors of 60: 2, 3, 4, 5, 6, 10, 12, 15, 20, and 30. This system was of importance for astronomy, principally in reckoning angles and subdivisions of circles: 30°, 45°, 60°, 90°, 120°, 180°, 240°, 300°, and 360°. It is also still used in measuring time: 60 seconds, 60 minutes, or 12 hours (as ⅕ of 60), or 24 hours (as ⅖ of 60).

22 The first use of '0' (*ṣifr*) as a placeholder within a positional base notation system is attributed by some to al-Khwārizmī, as noted in his *al-Ḥisāb al-hindī* (*Algoritmi*

م	ل	ك	ي	ط	ح	ز	و	ه	د	ج	ب	ا
40	30	20	10	9	8	7	6	5	4	3	2	1

Based on this system, the correspondence between numerals and alphabet letters is as follows:

1–10 [ا، ب، ج، د، ه، و، ز، ح، ط، ي]
20–90 [ك، ل، م، ن، س، ع، ف، ص]
100–900 [ق، ر، ش، ت، ث، خ، ذ، ض، ظ]
1000–10,000 by combinations [غ، بغ، جغ، دغ، هغ، وغ، زغ، حغ، طغ، يغ]

Of the sixteen numerical ranks (*marātib*) that the Ikhwān utilised, after the Pythagoreans, the largest set they noted in relation to usable quantities was that of quadrillions: 1,000,000,000,000,000 = 10^{15} (named *mārū ulūf ulūf ulūf ulūf ulūf*).[23] The sixteen ranks themselves started with units of 1 (*aḥād*) and ended with quadrillions, as follows: 1, 10, 10^2, 10^3, 10^4, 10^5, 10^6, 10^7, 10^8, 10^9, 10^{10}, 10^{11}, 10^{12}, 10^{13}, 10^{14}, 10^{15}.

Regarding fractions (*al-kusūr*), the Ikhwān noted that these started with the largest quantity and proceeded by way of division; hence, fractions point to 'the one' as a part within an integer (w) that is greater than 'one' ($w > 1$): e.g., 1 in 2 is a half ($\frac{1}{2}$; *niṣf*); 1 in 3 is a third ($\frac{1}{3}$; *thulth*); 1 in 4 is a quarter ($\frac{1}{4}$; *rubʿ*). Moreover, fractions were divided into 'fractional portions', like, for instance, $\frac{1}{15}$, which is referred to as *thulth al-khums* (a third of a fifth; which can be notated as: $\frac{1}{3}$ of $\frac{1}{5}$ = $\frac{1}{15}$). The *kusūr* were also represented by way of Arabic alphabet letters (the 'abjad' sequence; *ḥurūf al-jumal*) as follows:

ي	ط	ح	ز	و	ه	د	ج	ب
عشر	تسع	ثمن	سبع	سدس	خمس	ربع	ثلث	نصف
$\frac{1}{10}$	$\frac{1}{9}$	$\frac{1}{8}$	$\frac{1}{7}$	$\frac{1}{6}$	$\frac{1}{5}$	$\frac{1}{4}$	$\frac{1}{3}$	$\frac{1}{2}$

de numero Indorum). Throughout the tenth century, finger-reckoning arithmetic was the calculation system deployed by the business community, based on *ḥurūf al-jumal* (i.e., the 'abjad' numeric system), with varying calligraphic forms.

23 *Rasāʾil*, pp. 51, 55–56.

يه	يد	يج	يب	يا
ثلث الخمس	نصف السبع	جزء من	نصف السدس	جزء من
a third of one-fifth	half of one-seventh	a part of thirteen	half of one-sixth	a part of eleven
1/15	1/14	1/13	1/12	1/11

For instance, the *alif* represents 1; the *bāʾ*, ½; the *jīm*, ⅓; *dāl*, ¼; *hāʾ*, ⅕; *wāw*, ⅙; *zayn*, ⅐; *ḥāʾ*, ⅛; *ṭā,ʾ* ⅑; *yāʾ*, ⅒; or, combining *yāʾ* and *bāʾ*, as in *yab*, the resulting quantity is a fraction: ½ × ⅙ = 1/12 (half of one sixth; *niṣf al-suds*).[24]

Arithmetic IV

As regards the attributes of integers (*khawāṣṣ al-ʿadad al-ṣaḥīḥ*), the Ikhwān followed the Greeks in stating that 2 was the first of all numbers, as well as being the original numerical quantity initiating the series of even numbers (*azwāj*; *arithmos artios*), and the basis of their generation (hence, producing half of all numbers). After all, the number 2 was construed as being 'the first of the numbers' given that 1 is a unit (*waḥda*; *monad*) used in counting numbers but is not itself considered a *number* (*ʿadad*) as such. As for 3, it is the first odd number (*fard*; *arithmos perissos*), and supposedly acts as the basis for the generation of one third of all numbers; while 4 is the first perfect square number (*awwal ʿadad majdhūr*), since 4 = 2 × 2 (in other words, 4 being the first number that results from the multiplication of another number by itself). Moreover, 5 is the first 'automorphic' number (*ʿadad dāʾir*; it is also referred to as: a 'round', 'circular', 'cyclical' or 'spherical' number); since whenever 5 is exponentiated, it reappears at the end of the resulting quantity, as in: 25, 125, 625, 3125, 18625, 93125, etc. Similarly, 6, which is the first perfect number (*tāmm*; *teleios*), is also an automorphic number, recurring in multiples of 6 as: 36, 216, 1296,

24 Ibid., pp. 51, 56. In a fraction: the *monad* is the numerator (top number in the fraction) with the *integer* ($w > 1$) as the denominator (lower numbers in the fraction); as for the factor, it is a number that divides into another number without resulting fractions (so, for example, the factors of 10 are: 1, 2, and 5; or those of 12 are: 1, 2, 3, 4, 6). Decimals are contrasted here with 'duodecimals', which are based on sets of 12, and have more divisors than the sets of 10.

7776, etc.[25] The Ikhwān also held that 7 is the first 'complete number' (*'adad kāmil*), since, arguably, it carries the properties of all the numbers preceding it, and is equal to the first odd number added to the second even number (i.e., 3 + 4), and is also equal to the first even number added to the second odd number (i.e., 2 + 5), and is furthermore equal to the unit 1 added to the first perfect number (i.e., 1 + 6). As for the number 8, it is the first cubic number (*muka''abb*; 'cube', or, 'perfect cube'), since it results from the multiplication of a *majdhūr* (i.e., a 'perfect square' like 4) with its *jadhr* (i.e., its 'square root' like 2). It is also considered by the Ikhwān as being the first figurate solid number (*'adad mujassam*).[26] Regarding the number 9, it is the first odd perfect square (*fard majdhūr*), while 10 is the first in *al-'asharāt* (the series 10–90). The number 11 is considered by the Ikhwān as being the first 'deaf' number (*'adad aṣamm*; 'irrational number'),[27] because it does not have a fractional part that carries a name of its own; they then listed other 'deaf' numbers: 13, 17, 19, 23, 29, 31, 37, 41, 43, 47, 53, 59, 61, 67, 71, 73, 79, 83, 89, 91, etc. However, they did not indicate how a 'deaf' number differs from a 'prime' number (i.e., one that has divisors of only the unit one and itself). They also omitted the numbers 3, 5, and 7 from the list of what they designate as 'deaf numbers'; perhaps because they preferred to group these according to other specific properties that they gave them (namely, that 3 is the first odd number; 5, the first automorphic number; and 7, the first complete number). But if it is the case that the Ikhwān identified 'deaf' numbers as being 'irrational' numbers and not 'prime' numbers per se, then it is unclear why they did not consider the numeric values of irrational numbers like π or $\sqrt{2}$.

Moreover, the Ikhwān took the number 12 to be the first abundant/

25 *Rasā'il*, p. 58.

26 See *Rasā'il*, p. 59. We could also note that 8 is a power of the base 2 raised to an exponent 3, i.e., $8 = 2^3$. This describes what came to be known later as a *logarithm*, in the sense that it refers to the power to which a fixed number (the base) must be raised in order to equal a given number.

27 A *deaf* number (*'adad aṣamm*; *arhêtos/alogos*; or, in Latin, *surdus*) is an *irrational* number, in the sense that it cannot be expressed as a *ratio* of two integers, unlike a *rational* number (*'adad munṭaq*; *rhêtos*). It is believed that the appellation 'deaf number' (*'adad aṣamm*) was used to render Euclid's *arhêtos*. See Sabra, "Ilm al-ḥisāb'; see also *Rasā'il*, p. 60.

excessive number (*ʿadad zāʾid; hupertelês arithmos*), given that the sum of its divisors (*qua* factors; *ajzāʾuh*) results in a quantity larger than the number itself: i.e., $6 + 4 + 3 + 2 + 1 = 16 > 12$. In contrast, with any number called deficient/defective (*ʿadad nāqiṣ; ellipês arithmos*), the sum of its divisors (*ajzāʾuh*) produces a quantity that is less than it, such as 4, 8, or 10 (i.e., $2 + 1 = 3 < 4$; $4 + 2 + 1 = 7 < 8$; $5 + 2 + 1 = 8 < 10$).

Arithmetic V

The Ikhwān divided even numbers into three categories: (i) powers of two (*zawj al-zawj; artiakis artios*); (ii) pairs of odd numbers (*zawj al-fard; artio perissos*); (iii) couples of pairs of odd numbers (*zawj zawj al-fard; perissartios*).[28] Following this classification, a *zawj al-zawj* takes the form 2^n, and it is always an even number that is divisible into two even integers, which can, in turn, be split into two equal halves that are integers, e.g., $64 = 2^1 \times 32 = 2^2 \times 16 = 2^3 \times 8 = 2^4 \times 4 = 2^5 \times 2 = 2^6$. Moreover, in the series: 1, 2, 4, 8, 16, 32, 64, multiplying the numbers from the outer margins with each other results in equal quantities as we move towards the middle of the series: $1 \times 64 = 2 \times 32 = 4 \times 16$. As for a *zawj al-fard*, it takes the form: $2(2p + 1)$, meaning that each can be expressed as a pair of odd numbers. For instance, each of these numbers: 6, 10, 14, 18, 22, 26, is divisible into two equal half quantities only once, resulting in the respective values: 3, 5, 7, 9, 11, 13, which are, in their turn, indivisible into equal halves as integers (since the divisions of: 3, 5, 7, 9, 11, and 13 result in fractions). As regards a *zawj zawj al-fard*, it takes the form: $2^n(2p + 1)$, where $n \geq 2$ and $p \neq 0$; and it refers to numbers like 12, 20, 24, 28, 36, 44, 52, 60, 68, whereby each is divisible into equal quantities twice, as with: $20 = 2 \times 10 = 2\ (2 \times 5)$.

Furthermore, the Ikhwān considered the odd number (*ʿadad fard*) as either a prime number (*ʿadad awwal; prôtos arithmos*) or a composite (*ʿadad murakkab; sunthetos arithmos*). The former type refers to numbers like 3, 5, 7, 11, 13, 17, 19, 23, that are only divisible (without fractions) each by itself and by the unit 1; this definition was inspired by Book VII of Euclid's *Elements*, wherein it was noted that a *prime number* is that which is measured by a unit alone (Def. 11), and that numbers are *relatively prime* (*mutabāyina*) if they get measured by a common scale (Def. 12), like, for example, 9 and 25, where 3 generates 9 but does not produce 25, and 5 generates 25 but does not produce 9. The Ikhwān also referred to *associated* numbers (*'adad mushtarak*), like 9, 15, and 21, whereby each is counted by the unit 1 and the common denominator 3. Ultimately, a 'prime number' is a 'natural number' (a positive integer $w > 1$) that is divisible only by itself (w) and by the unit 1.

Regarding composite (*murakkab*) odd numbers, the Ikhwān highlighted that these can be counted with the unit 1 as well as numbers, like 3, 5, 7, 9 in reference to the corresponding numbers 9, 25, 49, 81 — whereby 9 is counted by 1 and by 3; 25 is counted by 1 and 5; 49 by 1 and 7; 81 by 1 and 9. This statement echoes Book VII of Euclid's *Elements* where it was first noted that a *composite* is a number that is measured by some other number (Def. 13), and that numbers are *relatively composite* if they are measured by some number as a common measure (Def. 14).

Arithmetic VI

In their attempt to define the 'perfect number' (*al-'adad tāmm*), the Ikhwān were inspired by Proposition 36 of Book IX of Euclid's *Elements*, whereby it is noted that a perfect number is equal to the sum of its proper divisors, and can be expressed in the form, $2^{p-1}(2^p - 1)$ where $p > 1$ is a prime number and $(2^p - 1)$ is a prime.[28] For example: 6, 28, 496, 8128 are perfect numbers, with their corresponding primes for $p > 1$ being respectively: 2, 3, 5, 7; these result in the correlative primes for $2^p - 1$, respectively: 3, 7, 31, 127.[29] The Ikhwān also noted that only one of the perfect numbers is to be found in a rank of numbers (*martaba*); like 6 within the series 1–9, 28 in the range 10–90, 496 in 100–900, and 8128 in 1000–9000.[30]

28 Ibid., pp. 60–64.

29 The quantity: $S = 2^p - 1$, where p is a prime number, refers to what came to be known later as a 'Mersenne prime' (named after the mathematician Marin Mersenne, d. 1648).

30 See also Roshdi Rashed, 'Analyse combinatoire, analyse numérique, analyse diophantienne et théorie des nombres', in *Histoire des sciences arabes*, ed. Roshdi Rashed with Régis Morélon, vol. 2 (Paris: Editions du Seuil, 1997), p. 87.

As for the 'perfect square' (*al-ʿadad al-majdhūr*),[31] it was defined by the Ikhwān as being a number having the quantity y^2, which, when added to twice its square root, i.e., $2y$ (*jadhrayh*), then added to the unit 1, results in a perfect square, n. The expression can be simplified to: $n = y^2 + 2y + 1$. They also noticed that if a perfect square, y^2, is diminished by twice its square root, $2y$, then added to 1, the remainder is a perfect square, m: $y^2 - 2y + 1 = (y - 1)^2 = m$.

Arithmetic VII

Following the Graeco–Arabic mathematical tradition, the Ikhwān considered two integers as 'amicable' numbers (*mutaḥābān*) if the sum of the aliquot parts (i.e., divisors) of one is equal to that of the sum of the aliquot parts of the other. Indeed, the mathematician Thābit ibn Qurra (whose work predates that of the Ikhwān, even allowing for chronologies situating their *Rasāʾil* at the earliest possible date) studied amicable numbers in the *lemmas* of his *Maqāla fī istikhrāj al-aʿdād al-mutaḥābba* (Treatise on the Determination of Amicable Numbers), which he composed in response to the legacies of Pythagoras and Euclid; based on his theory, the *mutḥābbān* (a pair of amicable numbers) are two numbers each of which equals the sum of the other's aliquot parts. So, if we take S_n to be the sum of the aliquot parts of the dividend number n (in other words, the sum of the positive integer divisors of n, excluding n itself) then, $S_{220} = 1 + 2 + 4 + 5 + 10 + 11 + 20 + 22 + 44 + 55 + 110 = 284$ (where 284 is a deficient number), and

31 *Rasāʾil*, pp. 64–65. A 'perfect square' is by mathematical convention also any 'square' (though this usage is rarer). However, in the case of offering an English rendering of '*majdhūr*' in this context, the use of 'perfect square' would allow for a distinction to be made with a '*murabba*ʿ' (*qua* 'square'). The algebraists, following Khwārizmī, used an additional term to designate a 'square', namely the technical term: '*māl*'. These variations reflect the distinctions between arithmetic, geometric, and algebraic definitions, which in that period already had cross-disciplinary applications unto each other in mathematics. The use of these terms is more closely examined in the *Arithmetic IX* section below. See also *Rasāʾil*, p. 72; al-Kwārizmī, *al-Jabr waʾl-muqābala*, ed. and trans. into French by Roshi Rashed as *Le commencement de l'algèbre* (Paris: Librairie A. Blanchard, 2006).

$S_{284} = 1 + 2 + 4 + 71 + 142 = 220$ (where 220 is an abundant number).[32]

Arithmetic VIII

The Ikhwān asserted that the multiplication of numbers, or what may be referred to as 'duplication'/'duplation' (*taḍʿīf al-ʿadad*), increases *ad infinitum* (*bilā nihāya*). This process is evoked by them in reference to (the Greek) 'natural progression' (*al-naẓm ṭabīʿī*), as in the series: 1, 2, 3, 4, 5, 6, 7, 8, 9, 10, 11, 12, 13, 14, 15, . . . , *n*.[33] It is, moreover, reflected in the progression of even numbers (*naẓm al-azwāj*), as in: 2, 4, 6, 8, 10, 12, 14, etc.; or through the progression of odd numbers (*naẓm al-afrād*), in the form: 1, 3, 5, 7, 9, 11, 13, 15, etc. Furthermore, this process involved a more direct form of 'multiplication' (*ḍarb*) that is of the type: $(a \times b)$, wherein the increase of the quantity a is produced through b iterations.

To elaborate on 'natural numbers' (positive integers), we note that these are arranged in an ascending series, where each term is greater than the one preceding it by a unit of 1, and the sum of such a set of numbers with an nth term would be:

$$S = n/2 \times (n + 1) = 1 + 2 + 3 + \ldots + (n - 1) + n.$$

Moreover, progressions can be arithmetical or geometrical;[34] whereby the former refers to a series of numbers each differing from its predecessor by a constant quantity of addition or subtraction d, while the latter designates a series of numbers, which stand with respect to each other in a constant ratio of multiplication or division r. An arithmetical progression takes the form:

$$S = n/2 \times [2a + d\,(n - 1)],$$

where S is the sum of the series, a its first term, and d the constant

32 In contrast with abundant/excessive numbers and deficient/defective numbers, equivalent numbers (*mutaʿādilān*) consist of two numbers where the sum of the divisors of each is the same, e.g., 39 and 55, since $S_{39}\,(1 + 3 + 13) = S_{55}\,(1 + 5 + 11)$. It is perhaps worth noting that the 'discovery' of the pair of amicable numbers 220/284 was attributed by Iamblichus of Chalcis (d. ca. 326) to Pythagoras.

33 Rashed, 'Analyse combinatoire', p. 87. We would note that if $p < (2^n - 1)$, where $(2^n - 1)$ is prime, then $p\,(2^n - 1)$ is an abundant number; we would also add that if $p > (2^n - 1)$, where $(2^n - 1)$ is prime, then $p\,(2^n - 1)$ is a deficient number.

34 *Rasāʾil*, p. 66.

difference of addition or subtraction. The geometrical progression takes, in its turn, the following form:

$S = a + a(r) + a(r^2) + \ldots + a(r^{n-2}) + a(r^{n-1}) \ldots = a(r^n - 1) \div (r - 1)$,

where S is the sum of the series, a its first term, and r the constant ratio of multiplication or division.

Arithmetic IX

Concerning their reflections on square numbers (*al-aʿdād al-murabbaʿa; tetragônos arithmos*), the Ikhwān's definitions of the properties of 'multiplication', of 'roots' and 'cubes' (*ḍarb*, *jadhr*, and *mukaʿʿabāt* respectively) rested on Def. 16–19 of Book VII of Euclid's *Elements*. Therein, it was noted that two numbers multiplied by one another produce a plane number (*ʿadad muṣaṭṭaḥ; epipedos arithmos*), *c*. As in: $a \times b = c$, where $a \neq b$ (Def. 16).

Whereas, three numbers multiplied by each other result in a solid number (*ʿadad mujassam; stereou arithmou*): $x \times y \times z$ (Def. 17). Moreover, a square number (*ʿadad murabbaʿ; arithmos tetragônos*) is taken to be the product of a number multiplied by itself: x^2 (Def. 18); while a cube number (*mukaʿʿab*) is the result of a number multiplied by itself and then multiplied by itself again: x^3 (Def. 19).

The Ikhwān also added that when a square number (*al-ʿadad al-murabbaʿ*), be it a perfect square (*murabbaʿ majdhūr*) or not, is multiplied by any number whatsoever, the product is a solid number (*ʿadad mujassam*). If a perfect square (*murabbaʿ majdhūr*) is multiplied by a number less than its square root (*aqall min jadhrih*), the product is called a diminished solid number (*ʿadad mujassam libnī*); while, if a perfect square (*murabbaʿ majdhūr*) is multiplied by a number greater than its square root (*akthar min jadhrih*), its product is an augmented solid number (*ʿadad mujassam bīrī*). In addition, they indicated that if a rectangular number (*ʿadad murabbaʿ ghayr majdhūr*; imperfect square) was multiplied by its shorter side (*ḍilʿih al-aṣghar*) the product is a diminished solid (*mujassam libnī*), and if multiplied by its longer side (*ḍilʿih al-aṭwal*) the product is an augmented solid (*mujassam bīrī*); while if it was multiplied by a number smaller or greater than both of them, the product will be called a free solid (*mujassam lawḥī*).

Arithmetic X

Although the Ikhwān affirmed that their 'number theory' was principally based on the mathematical traditions of Pythagoras and Nicomachus, it is also the case that their arithmetic was Euclidean, as is explicitly manifest in their reflections on the first ten Propositions of Book II of the *Elements* (*al-Maqāla al-thāniya min Kitāb Uqlīdis fī al-uṣūl*), which may be summarised in the more modern algebraic form by the following quadratic equations (namely equations containing a coefficient of x^2):[35]

II.1: $a(b + c + d) = ab + ac + ad$ (i.e., illustrating 'distributivity');[36]
II.2: $(a + b)(a + b) = (a + b)a + (a + b)b$;
II.3: $(a + b)b = ab + b^2$;
II.4: $(a + b)^2 = a^2 + b^2 + 2ab$;
II.5: $[(a + b)/2]^2 = ab + [(a - b)/2]^2$;
II.6: $(x + a)a + (x/2)^2 = [(x/2) + a]^2$;
II.7: $(a + b)^2 + b^2 = 2(a + b)b + a^2$;
II.8: $(2a + b)^2 = 4a(a + b) + b^2$;
II.9: $a^2 + b^2 = 2[(a + b)/2]^2 + 2[(a - b)/2]^2$;
II.10: $(a + x)^2 + x^2 = 2[(a/2 + x)^2 + (a/2)^2]$.

Arithmetic XI

Following the Greek traditions of the Pythagoreans and the Platonists, the Ikhwān pointed (*tanbīh*) to the entanglement of arithmetic with 'psychology' (*'ilm al-nafs*; *de anima*).[37] As noted earlier, they held that the mathematical disciplines (*riyāḍiyyāt*) ought to be acquired as a

35 Ibid., pp. 70–71.

36 For further particulars, see Euclid's *Elements*, Book II (1–10); see also Bernard R. Goldstein, 'A Treatise on Number Theory from a Tenth-Century Arabic Source', pp. 154–157; *Rasā'il*, pp. 72–75.

37 To give a geometric flavour of how this proposition (II.1) reads in Euclid's *Elements*: 'If there are two straight lines, and one of them is cut into any number of segments whatever, then the rectangle contained by the two straight lines equals the sum of the rectangles contained by the uncut straight lines and each of the segments.' An algebraic illustration of this proposition can be also expressed as follows: $x(y_1 + y_2 + \ldots + y_n) = xy_1 + xy_2 + \ldots + xy_n$.

foundation for studies in natural philosophy (i.e., physics; *ṭabīʿiyyāt*) as well as psychical sciences (*nafsāniyyāt*), which would ensure a learned approach to the treatment of theological topics (*ilāhīyyāt*; *nāmūsiyyāt*). Consequently, enquiring about the substance/essence of the soul (*jawhar al-nafs*) is a necessary stage on the path to studying theology.

After all, as noted in the introduction to the mathematical part of the *Epistles*, the philosophical aim behind the study of arithmetic and geometry was to gain access to natural philosophy (*ṭabīʿiyyāt*), and, through them, to an elucidation of 'the essence of the soul' (*jawhar al-nafs*) by way of preparation for investigating the principal problems of theology.[38]

Moreover, any enquiry into arithmetic points to the positing of a computing person, whereby numbers are grasped as being the quantity of the forms of things in the soul of the one who performs the counting or numbering (*kammiyyat ṣuwar al-ashyāʾ fī nafs al-ʿādd*).[39] The Ikhwān's observations in this regard were also reminiscent of what Aristotle noted regarding time in Book Delta (specifically, Chapters 10–14) of the *Physics*, whereby he defined *khronos* as a particular kind of *arithmos* (sections 219b1–2, 7–8), and, hence, as countable — for, no numbering or counting can take place unless a soul or intellect undertakes it (223a21–28). In addition, the generation of the 'soul' (*nafs*; *psukhê*; *anima*) was understood by Plato in the *Timaeus* (35b–36c) as being constituted out of numerical series (multiples of 2 and of 3). This opinion was explored in a quasi-arithmetic fashion by way of the following schema of portions, which was established in reference to seven numbers:

			1			
		2		3		
	4				9	
8						27
(Multiples of 2)						(Multiples of 3)

38 The Pythagorean Philolaus (who was mentioned by Plato in the *Phaedo*, and by Aristotle in the *De Anima*) held that the soul was 'the *harmony* of the body'.

39 *Rasāʾil*, pp. 75–77. The Ikhwān seem to have echoed the Socratic injunction: 'Know thyself!'. They also affirmed the associated maxim: 'If thou knowest thine essence [or 'knowest thyself'], then thou knowest God.'

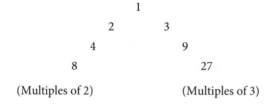

Geometry[40]

Geometry I

The second epistle of the *Rasā'il Ikhwān al-Ṣafā'* on geometry (*Risāla fī al-jūmaṭrīyā*; transliterated from the Greek '*geômetria*') was inspired by the assimilation of Euclid's *Elements* within the mathematical circles of mediaeval Islamic civilisation.[41] Euclid's objectives in the *Elements* (Books I–XIII)[42] were determined primarily with a view to systematising

40 Ibid., pp. 49, 75.

41 The epistle on geometry is divided into twenty-one chapters (*fuṣūl*), as follows (with pagination references based on the Dār Ṣādir Beirut edition of 1957): 'Untitled' preamble (pp. 78–81); 'Types of Lines' (p. 81); 'Epithets of Rectilinear Segments' (pp. 82–83); 'Names of Rectilinear Segments' (pp. 83–84); 'Types of Angles' (p. 85); 'Types of Planar Angles' (pp. 85–86); 'Types of Curvilinear Segments' (pp. 86–87); 'Surfaces' (pp. 87–88); 'Rectilinear Polygons' (pp. 88–89); 'Visual Points Figurations' (pp. 89–90); 'Demonstration that the Triangle is the Origin of All Shapes' (pp. 91–92); 'Types of Surfaces' (pp. 92–93); 'Surveying' (pp. 97–99); 'The Human Need for Co-operation' (pp. 99–100); 'Mental [Intellective] Geometry' (p. 101); 'On Imagining Distances' (pp. 101–103); 'The Reality of Distances in Mental Geometry' (pp. 103–104); 'Properties of Geometric Figures' (pp. 104–106); 'Demonstrating these Properties' (pp. 106–113); 'The Benefits of this Art' (p. 113). In this chapter, we shall follow an approximately parallel subdivision of the section on geometry, suitable for the purposes of this study, to be noted with the sub-title '*Geometry*' followed by a roman numeral.

42 According to the *Fihrist* (Index) of the bio-bibliographer Ibn al-Nadīm, the first recorded translation of Euclid's *Stoikheia* into Arabic (*Kitāb Uqlīdis fī al-uṣūl*; also known as *Kitāb al-Arkān*) was prepared by al-Ḥajjāj ibn Yūsuf ibn Maṭar under the patronage of the caliph Ḥārūn al-Rashīd (r. 786–809). Another rendition was later commissioned to Ḥunayn ibn Isḥāq (known in Latin as Johannitius; ca. 808–873) by the caliph al-Ma'mūn. The first translation was known as *al-Hārūnī*, and the second was titled *al-Ma'mūnī*; a third version based on a revision of the latter was established by Thābit ibn Qurra, who also worked on Euclid's *Dedomena* (*al-Muʿṭayāt* — a text connected mainly to Books I–VI of the *Stoikheia*). In addition, it is believed that the Latin translation of the *Elements* by Gerard of Cremona was based on Isḥāq's and Thābit's Arabic versions of the text. In the Greek tradition, the most notable commentary on the *Elements* was by Pappus of Alexandria (fl. fourth century). For particulars, see Euclid, *The Thirteen Books of the Elements*, ed. Thomas L. Heath, 3 vols. (Cambridge: Cambridge University Press, 1925; repr., New York: Dover, 1956); Benno Artmann, *Euclid, the Creation of Mathematics* (New York: Springer-Verlag, 1999); Ian Mueller, *Philosophy of Mathematics and Deductive Structure in Euclid's Elements* (Cambridge, MA: MIT Press, 1981); Carmela Baffioni, 'Euclides in the *Rasā'il* by Ikhwān al-Ṣafā'', *Études Orientales*, 5–6 (1990),

the multiple relationships between figures into axiomatic forms,[43] which he regarded as the ideal way to represent concrete physical bodies. The Euclidean axioms (*koinai doxai*; literally, 'common notions') and postulates, along with their supporting technical definitions, were presented as the foundational [true] premises-cum-general initial statements for the system of deductive logical inference, by virtue of which theorems were derived.

The Euclidean axioms may be designated as follows: Axiom 1 — Things which are respectively equal to the same thing are necessarily also equal to each other (i.e., if $x = a$ and $y = a$, then $x = y$). Axiom 2 — If equals are added to equals, the resulting wholes are equal (i.e., if $x = y$, then $x + z = y + z$). Axiom 3 — If equals are subtracted from equals, the resulting remainders are equal (i.e., if $x = y$, then $x - z = y - z$). Axiom 4 — Things that coincide with each other are equal (e.g., triangle ABC is equal to triangle XYZ if points A, B, C are superposed respectively on points X, Y, Z). Axiom 5 — A whole is greater than any of its parts (i.e., if $x = y + z$, then $x > y$ and $x > z$). Moreover, the five postulates of Euclid's plane geometry can be summarised as follows: Postulate 1 — A straight line segment can be drawn to join any two points. Postulate 2 — Any straight line segment can be extended indefinitely in a straight line. Postulate 3 — Given any straight line segment, a circle can be drawn that has the segment as its radius and one end-point as its centre. Postulate 4 — All right angles are congruent.

pp. 58–68; G. de Young, 'The Arabic Textual Traditions of Euclid's *Elements*', *Historia Mathematica*, 11 (1984), pp. 147–160.

43 The table of contents of Books I–XIII of Euclid's *Elements* may be summarised as follows: I 'Fundamentals of Plane Geometry: Theories of Triangles, Parallels, and Areas'; II — this book is titled inaccurately by some modern scholars as 'Geometric Algebra', but, rather, it interprets material attributable to algebra through geometry; III 'On Circles and Angles'; IV 'Constructions for Inscribed and Circumscribed Figures (Regular Polygons)'; V 'Theory of Abstract Ratios and Proportion'; VI 'Similar Figures and Geometric Proportions'; VII 'Fundamentals of Number Theory'; VIII 'Continued Proportions in Number Theory (Geometric Progressions)'; IX 'Number Theory Propositions'; X 'Classification of Incommensurables (Irrational Magnitudes)'; XI 'Solid Geometry'; XII 'Measurement of Figures'; XIII 'Regular Solids (On Constructing Regular Polyhedrons)'. The definitive edition of Euclid's *Elements* is preserved by the Teubner Classical Library in 8 volumes, with a supplement entitled *Euclides opera omnia*, ed. J. L. Heiberg and H. Menge (Leipzig: Teubner, 1883–1916).

Postulate 5 — If two lines are drawn which intersect a third in such a way that the sum of the inner angles on one side is less than two right angles, then the two lines inevitably must intersect each other on that side if extended far enough.

The 'Fifth Postulate', also known as the 'parallel postulate' (*qaḍiyyat al-mutawāziyāt*), was problematic for geometers, given that it could not be proved as a theorem despite the attempts of many polymaths throughout history who believed that it could be decisively established on the basis of the other four postulates.[44] It is, however, worth noting in this context that the Ikhwān did not preoccupy themselves with demonstrations related to this question in their epistle on geometry; perhaps this was because of the complex and specialist mathematical content of this problem, which nevertheless had already been treated with great care and detail both by their predecessors and their contemporaries amongst the mathematicians.

In conjunction with the 'Fifth Postulate', Definition 23 in Book I of the *Elements* also addressed the question of 'parallelism' by stating that 'parallel lines are straight lines which, being in the same plane and being produced *indefinitely* in both directions, *do not* meet one another in either direction' [italics added]. Nevertheless, this Euclidean definition would not have satisfied mathematicians and logicians, given that it proceeded by way of negation ('*do not* meet one another'), as well as lacking notional definiteness ('produced *indefinitely*').[45] To illustrate

44 The 'relationships' between geometric figures are not established in this context in the so-called 'Euclidean space', given that such a concept and appellation was coined in relatively modern times, and is historically posterior to the geometry of figures as embodied in Euclid's *Stoikheia* (*Elements*). As noted in Euclid's *Data*, Proposition 55 (in correspondence with Proposition 25 of Book VI of his *Elements*), *khôrion* (which derives from *khôra*, approximating a designation of space in Greek) refers to '*area*'— as in, 'if an area [*khôrion*] is given in form and in magnitude, its sides will also be given in magnitude'. The emergence of spatiality *qua* extension in geometry originated in the epistemic breakthrough associated with Ibn al-Haytham's (also known as Alhazen) mathematical definition of place in his *Qawl fī al-makān*; the Arabic critical edition and annotated French translation of this remarkable tract are both presented in Roshdi Rashed, *Les mathématiques infinitésimales du IXe au XIe siècle*, vol. 4 (London: al-Furqān Islamic Heritage Foundation, 2002), pp. 666–685.

45 In the early 1700s, the Italian scholar Girolamo Saccheri (1667–1733) proposed negating the parallel postulate and then probing whether this results in

the content of 'Postulate 5' further, let us construct two straight lines, *AB* and *CD* (Figure 1), and let a third straight line, *EF*, intersect *AB* and *CD* respectively at points *G* and *H*; if the angle $\angle AGH$ added to the angle $\angle GHC$ resulted in a value < 180°, then lines *AB* and *CD* will meet if they were respectively extended far enough in the direction of *A* and *C*.

Figure 1

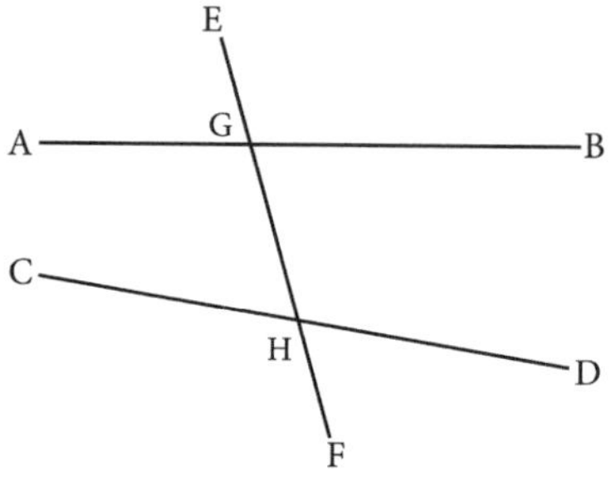

Geometry II

Following Euclid, the Ikhwān held that the art of geometry begins with the point (*nuqṭa; sêmeion*), which is without parts and is one of the extremities of the line (*Elements*, Book I, Def. 1, 3). The line (*al-khaṭṭ; grammê*), which carries just one property, namely length (*ṭūl*), was conceived as being a segment generated between two points, acting as the origin of the surface (*al-saṭḥ; epiphaneia*). The surface, in its turn, has two properties, length and width, and produces the geometric solid (*al-jism; stereos*) that carries the three properties of length, width, and depth (*dhū thalāthat ab'ād; trikhê diastaton — Elements*, Book XI, Def. 1).

The Ikhwān classified lines as being straight (*mustaqīm*), curved/arched (*muqawwas*), or bent (*munḥanī*; namely, a line consisting of a straight and a curved segment). Moreover, they noted that lines

a contradiction. However, the Russian mathematician Nicolai Lobachevsky, who developed the rudiments of 'non-Euclidean' geometry, was the one who accomplished the bold departure from Euclid's tradition; and his efforts were also anticipated by the Hungarian mathematician Janos Bolyai, and reinforced by the German mathematician Bernhard Riemann.

could be equal (*mutasāwiyya*), parallel (*mutawāziyya*),[46] convergent (*mutalāqiyya*), in tangential contact with each other (*mutamāssa*), or intersecting (*mutaqāṭiʿa*). Lines could also have the additional properties of being perpendiculars, base-lines, chords, sides of polygons, hypotenuses, or diagonals.[47]

Progressing from lower-dimensional geometric entities, like lines, to higher-dimensional entities, like surfaces, the Ikhwān considered the variegated types of polygons that are generated from linear configurations, such as the triangle (trilateral figure), the square (the quadrangle, along with quadrilateral figures), the pentagon, hexagon, heptagon, octagon, nonagon, decagon, etc. It was also believed that, hypothetically, more shapes could be generated *ad infinitum*, in correspondence with the infinite progression of numbers in arithmetic.[48]

Furthermore, the Ikhwān classed surfaces (*suṭūḥ*) into three kinds: the planar (*musaṭṭaḥ*), the concave (*muqaʿʿar*), and the convex (*muqabbab*) — the latter two derived from conics. Moreover, surfaces enveloped solids as well as being the lower-dimensional entities from which the latter are generated; hence, a figure like the sphere is bound by one surface; the half-sphere is delimited by two surfaces; a quarter of a sphere, by three; a fiery solid (like a tetrahedron) by four triangular surfaces — or a solid (like a pyramid) delimited by four triangles and a square base; and the cube (*mukaʿʿab*) by six square-shaped surfaces.

The Ikhwān also made reference to four types of right-angled parallelepiped (*mutawāzī al-suṭūḥ*; i.e., a solid bound by parallelograms): 1 — The cube (*mukaʿʿab*) was defined by them as a right-angled parallelepiped with three equal sides, hence having its length, equal to its height, equal to its width. 2 — The 'well-like shape' parallelepiped (*biʾrī*) has a height equal to its width, with each less than its length. 3 — The 'brick-like shape' parallelepiped (*libnī*) has a height equal to its width, with each greater than its length. 4 — The 'board-like shape'

46 It is worth noting here that the Ikhwān did not elaborate on Euclid's Fifth Postulate, which preoccupied the complex mathematical efforts of the polymaths of their age, as noted above in our section *Geometry I*.

47 *Rasāʾil*, pp. 81–84.

48 Ibid., pp. 88–89.

parallelepiped (*lawḥī*) has a height less than its width, which is less than its length.[49]

Geometry III

As stated in Plato's *Timaeus* (53a–55d), the four primary bodies — fire (plasma), air (gas), water (liquid), and earth (earth) — were all composed out of 'sub-atomic particles' having the shape of two right-angled triangles; namely, the scalene half-equilateral or the isosceles. Following this Platonic thesis, the trilateral figures were grasped by the Ikhwān as being the primary constituents of all figures (*al-muthallath aṣl kull al-ashkāl*);[50] it was similarly the case with the five Platonic solids: the *tetrahedron* (corresponding with fire), the *octahedron* (corresponding with air), the *icosahedron* (corresponding with water), the *hexahedron* (corresponding with earth), and the *dodecahedron* (corresponding with the visible cosmos).

In attempting to elucidate the properties of triangles, the Ikhwān based their definitions of angles (*zawāyā*) on Book I of Euclid's *Elements* (Def. 10–12). Thus, they noted that an angle could be planar (*muṣaṭṭaḥ*) or solid (*mujassam*), and that it could be right-angled (*zāwiya qā'ima*), acute (*zāwiya ḥādda*), or obtuse (*zāwiya munfarija*). The Ikhwān then listed seven types of trilateral figures, following Def. 20–21 of Book I of Euclid's *Elements*, which may be summed up as follows: 1 — The acute-angled equilateral triangle (*al-ḥādd al-zawāyā al-mutasāwī al-aḍlā'*). 2 — The acute-angled isosceles triangle (*al-ḥādd al-zawāyā al-mutasāwī al-ḍil'ayn*). 3 — The acute-angled scalene triangle (*al-ḥādd al-zawāyā al-mukhtalif al-aḍlā'*). 4 — The right-angled isosceles triangle (*al-qā'im al-zāwiya al-mutasāwī al-ḍil'ayn*). 5 — The right-angled scalene triangle (*al-qā'im al-zāwiya al-mukhtalif al-aḍlā'*). 6 — The obtuse-angled isosceles triangle (*al-munfarij al-zāwiya al-mutasāwī al-aḍlā'*). 7 — The obtuse-angled scalene triangle (*al-munfarij al-zāwiya al-mukhtalif al-aḍlā'*).[51]

On the different properties of trilateral figures, the Ikhwān evoked

49 Ibid., pp. 94–95.
50 Ibid., p. 91.
51 Ibid., pp. 104–106.

Pythagoras' theorem (without supporting explications), in noting that: within any right-angled triangle, the square of the hypotenuse is equal to the sum of the squares of each of its two other sides. Hence, with a right-angled triangle ABC, where $\angle BAC = 90°$, we have: $BC^2 = AB^2 + AC^2$. The Ikhwān also indicated that for any triangle that has two acute angles, its third angle must be either 90° (right-angled; *zāwiya qā'ima*) or obtuse (*zāwiya munfarija*), because the sum total of all its angles is 180°.

Geometry IV

Regarding the definition of quadrilateral figures, the Ikhwān followed what was noted in Book I, Def. 22 of Euclid's *Elements*. Hence, they pointed out that a square (*al-murabbaʿ*) is a quadrilateral figure that is equilateral and right-angled; that the oblong *qua* rectangle (*al-mustaṭīl*) is a quadrilateral figure that is right-angled but not equilateral; that the rhombus (*al-muʿayyan*) is a quadrilateral figure that is equilateral but not right-angled; that the rhomboid parallelogram (*al-shabīh bi'l-muʿayyan*) is a quadrilateral that is neither equilateral nor right-angled, and whose opposite sides and angles are equal; and that the trapezoid (*mukhtalif al-aḍlāʿ wa'l-zawāyā*) refers to the remainder of quadrilaterals.[52]

In reference to the properties of polygons, the Ikhwān noted that the side (*ḍilʿ*) of a hexagon (*musaddas*) is equal to half the diameter (i.e., equal to the radius) of the circle that contains it.[53] However, they did not provide a demonstration of this statement, which could be shown by way of a geometric construction (partly based on Proposition 1 of Book I of Euclid's *Elements*) as follows: let us construct two equal circles, C_1 and C_2 (with respective radii of R_1 and R_2, as shown in 'Figure 2' below), and let their circumferences pass through each others' centres, O_1 and O_2, and also intersect at a point, Z; then, let each of the circles contain a hexagon, such as C_1 encircles hexagon H_1 with a corresponding side S_1, and C_2 encircles hexagon H_2 with a correlative side S_2. This construction means that:

$$R_1 = R_2 = O_1O_2 = O_1Z = O_2Z = S_1 = S_2$$

52 Ibid., pp. 107–108.

53 Ibid., p. 108.

where S_1 and S_2 are each subtended from centres O_1 and O_2 respectively at an angle of 60° to the line O_1O_2.

Figure 2

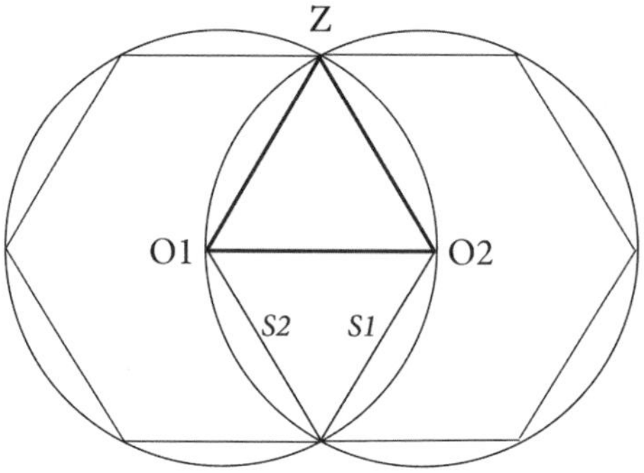

Geometry V

Based on the Pythagorean tradition related to figurate numbers (namely, numerical quantities represented by regular geometric configurations of equally spaced points),[54] the Ikhwān illustrated triangular numbers with ten marker dots (*al-muthallath min ʿasharat ajzāʾ*) as follows:

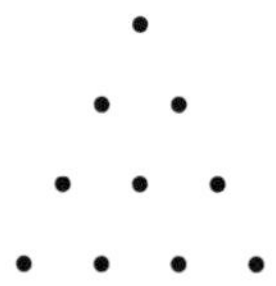

54 For instance, the series of triangular numbers (*ʿadad muthallath*; *arithmos trigônos*) begins with the following: 1, 3, 6, 10, 15, 21, 28, 36, 45, etc. In contrast, the so-called square numbers (*murabbaʿāt*; *arithmos tetragônos*) proceed in the form: 1, 4, 9, 16, 25, etc. Both of these series (triangular and square) increase following a pattern akin to the natural progression of numbers. Consequently, triangular numbers can be expressed with reference to the nth term as: $n(n + 1) \div 2$. While square numbers, which are equal to the sum of two successive triangular numbers, are of the form n^2. In addition, we could note the following figurative numbers: Pentagonal numbers [like 1, 5, 12, 22, 35, 51] have the form: $n(3n - 1) \div 2$. Hexagonal numbers [like 1, 6, 15, 28, 45, 66] take the form: $n(2n - 1)$. Heptagonal numbers [like 1, 7, 18, 34, 55, 81], are of the form: $n(5n - 3) \div 2$. Octagonal numbers [like 1, 8, 21, 40, 65, 96] are of the form: $n(3n - 2)$. See J. H. Conway and R. K. Guy, *The Book of Numbers* (New York: Springer-Verlag, 1996), pp. 30–62.

This diagram represented the so-called *tetraktys*,[55] which was a 'holy' symbol for the Pythagorean *mathêmatikoi* (i.e., Pythagoras' disciples). According to the Ikhwān, this figure described the procession of all numbers and corresponded with an onto-theological account of the manner in which God fashioned things (*al-ashyā'*) in the intellect (*al-'aql*), and how He manifested them in the soul (*al-nafs*) and in matter (*ṣawwarahā fī al-hayūlā*). Likewise, they held that this figure applied to the linguistic formation of letters (*al-ḥurūf*), and pointed to the theosophical significance of unity in arithmetic as a sign of God's Oneness (*waḥdāniyya*).[56]

Geometry VI

The Ikhwān presented highlights from Euclid's *Elements* (Book I, Def. 15–18) regarding geometric entities associated with circles, such as the centre (*markaz*), diameter (*quṭr*), chord (*watar*), sagitta (*sahm*), arc (*qaws*), sine (*jayb mustawī*), and versine, (*jayb ma'kūs*), as well as references to the positional relations between circles (concentric, tangential, intersecting).[57]

For instance, *al-watar* (the chord) is the straight line joining the two extremities/vertices of an arc (*al-khaṭṭ al-muqawwas*); as for *al-sahm* (the sagitta), it is the axis cutting both the chord and the arc into two equal segments (by being extended perpendicularly to the mid-point of

55 Marquet addresses the mystical significance of this figurate number in *Les 'Frères de la Pureté', pythagoriciens de l'Islam*, pp. 163–168.

56 *Rasā'il*, p. 54.

57 Ibid., pp. 85–88. In addition, the Ikhwān named a variety of ellipsoids (derived as conic sections) that differ according to the proportions between their minor and major axes (carrying the following appellations: *bayḍī, hilālī, makhrūṭ ṣanawbarī, ihlīlajī, nīm khānjī, ṭablī, zaytūnī*); on this, see ibid., pp. 92–93. Besides the geometrical tradition of Euclid, and the editions of Theon of Alexandria, Pappus, and Diophantus, an influential line of transmission into Arabic geometry is also linked to the *Conica* of Apollonius of Perga (d. ca. 190); on this, see Apollonius of Perga, *Les coniques d'Apollonius de Perge*, trans. Paul ver Eecke (Bruges: Desclée De Brouwer, 1923). However, it is unclear whether the Ikhwān were aware of this tradition, or of the *Sphaerica* of Theodosius of Bithynia (d. 90 BCE) and that of Menelaus of Alexandria (d. 130), or even, whether they knew of the groundbreaking mathematical research of tenth-century polymaths like Ibn Sahl, al-Qūhī, and al-Sijzī.

the chord from the furthest point away from the chord situated on the arc). The sagitta is also the resultant of subtracting the apothem (i.e., the perpendicular dropped from the centre of a circle to the mid-point of a chord) from the radius of the circle, of which the arc is a segment. If the sagitta (*sahm*) relates to half the arc (*qaws*) it will be an inverted sine, namely a *versine* (*jayb maʿkūs*); while, if half the arc (*qaws*) relates to half the chord (*watar*), it will then be a 'straight' sine (*jayb mustawī*).[58]

To illustrate these geometric relationships, let us consider a unit circle with a centre O (radius = 1) and then introduce a vertical sector chord AB that delimits the arc *AB* on that circle ('Figure 3' below). Let us then project a perpendicular chord from O to chord AB such that it intersects with its mid-point D, as well as intersecting with the arc *AB* at point C that is furthest away from the chord AB. Based on this figure, the radius of the circle is OC = 1, the apothem is OD (namely, the perpendicular distance from the mid-point of the chord AB to the centre of the circle, O; this is also known as 'short radius' or 'in-radius'), and the sagitta (*sahm*) is DC (namely the radius, OC, minus the apothem, OD). Let us also consider the angle $\angle AOC$ with a value (θ); then, the ordinary sine of ∠θ (i.e., vertical sine; *jayb mustawī*, or in Latin, *sinus rectus*) would be:

$$\sin(\theta) = \frac{\tfrac{1}{2}\text{ chord AB}}{\text{radius OA}} = \text{AD. (since radius OA} = 1)$$

While the cosine would be:

$$\cos(\theta) = \frac{\text{apothem OD}}{\text{radius OA}} = \text{OD.}$$

And the *sinus versus* of ∠θ (also called flipped sine, versed sine, or versine; *jayb maʿkūs*) would then be:

$$\text{versin}(\theta) = 1 - \cos(\theta) = \text{OC} - \text{OD}$$
$$\text{versin}(\theta) = \text{sagitta DC.}$$

Consequently, in a unit circle (with radius = 1), the sine is equal to half the chord; the cosine to the apothem; and the versine to the sagitta.[59]

58 *Rasāʾil*, p. 86.

59 Just as a reminder concerning trigonometric ratios and reciprocals in reference to triangles: if we consider a right-angled triangle *ABC*, such as angle $ACB = \angle C = 90°$, we obtain the following trigonometric ratios: $\sin B = AC/AB$, $\cos B = BC/AB$, $\tan B = AC/BC$. This also results in the reciprocal ratios: $\csc B = AB/AC$, $\sec B = AB/BC$, $\cot B = BC/AC$.

Figure 3

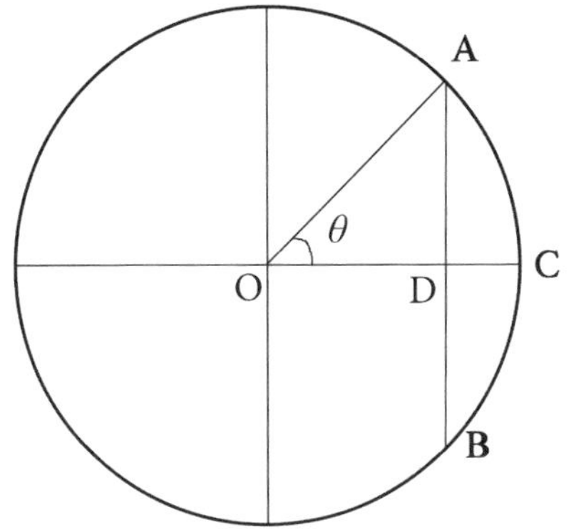

Geometry VII

The Ikhwān conceived geometry as a discipline that entailed knowing magnitudes (*maqādīr*) and dimensions (*abʿād*; literally 'distances'), along with their properties. They also divided this mathematical science (*ʿilm*) into two branches: sensible geometry (*handasa ḥissiyya*) and intelligible geometry (*handasa ʿaqliyya*). The former referred to sense perception, while the latter designated what is abstract and conceivable only by the intellect. Furthermore, the Ikhwān held that intelligible distances (*al-abʿād al-ʿaqliyya*) were attributes of the (quantitative) sensory magnitudes (*al-maqādīr al-ḥissiyya*).[60] They also noted that the sensible (*ḥissiyya*) applications of this art were of expedient use to artisans, whilst the theoretical side assisted in understanding the influence that heavenly bodies and musical harmonics exercised on souls, as well as furnishing significant clues concerning the impact that the separate souls/intellects (*al-nufūs al-mufāriqa*) may have on corporeally embodied souls (*al-nufūs al-mutajassida*).[61]

Although the Ikhwān did occasionally conflate abstract mathematical propositions with empirical observations, or confuse them with the metaphorical and symbolic constructs of analogy, they nonetheless differentiated between the practical applications of geometry and its theoretical forms. This matter was articulated mainly in terms of the

60 *Rasāʾil*, pp. 79–81.
61 Ibid., p. 113.

role assigned to the 'imagining' (postulating) of distances (*tawahhum al-abʿād*).[62]

The phenomena of the sensible environment present geometers with various figures associated with material quantities. In this, sensible geometry, which is attested by means of physical objects, appears as an approximation of structurally abstract geometrical figures taken as invariant *idealities* (mathematical objects; *mathêmata*).[63]

In highlighting the practical merits of sensible geometry, the Ikhwān emphasised the connection between abstract geometry and its applications in the art of surveying (*al-misāḥa*). However, they did not develop a method by which geometry could offer a theoretical grounding for the science of measurement, but, rather, they focused primarily on listing five scales or magnitudes (*maqādīr*) of linear measure (some being anthropometric), which (according to them) were deployed in Iraq by their contemporaries.[64]

In the attempt to realise an abstractive imagining (*tawahhum*) of pure mathematical figures, the Ikhwān made use of the application of (mechanical) motion to geometrical entities; perhaps this fact is an indication of the broad epistemic reception of the developments in mathematics that were realised through the research of mathematicians in ninth- and tenth-century Islamic civilisation.[65] In respect of this, the

62 Ibid., pp. 101–103. It is worth noting that the process of abstraction in mathematics by means of imagining (postulating) proved to be a decisive factor in the geometrisation of place by Ibn al-Haytham (Alhazen), in his *Qawl fī al-makān*, and its definition as *khalāʾ mutakhayyal* (imagined/postulated void).

63 Geometry opened up a growing stratification of mathematical primal *idealities* ('ideal' *qua* 'notional' objects), which were defined by their trans-temporal constancy, their singularity, universality, exactness, and objectivity; as Hume observed: 'Although there never was a circle or a triangle in nature, the truths demonstrated by Euclid would forever retain their certainty and evidence.' See David Hume, *An Enquiry Concerning Human Understanding*, ed. L. A. Shelby-Bigge (Oxford: Oxford University Press, 1972), Section IV, Part 1, p. 25.

64 *Rasāʾil*, pp. 97–99. These scales included: (i) the finger (*al-iṣbaʿ*) = 6 capillaries (*shuʿayrāt*); (ii) the palm (*al-qabḍa*) = 4 fingers = 24 capillaries; (iii) the cubit (*al-dhirāʿ*) = 8 palms = 32 fingers = 192 capillaries; (iv) the doorway (*al-bāb*) = 6 cubits = 48 palms = 192 fingers = 1152 capillaries; (v) the cord or rope (*al-ashl* or *al-ḥabl*; equivalent to the Greek *skhoinion*) = 10 doorways = 60 cubits = 480 palms = 1920 fingers = 11520 capillaries. These measurements were also supplemented with fractional scales named *juraybāt*, *qufayzāt*, and *ʿushayrāt*.

65 Euclid avoided the application of motion in theoretical geometry, given that

Ikhwān observed that a displaced point generates an imaginary line, and that a mobile line defines an imaginable surface, while a surface set in motion generates an imagined solid by its inclination in depth. And they showed that mathematical points, lines, surfaces, and solids are thus amenable to transformation via movement.

The imagined dimensions of height, length, and width are fully abstracted from physical entities in the intelligible type of geometry, unlike the application and manifestation of geometry in the sensible realm. Imagination (*tawahhum*) is considered to be the source of the effective processes of abstraction that ground the study of theoretical geometry. This state of affairs describes an epistemic procession from focusing on knowable entities (*al-maʿlūmāt*) as objects of our sense faculties (*al-quwā al-ḥāssa*), to accounting for them by way of our imaginative faculty (*al-mutakhayyila*), and, through the imaginative faculty, grasping them via our cognitive capabilities (*al-mufakkira*), as well as preserving our knowledge of them in the memory (*al-ḥāfiẓa*).[66]

The shift from sensory geometry to an intellective one constitutes a transition from practical arts (*ṣanāʾiʿ ʿamaliyya*) to theoretical ones (*ṣanāʾiʿ ʿilmiyya*),[67] as well as describing an epistemic shift from focusing on the sensible to placing emphasis on the intelligible (*min al-maḥsūsāt ilā al-maʿqūlāt*). Consequently, theoretical geometry studies the distances (*abʿād*) of solid figures as they are abstracted from corporeal entities and their sensible magnitudes (*maqādīr ḥissiyya*). It is thus a discipline that cultivates a capacity for abstraction and develops an ability to comprehend the intellective without recourse to the sensible. Even though it remained unclear, based on the Ikhwān's opinions, whether they believed that this 'movement of abstraction', which is attested with a shift from a theoretical-cum-intellective geometry to a practical-cum-sensible geometry, was ultimately coupled with a 'movement of concreteness' that was articulated in terms of the generation of tools or instruments; whereby, theoretical geometry grounds the practical applications of geometry.

motion was primarily a phenomenon of physics.

66 *Rasāʾil*, pp. 103–104.

67 The Ikhwān addressed the practical arts in Epistle 7, and the theoretical sciences in Epistle 8; see respectively, ibid., pp. 258–275 and 276–295.

The apprentice of mathematics is progressively trained to minimise reliance on the senses in conceiving entities that can be known (*al-maʿlūmāt*) or in attempting to contemplate the intelligible (*al-maʿqūlāt*). According to the Ikhwān, this pedagogical exercise opens up possibilities for acquiring 'higher forms of knowledge' that ultimately facilitate the study of theology, as well as training the intellect to hypothetically exit the realm of matter (*al-hayūlā*) in its reflections.[68] It is in this regard that arithmetic and geometry prime the apprentice for speculating on nature and the soul for studying theology. For this reason, the Ikhwān believed that arithmetic and geometry, along with their related measured proportions,[69] (theoretically) may all contribute to the constructive cultivation of the soul and the reformation of its ethics (*tahdhīb al-nafs wa iṣlāḥ al-akhlāq*).[70] However, the significance of mathematics for the Ikhwān was principally pedagogic rather than epistemic per se. Unfortunately, by not exploring how mathematics assists in solving the problems of theoretical philosophy, they failed to

68 The abstractive capabilities of the imagination were elaborated by Ibn al-Haytham (Alhazen), who established an ontological distinction between sensible beings (those apparent by means of *al-ḥiss*) and entities that exist through *al-takhayyul* ('the imagination'). Moreover, he held that the latter class exists by way of an ascertaining *taḥqīq* (in Latin, '*certificatio*'), while the perception of sensible entities is prone to error. He also added that the sensible does not exist in reality (*laysa huwa mawjūd ʿalā al-ḥaqīqa*), and that it is corruptible and unstable, while the imagined form is grasped according to its truth, and does not continuously change with the variation of whomsoever imagines it (*lā tataghayyar bi-taghayyur al-mutakhayyil lahā*). For an explanation of the role of imagination in mathematics in reference to Ibn al-Haytham's tract *Fī ḥall shukūk Kitāb Uqlīdis fī al-uṣūl*, see Roshdi Rashed, *Les mathématiques infinitésimales du IXe au XIe siècle*, vol. 4, pp. 8–10.

69 The Ikhwān classed arithmetic proportion (*nisba ʿadadiyya*) as quantitative (*nisba bi'l-kammiyya*), while they took geometric proportion (*nisba handasiyya*) to be qualitative (*nisba bi'l-kayfiyya*).

70 Ethics was also treated by the Ikhwān in Epistles 6 and 9; see respectively, *Rasāʾil*, pp. 242–257, 296–389. However, it is worthy of note that their reflections on the importance of mathematics in cultivating a moral inclination in their apprentices diverged from the classical precedents associated with Porphyry's *Sententiae ad Intelligibilia Ducentes* (*Aphormai pros ta noêta*), Plotinus' *Institutio Theologica* (*Stoikheiôsis theologikê*), or what centuries later emerged with Spinoza's *Ethica Ordine Geometrico Demonstrata*, whereby a philosophical tract is composed in a quasi-geometric fashion: stating definitions and axioms that carried 'metaphysical' content, followed by the deductions of 'demonstrative reasoning'.

give their epistemology a proper grounding in a mathematical-logical rationale, and, furthermore, by undermining the empirical impetus in research due to their mistrust of the senses, they mistook *mystical analogism* for *rigorous explanation* and *sound acquisition of scientific knowledge* (*ʿilm*).[71]

71 I am deeply grateful to Professors Roshdi Rashed and Wilferd Madelung for their comments on an earlier draft of this chapter, which assisted me in refining its content.

EIGHT

Music and Musicology in the Rasā'il Ikhwān al-Ṣafā'

Owen Wright

Preamble

This study has no pretensions towards giving a general account of the philosophical framework within which the *risāla* on music is set, nor does it attempt to discuss in any detail the origins and evolution of the themes it treats.[1] Rather, the aim is the much more modest one of providing sufficient contextualisation to make its contents readily accessible to both musicologists and general readers, especially those approaching it in isolation from the work as a whole. To begin with,

1 In addition to the coverage in the present volume, an excellent general introduction is: Alessandro Bausani, *L'enciclopedia dei Fratelli della Purità* (Naples: Istituto Universitario Orientale, 1978); this includes a lucid summary of the main contents of the *risāla* on music (as well as of all the other *rasā'il*). For specific issues, valuable references are provided, particularly on antecedents and parallels, in the footnotes of 'L'Épître sur la musique des Ikhwān al-Ṣafā'', ed. and trans. Amnon Shiloah, *Revue des Études Islamiques*, 32 (1965), pp. 125–162; ibid., 34 (1967), pp. 159–193. For the Arabic text of the Ikhwān's *risāla* on music, see *Rasā'il Ikhwān al-Ṣafā'*, ed. Buṭrus Bustānī (Beirut: Dār Ṣādir, 1957), Epistle 5, vol. 1, pp. 183–241.

it would obviously be in order to outline the particular approach the Ikhwān take to music, to account for the nature of their treatment of it, and, indeed, to ask what they themselves consider it to be. But a better initial question to ask is what they think it *does* and what function it performs. The intellectual framework within which the Ikhwān operate, in common with that of mediaeval Europe, is marked by an omnipresence of affinities, and by networks of equivalences or similarities that point both to the fundamental unity of Creation and also to relationships and pathways between discrete positions within the hierarchical order imposed by its Creator.[2] Accordingly, they consider music to represent, albeit in imperfect imitation, phenomena belonging to a higher plane, namely, the eternal domain of the celestial; indeed, the two share a structural parallel of common reliance on certain fundamental mathematical relationships. In particular, we find an emphasis on those that can be expressed as simple ratios, considered to exhibit ideal proportions. At the same time, this seemingly static model yields to the instrumental, so that music becomes a dynamic means of effecting change: by manifesting (and being governed by) harmonious proportions, it has the power to act upon the emotions, and can also be used to counteract physical disharmony and restore balance to the humours when disturbed by illness. Above all, music has the power to affect the soul during its temporary sojourn in this world (the transient domain of 'generation and corruption'), inspiring in it the desire to achieve salvation and ascend to the celestial realm. Thus, although the *risāla* contains a certain amount of practical information and scientific observation, its primary concern is with the way music is structured according to Pythagorean concepts of numerical relationships. It is, indeed, on account of this fundamental structural property of music that it can be taken as a profound analogy for the structure of the cosmos as a whole, and, in this way, as a means by which the human soul can glimpse the beauties of a higher world and become filled with a yearning to be present with them. Given this implicit theme, it is not surprising to find an explicit disclaimer concerning music as manifested

2 For an account of the prolongation of this world-view in Europe until the sixteenth century, see, e.g., Michel Foucault, *Les mots et les choses* (Paris: Gallimard, 1966; repr. 1992), pp. 32–55.

in human and mechanical agents. Within this *risāla*, the aims of the Ikhwān are not to give a descriptive account of local phenomena (i.e., primary musicological concerns such as particular modal and stylistic features, the training of musicians, instruments and their techniques), but rather, to find a model of general applicability from observing the universal patterns underlying the production of musical sound, and at the same time to provide moral motivations and urge specific spiritual goals. This emphasis is made manifest in the admirably succinct yet telling summary given in the introductory outline of one manuscript,[3] which informs us that this *risāla* provides:

> A clarification that notes and rhythmical melodies have effects on the souls of listeners like those of medicines, potions and antidotes on the physical body, and that the movements of the celestial spheres in their orbits and their friction produce sweet notes and rhythmical melodies like the notes of lute strings and wind instruments; the purpose of all of these being to arouse in the angelic human soul a desire to ascend there after that parting from the body we call death.

Outline of the Contents of the Epistle on Music

Correspondingly, in the introductory section we find a statement and elaboration of the central claim that music affects the soul.[4] Of all the arts, the Ikhwān hold that music is the one that contains inherently a spiritual element and, unlike the others, which all work through various material mediations, the substance music manipulates is the human soul. Relying upon a command of the proportional relationships that govern melodic and rhythmic structures, through music it is thus possible to influence the psyche, inspiring it to acts of endurance and courage, and comforting it in times of distress. Mention is also made of the creation of music by the sages, whose purpose was religious, using it to inspire

3 MS Esad Efendi 3637 (ca. thirteenth century, Istanbul).

4 For an alternative general account of the contents of this *risāla*, see Fadlou Shehadi, *Philosophies of Music in Medieval Islam* (Leiden: Brill, 1995). For the sake of clarity, Shehadi presents the material under thematic heads, rather than, as herein, in the form of a strictly linear narrative.

humility and obedience. (In the later development of this important thematic area, the sages are usually coupled with the prophets, the former representing the truths of ancient wisdom and the Greek philosophical tradition; the latter, the congruent truths of the monotheistic tradition of divine revelation.) From such psychological and ethical principles, the focus of the *risāla* moves to the physical world, with music being defined in terms of harmoniously ordered combinations of sounds. Types of sound are distinguished and its propagation and transmission elucidated, with a disquisition on the causes of the loudest sound, thunder, being followed by a discussion of the ways in which sounds are produced by other natural phenomena, as well as animals and instruments. The definition of music is then revisited, and the topic shifts in an apparently haphazard way to the concepts of motion and rest. The reason is, however, not far to seek: as with other theorists, the analytical paradigm for the subsequent discussions of rhythm is imported from the science of prosody,[5] in which the distinctive patterns of long and short syllables that characterise Arabic metre are articulated in terms of properties of the Arabic script, a basic distinction being drawn between letters categorised as 'moving' (*mutaḥarrik*) and others as 'still' (*sākin*). The concept of motion leads to the comparative terms fast and slow, and thence to a consideration of how to assess this and other forms of contrast, of which the least obvious is the distinction between continuous and discontinuous, illustrated by reference to instruments and their methods of sound production.

In the next chapter of the *risāla*, there is a smooth transition to the factors that determine relative pitch in strings: length, thickness and tension; and thence, to consonant and dissonant intervallic relationships. These, however, are neither defined nor discussed in terms of the physics of sound: what is referred to, rather, is the effects they have, and in particular, the way in which they fit into the humoral system, so

5 Not for nothing is al-Khalīl ibn Aḥmad (d. ca. 791), to whom the codification of poetic metres is attributed, considered to be the first theoretician of musical rhythm also; see Eckhard Neubauer, 'Al-Khalīl ibn Aḥmad und die Frühgeschichte der arabischen Lehre von den "Tönen" und den musikalischen Metren, mit einer Übersetzung des *Kitāb al-Nagham* von Yaḥyā ibn 'Alī al-Munajjim', *Zeitschrift für Geschichte der Arabisch-Islamischen Wissenschaften*, 10 (1995–1996), pp. 255–323 (henceforth cited as 'Al-Khalīl ibn Aḥmad').

that they can be applied therapeutically to redress loss of equilibrium. In order to be effective, then, music must be attuned to the temperament of the listener, whether an individual or a group, and recognition is given to the existence of differing styles and tastes among a variety of ethnic groups, although none is illustrated.

The next transition is more abrupt, with the *risāla* now proceeding to a consideration of rhythmic structures. Not unexpectedly, the approach is via an outline of prosodic principles and patterns, by analogy with which a number of possible rhythmic sequences are proposed. These are purely theoretical entities, and discussion of the eight existing rhythmic cycles is deferred: here they are merely listed, and attention is drawn to the importance of the number 8, and thence to the symbolic importance of numerical associations in general. The vital importance of this area results from the convergence of two traditions. In the Pythagorean tradition, emphasis is laid upon intervals of the form: $(n + 1):n$. Ratios of this form of especial importance are: 2:1, the octave; 3:2, the fifth; 4:3, the fourth; and 9:8, the whole tone, from which diatonic scales can be constructed. Such intervals are associated with the pure, harmonious sounds of the music of the spheres that human music distantly echoes. While in the Neoplatonic tradition, the numbers 1 to 4, from which all others can be generated,[6] represent the first layers within the theory of emanations associated particularly with Plotinus: 1 is equated with God the Creator; 2, with the intellect; 3, the soul; and 4, matter. At this point, it is only the Neoplatonic strand that is developed by the Ikhwān, parallels being drawn with other disciplines also (1, being prior in arithmetic, corresponds with the point in geometry, the sun in astronomy, and substance among logical categories), and, not surprisingly, it is with praise of the *one* God that this excursus ends.

Although it reverts to rhythm, the ensuing discussion still postpones consideration of the structure of the eight rhythmic cycles, preferring to address variations of tempo. Four bands are distinguished and, although the labels attached to them partake of the vocabulary of the cycle names, it is clear that the discriminations made are abstract ones, even if they do expressly involve considerations of perception,

6 E.g., 7 = 4 + 3, 10 = 4 + 3 + 2 + 1.

the argument being that the duration between attacks should not be so long as to blunt awareness of the meaningfully patterned relationship that subsists between them.[7] From attacks and their relative durations we move to the means by which they are produced, with an initial, brief survey of types of instrument being followed by a more extended discussion of the one considered the most perfect, the lute (*al-ʿūd*). Here, for the first time, we touch upon the practical, with an account of dimensions and materials, and a definition of tuning and fretting.[8] However, neutral intervals are excluded, and the Pythagorean fretting given is such as to produce, on any one string, a whole tone followed by three semitones,[9] and all four strings are tuned a perfect fourth apart. But no mention is made of the melodic uses that might have been made of such tonal resources, and the emphasis throughout is on the 'perfect proportions',[10] that is, the ratios 2:1, 3:2, 4:3, and 9:8 that variously govern not only the positioning of the frets, but also the relative thickness of the strings, the very proportions of the instrument and, beyond that, the correspondence between the music of this world and that of the realm above. With the observation that among the various notes the lute can produce, the relationship between the low, slow ones and the high, fast ones is like that of body to spirit, this core theme of the coupling of the material and spiritual comes to the fore. It is then developed towards the proposition that the temporal organisation of such notes imitates the vastly distended dimensions of siderial time as manifested in the music of the cosmos. Crucially, and here we reach the central salvific concern of the whole *risāla*, awareness of this relationship, according to the Ikhwān, inspires in

7 'Attack', here and below, refers to 'the onset of the sound' (for example, from the blow of a drumstick, the plucking of a plectrum, or the setting of bow to string).

8 For a full presentation and discussion of this material, see Eckhard Neubauer, 'Der Bau der Laute und ihre Besaitung nach arabischen, persischen und türkischen Quellen des 9. bis 15. Jahrhunderts', *Zeitschrift für Geschichte der Arabisch-Islamischen Wissenschaften*, 8 (1993), pp. 279–378 (henceforth cited as 'Der Bau der Laute').

9 Actually, these three 'semitones' differ slightly in size: more precisely they consist of a *limma* (90 cents), an *apotome* (114 cents), and another *limma*.

10 The adjective used by the Ikhwān is *sharīf*, a more literal translation of which would be 'noble'.

the individual soul a yearning to be united in celestial bliss with the souls of the saved.

The solemnly exalted tone is then suspended for a curious objection concerning whether the nature of the celestial spheres might not in fact prevent the celestial bodies within them from emitting sounds. This is disposed of, first, by reaffirmation and assertion rather than logical demonstration, but then, in the next chapter, by the delightful argument, developed with further ramifications, that if they were silent there would be no point to the inhabitants of the celestial realm possessing the power of hearing. But the celestial realm is itself stratified, for just as musical tones in the world below remind the soul of the joys of the celestial realm, so the music of the spheres recalls the joys of the world of the spirits above. Within this hierarchy, the lower imitates the pre-existing higher, just as children and pupils imitate parents and teachers and aspire to be like them.

It was awareness of these relationships and aspirations, we are then informed, which inspired the sages to have recourse to music. Although Pythagoras is mentioned as the first to enquire into music as a science, the insight of the sages that is emphasised, rather, is music's capacity to stimulate devotion, repentance, spiritual enlightenment and a wish in the soul to ascend to the celestial realm. To elucidate this, specifically Islamic parallels in which the theme is salvation through self-sacrifice are inserted, but the argument reverts immediately to the exhortations of the sages, and it is in this context, rather surprisingly devoid of any explicitly Islamic formulation, that the controversy about the ethical status of music is alluded to. Music has been proscribed at times, so the argument goes, because it has been perverted for frivolous and idle purposes, used to accompany verse celebratory of this world and its transient pleasures but dismissive of the next; the reader is sternly reminded of the need to avoid such error, and to understand the true purpose of the teachings and laws of the prophets and sages.

Beginning with the analogy of an enduring mental image prompted by evanescent musical sounds, the soul within the body is delineated as something precious (pearl, foetus, seed, fruit) that grows to maturity within its container (shell, womb, calyx, rind) and is then extracted from it to begin a new life. (This process, it may be noted, is also held to be valid for sacrificial animals.) In reaching this culminating point the *risāla* has, in effect, completed its thematic journey, even though it has barely passed its half-way point, and that which remains is, essentially, amplification and illustration.

The first theme to be developed further is that of numerical correspondences. Starting with the four strings of the paradigmatic lute, we are led to the four elements and the four humours, emphasis being laid on the power of the notes of the various strings to affect bodily humours, and hence, when suitably combined, to be used in hospitals to alleviate suffering. From the strings we pass to the 4:3 ratio governing their relative thicknesses, and are presented with parallels in the relationships between the various celestial spheres. The next ratio to be discussed is the 9:8 of the whole tone, deemed preferential because of the singular importance of 8, the first cube number, from which we move to a discussion of the numerical virtues of the cube itself, and thence, via the 9:8 relationship of the radii of the spheres of the air and the earth, to a listing of the dimensions of the diameters of the celestial spheres and, selectively, of their relationships to each other.

At this point, we arrive at a form of the argument from design, the harmonious ordering of the cosmos being attributable to a single, unique creator. But we have not yet finished with the virtues of 8, it being the number of prosodic feet, musical rhythms, and levels of paradise; and eightfold sets are esteemed as superior to others, especially in the domain of doctrine, where the views of the Ikhwān are deemed more rounded and complete than those of, e.g., dualists or trinitarians. The notion of completeness and perfection is then illustrated, yet again, by reference not only to the proportions and tuning of the lute, but also to correspondingly well-formed compositions, and the allusion to matters more obviously musical will be followed eventually by a further section on rhythm. That a discussion of prosodic patterns should be intercalated here is hardly unexpected, but rather less predictable is the insertion between them of further and extensive illustrations of the underlying theme of harmonious proportional relationships taken from the fields of calligraphy and anatomy, concluding with an impassioned passage of moral exhortation.

The calligraphic principle declared — and which is deemed valid not only for Arabic, but for all scripts — is that every letter shape derives from just two elements: a straight line of a given length (or fractions thereof), and the circumference (or segments thereof) of a (semi-)circle of which that line is the diameter. From the ensuing, detailed account of the individual letter shapes found in the Arabic script it becomes apparent, though, that this is a little too neatly reductive, for some are stated to be formed by compressing a segment of the circumference of the circle, others by curving the initial straight line or turning it into a circle. A final reference to 'perfect proportion' provides a transition to analogous patterns taken from the proportional relationships observable in the human body, whether of length, size, or shape. A more specific and quite detailed set of measurements, calculated in terms of handspans, is then given for the parts of the body of a healthy new-born, the body at one point having the limbs extended and then a circle inscribed around by a compass in a manner now reminiscent of Leonardo da Vinci's Vitruvian Man. There ensues a restatement of the argument from design, followed by an appeal for the purification of the soul so that, after death, it may return to the celestial abode whence it originated.

The ensuing discussion of rhythm finally addresses the eight rhythmic cycles, the species from which everything is derived. Each is defined in some (if not always sufficient) detail, by reference to the number and nature of the attacks and the durations between them. In line with the earlier recognition that different ethnic groups possess distinct musical styles, the point is further made that the cycles defined here are those fundamental to Arab music, and are not shared by Persians or Byzantines.

The next section revisits the theme of associated sets, this time the especially productive fourfold groupings. Taking the seasons as the axis, we find related to each a dizzying array of phenomena, from such obvious ones as segments of the zodiac and the celestial sphere through times of day, the life cycle, humours and faculties, to tastes, colours, and scents, not forgetting musical rhythms and lute strings. The efficacy of the harmonious conjunction of concordant elements is again touched upon, not only in medical therapy but also in the preparation of talismans, and there is then a further exploration of the relationship between music and psychological states, including its mood-altering ability. The capable musician will know how to match the setting to the sense of the poem, to vary the rhythm according to the mood of the moment (for example, to avert boredom), and, finally, to calm the rowdy.

The penultimate chapter of the *risāla* is devoted to an established literary genre, an anthology of wise sayings concerning music. The setting is explicitly Greek: a symposium to which various learned men have been invited, and whose profound observations are to be recorded. While not wholly eschewing the jocular, the tone is, for the most part, lofty, dwelling initially on the moral inspiration music can and should convey to the soul. The next theme to be elaborated is the uncanny ability of instruments, though not endowed with speech, to reveal the secret feelings of the heart, intimating a profound level of communication through structured sound that does not require the surface articulation of meaning in words. But stress is also laid on the need to protect the soul from the baser passions that music might provoke (although it is also pointed out that protection from such temptations is provided by the natural dimming of desire in old age). The central group of sayings is concerned with the competing claims of sight and hearing for superiority, preference being given, not surprisingly, to the latter. The final group nevertheless gives greater prominence to singing the praises of visual things, seeing in them, when perfectly proportioned, traces of celestial beauty. Gazing upon them, the individual soul is yet again inspired with a desire to ascend to the heavenly realm.

Here, as we approach the final chapter, we arrive once more at the central spiritual message with which the first half of the *risāla* ended, and, just as there, the tone is broken by the injection of a specious argument, indeed, by a dramatised variant of the same debate about the existence or non-existence of sound and hearers of sound in the celestial domain. Eventually, however, we reach a reaffirmation of the truth of the correspondences between this world and its superior, spiritual counterpart, accompanied by a vision of paradise as a distillation of sensory perfection. The *risāla* is then rounded off with expressions of that wisdom centred on the Sufi concept of ecstasy (*al-wajd*), in which the soul becomes detached from the world in its pursuit of union with the divine. But although earthly music may accompany these experiences, their fundamental orientation is now verbal. Indeed, in the final example, that of Moses, who sings in ecstasy upon hearing the Lord speak, human music is thereafter found wanting. Having inspired in the soul a love of the world beyond by reflecting its pure harmonious sounds, its task, it would seem, is complete: music is finally transcended, and the joy of souls in paradise comes not from song, but conversation with God.

Terminology and Concepts

Translation across cultures and language families is always problematic, and although the general absence of explanatory material from the above summary might be thought to imply the existence of a set of reliable terminological equivalences between Arabic and English, this is by no means the case: the match is frequently far from perfect, and several terms are sufficiently deceptive to warrant commentary. Detailed discussion, though, will have to be reserved for the volume devoted to the *risāla* on music within the series that is introduced by the present work; here the aim is the more eclectic one of giving a general indication of the nature of the problems involved, while at the same time touching upon further features of content, with the discussion now being organised by general areas or semantic fields.

Music and Sound

The most obvious place to start is with the very word 'music' and what it relates to in the original. In most cases it renders '*mūsīqī*', itself an evident loan-word, and one of the relatively recent introductions that, for the Ikhwān, requires some kind of elucidation or definition.[11] They

11 Its introduction may well be dated to the ninth-century translation movement; on which, see Dimitri Gutas, *Greek Thought, Arabic Culture* (London: Routledge, 1998; repr. 2002). It appears in none of the titles of works by authors preceding al-Kindī which are listed in Henry G. Farmer, *The Sources of Arabian Music* (Leiden: Brill, 1965). It will eventually become naturalised to the extent that its Greek origin is sometimes forgotten and folk etymologies are found for it.

offer two, the more succinct being simply a gloss of '*ghinā'*' ('song'). But, inevitably, these words are not fully synonymous: '*ghinā'*' points towards the practical, to music in performance, whereas '*mūsīqī*' tends to include (indeed, sometimes to emphasise more) the theoretical, the world of ideas surrounding the realised sound or even wholly independent of it. Thus, while 'music' serves well enough as a translation, it does not quite capture the bias '*mūsīqī*' often has, away from practice and towards the speculative end of the spectrum, to the extent that it may, in fact, contrast with or be complimentary to '*ghinā'*'.[12] It should be noted, further, that '*ghinā'*' is also not given full justice by being translated simply as 'song', for, beyond this basic sense, it may imply musical practice in general, instrumental sounds included.

Sound

An initial potential complication here is that the basic term '*ṣawt*' (plural, '*aṣwāt*') straddles both 'sound' and 'song' (in the sense of a specific vocal composition),[13] although the context usually decides clearly between them. But sound is also defined generically, and neutrally, as an acoustic phenomenon: a *qar'* ('knock', 'rap', 'thump') between two bodies produces shock waves in the air that are propagated spherically (the image used is that of a molten glass ball being distended by the breath of the glass-blower), and as they spread they gradually lose force until eventually the motion ceases. The *risāla* also distinguishes between various forms of sound production, presenting, for example, a typical branching set of discriminations based upon that favourite classificatory device, the binary opposition.

12 That one should be defined by the other here is thus not without interest, although it may simply be that if a one-word gloss for '*mūsīqī*' was required, nothing else came closer.

13 Elsewhere, and most particularly in the major contemporary non-theoretical source, the *Kitāb al-Aghānī* — on which, see George D. Sawa, *Music Performance Practice in the Early 'Abbāsid Era* (Toronto: Pontifical Institute of Medieval Studies, 1989) — '*ṣawt*' generally has the latter sense. It is possible that this meaning is implied in certain passages of the *risāla*, but the Ikhwān certainly do not so define it.

Figure 1

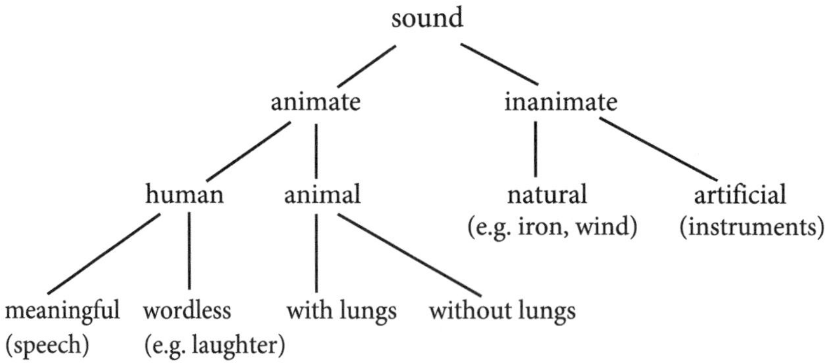

This gives two obvious locations where music could be appended, but instead of including it here we progress from sound to music by moving along a one-dimensional axis from simple to complex, with a clear hierarchy of elements. The first stage is the selection of a particular type of sound: a *naghma* ('note'; plural '*naghamāt*'), which is qualified by a term as implying a sung tone, and hence a sustained pitch, as contrasted with raw, non-musical noise.[14] Such notes may then follow one another in unbroken succession, thereby forming a melody. Finally, we reach *ghināʾ* ('song', 'vocal music', or simply, 'music'), here characterising melodies that are well-formed (*muʾtalif*, 'harmonious') with regard to their tonal relationships.

It should be noted, though, that the omission of any reference to the parameter of rhythm from this hierarchical definition is not wholly justified (as the section 'Melody versus Rhythm', below, makes evident). There is an explicit reference to it contained in a further definition of *ghināʾ*, which states that it consists of both 'harmonious melodies' (*alḥān muʾtalifa*) and 'rhythmically regulated notes' (*naghamāt muttazina*). Although 'notes' or 'tones' (*naghamāt*) first imply pitch, they must also be realised in time, and are qualified here as 'balanced', or

14 There is no equivalent to the precise definition provided by al-Fārābī, which, with minor variations, would be adopted by later theorists: 'A note is a sound sustained for a single perceptible duration as it resounds in the body where it occurs [*Al-naghma ṣawt lābith zamānan wāḥidan maḥsūsan dhā qadr fī al-jism al-ladhī fīhi yūjad*].' See *Kitāb al-Mūsīqī al-kabīr*, ed. Ghaṭṭās ʿAbd al-Malik Khashaba, with revision and introduction by Maḥmūd Aḥmad al-Ḥifnī (Cairo: Dār al-Kitāb al-ʿArabī li'l-Ṭibāʿa wa'l-Nashr, 1967), p. 214.

'regular' (*muttazin*), and elsewhere, by the related and more common term '*mawzūn*', which adds to the notion of balance that of metrical regularity. The necessary integration of the elements of pitch and duration that together make a melody, often obscured by the opposition of 'melodic' and 'rhythmic', is underlined here by the chiastic structure of this definition, with melodic qualification for a term which, as we shall see, frequently has a rhythmic bias, and *vice versa.*

Melody versus Rhythm: The Case of Laḥn

If it seems strange to conjoin these areas by opposing them, one of the main problems in translating this *risāla* is precisely that they are only rarely uncoupled and separated as decisively as the English terminology suggests. In particular, the key term '*laḥn*' (plural, '*alḥān*') carries rhythmic as well as melodic implications, and as a word to be translated it is rather slippery and not always best rendered (as it has been above) by 'melody', for, whereas this foregrounds the element of pitch organisation, frequently of at least equal importance as a component of *laḥn* is rhythmic structure. This is not to say that it never appears in contexts where the emphasis appears to be on the element of pitch (indeed, equipped with a different plural, '*luḥūn*', al-Kindī could use it as equivalent to 'scale' or 'mode').[15] But in others, it refers explicitly (and exclusively) to the way duration is organised, being, in the terminology of the Ikhwān, the technical equivalent to 'rhythmic cycle'.[16] When encompassing elements of both pitch and duration,

15 For use of '*luḥūn*', see al-Kindī, '*Risāla fī khubr ta'līf al-alḥān*', in *Mu'allafāt al-Kindī al-mūsīqiyya*, ed. Zakarīyā Yūsuf (Baghdad: Maṭba'at Shafīq, 1962), pp. 54–57. A more general, melodic sense attaches itself to such titles as '*ta'līf al-alḥān*' ('the composition of melodies'); both plurals, '*luḥūn*' and '*alḥān*', may convey this meaning. For examples, see Neubauer, 'Al-Khalīl ibn Aḥmad' (see above note 5), pp. 258–259. For al-Fārābī, too, the pitch organisation element of *laḥn*/*alḥān* is more prominent than temporal organisation. For a general historical survey of the uses of '*laḥn*', see Hūmān As'adī, '*Nigāhī-yi ijmālī bih sayr-i taḥawwul-i mafhūm-i "laḥn" dar mūsīqī-yi jihān-i Islām*', *Māhūr*, 13 (2001), pp. 57–68.

16 However, this does not appear to have been common usage. Elsewhere, one normally encounters '*īqā'*' (also the generic term for 'rhythm') or '*dawr*' ('cycle'). One other author to use '*laḥn*'/'*alḥān*' in the sense 'rhythmic cycle' is Sa'adya

it is thus perhaps most accurately rendered by the leaden 'measured melody' or the slightly less ungainly 'rhythmic melody'. Such semantic misalignments may be negligible when terms occur in isolation, or else they can be handled with a brief explanatory comment, but where terms accumulate they can be vexatious. For example, in one passage, '*mūsīqī*' (here, an adjective, 'musical') qualifies '*laḥn*', and the phrase 'musical *alḥān*' is defined as '*aṣwāt*' and '*naghamāt*'. The question here is whether the phrasing is simply an accumulation of near synonyms, the whole amounting to no more than a vacuous 'melodies consist of notes', or something both more specific and more complex. Translation here becomes an uncomfortable choice between interpretations, each potentially tendentious.

Composition/Harmony

The definition of '*mūsīqī*' mentioned above equated it with 'song' (*ghinā'*). The other definition the Ikhwān offer is that it is the 'art of composition' (*ṣināʿat al-taʾlīf*) and 'knowledge of proportions' (*maʿrifat al-nisab*). But the key term '*taʾlīf*', normally rendered baldly as 'composition', is difficult to capture exactly: it corresponds perfectly to the etymological sense of composition as 'putting together' (from the Latin, *componere*), suggesting, in addition, a good fit (hence the proposed rendering 'harmonious' for the related adjective '*muʾtalif*'). The emphasis here is on a key concept in the thinking of the Ikhwān, namely that the underlying principal of music should be an adherence to the ideal proportions (alluded to in the second part of the definition) which are manifest in the perfect ratios that govern the music of the spheres, and it is this aspect that Shiloah (see note 1) foregrounds in preferring to render '*taʾlīf*' as 'harmony'. It may be noted that, in the present context, the term 'harmony' (and likewise 'harmonious') has nothing to do with complexes of different pitches sounding at the same time. Although theorists certainly discuss two tones sounded simultaneously, and lute-players no doubt used lower-register notes to supply consonant intervals supporting the melody, in relation to

Gaon (892–942); see Henry G. Farmer, *Saʿadya Gaon on the Influence of Music* (London: Probsthain, 1943).

practice, consonance and harmony are to be thought of as primarily horizontal concepts, defining the nature of the relationship between successive tones.

Rhythm

Analytical Framework

When speaking of the specific rhythmic patterns or cycles used in practice, the term preferred by the Ikhwān, as noted above, is '*alḥān*', whereas in relation to the abstract concept, we encounter the term favoured by other theorists, '*īqā*°. However, '*īqā*° may also refer to an attack, while the plural, '*īqā'āt*', may imply specific structures, that is, rhythmic cycles. Also noted above is the term '*qar*°, defining a sound-producing impact, but in the specific context of musical rhythm, the preferred term for 'attack' or 'percussion' is '*naqra*', plural '*naqarāt*' (literally, a 'knock' or 'rap').[17] Despite such overlaps, these provide at least the nucleus of a specialist vocabulary, and it would have been logical for the Ikhwān to have developed around them an autonomous description of rhythmic phenomena drawing upon concepts in arithmetic (numerical proportions) or geometry (points and lines), but, in the event, they chose to follow al-Kindī, their principal source for the discussion of rhythm, and import the analytical framework used in the study of prosody, together with its related jargon. Musical rhythm is thus conceived as allied or analogous to poetic metre, a relationship strengthened by the fact that the names of some of the cycles also serve as names of metres (even if their structures do not appear to match).

Arabic prosody, for which the technical term is '*'arūḍ*',[18] uses a vocabulary reflecting phonological criteria as mediated by the Arabic script, so that, rather than distinguish various syllable types as abstract analytical categories to determine metrical patterns, it refers

17 One may be defined in terms of the other. For example, al-Fārābī says in his *Kitāb al-Mūsīqī al-kabīr*, p. 447, that: 'A percussion [*naqr*] is the impact [*qar'*] on one solid [*ṣulb*] object of another sharp-edged [*daqīq al-ḥarf*] solid object.'

18 For a general yet thorough account, see G. Meredith-Owens, ''Arūḍ', *EI2*, vol. 1, p. 667–677.

to combinations of letter (for consonant) and diacritic sign (for ± vowel).Transferring this approach to the analysis of rhythm, an attack (*naqra*) is equated with a letter representing a consonant that is initial in a syllable and, according to the way the script represents the syllabication rules of Arabic, must be followed by a short vowel (symbolised by a diacritic sign) which is termed a 'movement' (*ḥaraka*): together, then, 'letter + movement', termed 'moving letter' *(ḥarf mutaḥarrik)*. The musical equivalent of the short syllable, 'attack + (short) duration' (*naqra mutaḥarrika*), is viewed as the shortest perceptible separate sound (anything shorter, and the attacks would run together, forming a continuous sound). The Ikhwān symbolise it either by one of the short syllables ('*mu*', '*ta*', '*'i*') occurring in the abstract word shapes used to represent prosodic feet or, in corresponding syllable strings more closely related to percussion patterns, by '*ta*' or '*na*'. In effect, the attack itself is viewed as instantaneous, without duration, just as a point in geometry has no extent: it can only be perceived in time because of the following duration separating it from the next attack, so that the combination 'attack + (short) duration' is an indivisible entity. However, the gap between attacks can be long instead of short. This points to distinctions of note length (stateable in terms of the number of time-units between one attack and the next), and/or of tempo, and, in one passage, four length/tempo bands are discriminated.[19] Integral to the notion of a slowest possible band is that of a maximum possible duration (defined as eight time-units), beyond which the relationship between the preceding and following attacks can no longer be perceived as meaningful, thereby destroying the coherence of the rhythmic cycle in which they occur.

The Individual Cycles

The designations of two of the four bands are abstract, but the remaining two are given the names of cycles found in practice.[20] Whereas pitch

19 Al-Fārābī puts forward a parallel scheme; see Sawa, *Music Performance Practice in the early 'Abbāsid Era*, p. 39.

20 The first is also called 'the quickest possible', literally 'the lightest possible light' (*al-khafīf al-ladhī lā yumkin akhaff minh*), again including a term used in prac-

relationships are of interest to the Ikhwān solely for their cosmological affiliations, and no attempt is made to discuss modal structure or to name any of the melodic modes or classes in current use, considerable attention is given to the classification and description of rhythmic structures, which are both ordered into sets and analysed individually. There are eight in all, although, for some if not for all, there may have been a number of variant forms.[21] They are listed twice, with only negligible differences in some of the names, but a significant difference in one. They are first presented in the symmetrical form of a two-by-four arrangement, in which each of the four pairs is made up of an 'x' and its 'light' (*khafīf*) counterpart. But the second version offers as its final pair a 'light' counterpart of the 'light' (*khafīf al-khafīf*) and *hazaj*, and this discrepancy suggests that practice may not have been quite as neat and tidy as theory would wish. In all probability, the first list distorted matters by giving as a pair *hazaj* and a 'light' counterpart, that is, a 'light' counterpart for what was already a 'light' rhythm; while, in the second, a further asymmetry occurs in the absence of a 'light' entity to which there could be a light(er) counterpart.[22] The neat 'four-and-four' arrangement of the first set thus appears to have been an imposition on what, in practice, may have been a combination of three slower cycles and five faster ones or, indeed, a more fluid situation in which there may not even have been a consensus on a total of archetypal structures, let alone how they were grouped.[23]

---

tice, *khafīf*, while the second, likewise, is also called 'the second light' (*al-khafīf al-thānī*).

21 That the forms given are not the only ones occurring in practice is indicated clearly: following al-Kindī (in *Mu'allafāt al-Kindī al-mūsīqiyya*, p. 97) they are defined as genera (*ajnās*), from which all other forms derive (*wa-minhā yatafarra' sā'iruhā*). Also given is the analogy of the relationship between the prosodic feet and the metres compounded from them, which, if at all accurate, would point to the existence of quite complex cycles.

22 Furthermore, if 'light' were shorthand for the counterpart to either the first or second 'heavy' we would have a threefold stratification in contrast with binary oppositions elsewhere.

23 Al-Fārābī, for example, lists seven (see Sawa, *Music Performance Practice in the Early 'Abbāsid Era*, p. 41), but the 'four-by-four' arrangement is prefigured in al-Kindī's account (in *Mu'allafāt al-Kindī al-mūsīqiyya*, p. 97). There is an obvious parallel with the four-by-four mode classification attributed to Isḥāq al-Mawṣilī, but given its awkwardness it is difficult to see this four-by-four

The present introduction is not the place to attempt an interpretation of the definitions given for these cycles, for this would need to pay heed both to internal textual difficulties and to the discrepancies between the versions given by the Ikhwān and those of other authorities. But some account might be given of the nature of the problems encountered in the text, where we are sometimes confronted with what appear to be internal inconsistencies compounded by an occasional lack of specificity. These may be illustrated by looking at two representative cycles: one, the definition of which appears to be unequivocal, while the other is problematic.

Straightforward Definitions

Cycles are presented in two forms: a verbal description of the number and (sometimes) the relative position of the characteristic attacks, and a statement in prosodic and rhythmic syllables (which may be thought of as mnemonic definitions). In the first of these two cycles, the 'light of the light' (*khafīf al-khafīf*), verbal description and syllabic representation are congruent, yielding a cycle of three time-units. The verbal description specifies that it consists of two 'consecutive' (*mutawālī*) attacks; helpfully confirms that this means that they are not separated by a pause; and adds that each successive pair of attacks is separated by a pause, the duration of which is one time-unit (*zamān naqra*). We thus have /x x o / (where x = an attack; o = a pause, or the absence of an attack; and each symbol has the value of one time-unit).[24]

Problematic Definitions

Other definitions, however, are not so plain-sailing. The verbal description of the 'light' counterpart of the first 'heavy', for example, mentions seven attacks, but qualifies these in ways that make it difficult to establish the duration of the pauses, and hence the total number of time-units in the cycle. Two of the attacks (by implication, the first

organisation of the rhythmic cycles as anything other than an instance of *Systemzwang* (see Neubauer, 'Al-Khalīl ibn Aḥmad', pp. 283–284).

24 This definition is almost identical with that proposed by al-Kindī, in *Mu'allafāt al-Kindī al-mūsīqiyya*, p. 98.

two) are 'consecutive' and, as with the 'light of the light', the text then confirms that they are not separated by a pause. So far so good, but the third attack is then described as 'isolated' (*mufrad*) and 'heavy' (*thaqīl*), suggesting, in addition to a distinction of intensity and/or timbre, the possible presence of pauses both before and after, but without specifying their length; and there are then a further four attacks, considered as a group but characterised in such a way as to indicate that they may not have all been equally spaced. If we then turn for help to the mnemonic definitions, we find that the manuscripts of the *risāla*, unfortunately, fail to agree, although the most likely version yields a total of thirteen time-units, and by itself points to:

/x x o x x o x x x o x x o /

If we attempt to fit the seven attacks of the verbal definition into this framework we might arrive, just, at:

/x x o x o o x x o x o x o /

But this signally fails to convince. There is, in such cases, a discrepancy between what verbal definition and mnemonic representation suggest, and a further and more serious discrepancy between whatever one might tease out of them and what other authorities tell us. The most salient and surprising feature of the accounts given of the longer cycles is their relative complexity when compared with the versions offered by al-Kindī and largely confirmed by al-Fārābī. Thus although the Ikhwān follow al-Kindī programmatically in presenting a set of what are considered to be generic types, the forms taken by cycles such as the 'light' counterpart of the 'first heavy' are best understood as a combination of two cycles, the second a variant form involving the omission of an attack from the first, the primary matrix, and the addition of others.

Musicians/Instrumentalists

It is striking that vocal music is referred to primarily in the context of the morally uplifting, the world of the sages and prophets. The musician may be a singer (*mughannī*, a term that slips in almost inadvertently and is nowhere given prominence), but is rather viewed or, perhaps more precisely, tends to come into sharper focus, as an instrumentalist. In this respect it is noteworthy that the initial classification of music, despite the emphasis on its spiritual potential, is as an exception among the 'manual' arts (*bi-al-yadayn*). This is justified in so far as, having used his psychological intuition to sense what is appropriate to the situation, it is through the musician's manual dexterity in adjusting the tuning of his instrument and then performing on it within the appropriate modal and rhythmic range that he can achieve the desired result of creating a particular emotional response in the audience. The power of vocal music is certainly acknowledged, but it is the semantic charge of the text set that is foregrounded, not the charm of the melody; and in a key statement of the principle that music in this world reminds the soul of the joys of the world above, there is a striking absence of any reference to song: rather, it is specifically the sound 'produced by the movements of the musician' that is mentioned. This, we may assume, refers in the first instance to the potential of the fingers, as they move over the perfectly adjusted frets of a lute, to create melodies that embody the 'perfect proportions' common to microcosm and macrocosm and hence serve as a bridge between the two domains, activating in the soul reminiscence, nostalgia and yearning.

Given this bias, the general avoidance of the term '*mughannī*' is predictable, but instead of, say, '*ḍārib*' ('lute player'), we encounter for 'musician' the unusual '*mūsīqār*' obviously derived from '*mūsīqī*'.[25] It appears to be an early coinage that failed to establish itself as a standard term;[26] and its learned flavour serves, if anything, to distance it from the

25 It is, however, equated with '*mughannī*' upon its first appearance, in a terse set of lexical equivalences: this begins with the definition or explanation of *mūsīqī* as *ghinā'*, followed by that of *mūsīqār* as the related *mughannī*. Al-Khwārizmī says, 'the musician/singer [*muṭrib*] and composer [*mu'allif al-alḥān*] are called *mūsīqūr* and *mūsīqār*'; see Al-Khwārizmī, *Mafātīḥ al-'ulūm*, ed. Gerlof van Vloten (Leiden: Brill, 1895), p. 236. But that '*mūsīqūr*' was puzzling to later scribes is suggested by the existence of several manuscript variants (*mūsīqūn*, *mūsīqūrus*, *mūsīqiyyīn*, *mūsīqiyyūn*). The suffix '*-ār*' is not Arabic but a Persian agentive morpheme.

26 '*Mūsīqār*' appears in a treatise by al-Kindī, and may have been coined by him (see his *Mu'allafāt al-Kindī al-mūsīqiyya*, p. 83); it is subsequently used only by al-Khwārizmī and the Ikhwān, or, at least, no other source is cited by Lois I. al-Faruqi, *An Annotated Glossary of Arabic Musical Terms* (Westport, VA: Greenwood Press, 1981). But more frequent with al-Kindī are the more ponderous '*mūsīqārī*' (see *Mu'allafāt al-Kindī al-mūsīqiyya*, pp. 69, 72, 88) and the

world of everyday music-making, and makes of the musician a more august figure, worthy to be associated with the morally elevated and spiritually inspiring.

Instruments

A parallel and more *recherché* derivation is '*mūsīqān*', which does not appear to be attested elsewhere.[27] It is introduced along with '*mūsīqī*' and '*mūsīqār*', being defined as a 'musical instrument' (*ālat al-ghināʾ*). However, this time the common term '*āla*' is the one used in association with references to existing instruments, while the equally common '*malāhī*' appears just once, in conjunction with '*mughannī*', in a possibly derogatory reference to contemporary musicians who hold views contrary to those put forward by the Ikhwān.[28] In contrast, '*mūsīqān*' is reserved for passages more appropriate to its etymology and elevated tone, ones purporting to convey the views of ancient sages.

The description of the lute accords with, and may have been inflected by, the numerological and cosmological emphasis of the *risāla*, and no other instrument is discussed in similar detail.[29] Indeed, apart from the occasional explanatory comment, the other instruments are simply named, so that interest resides principally in the number selected and in the ways they are grouped: the text of the *risāla* eschews the descriptive precision that would allow us to solve the sometime thorny problems of identification that beset instrumental terminology. The Ikhwān focus

simpler '*mūsīqiyy*' (ibid., pp. 84, 93, 95); he also uses the circumlocution (ibid., p. 69): 'the users of this musical art [*mustaʿmilū hādhih al-ṣināʿa al-mūsīqiyya*]'.

27 Again, no other instances are cited in al-Faruqi, *An Annotated Glossary of Arabic Musical Terms*. This also includes '*mūsīqāt*', presumably a misreading. The manuscripts sometimes confuse '*mūsīqār*' and '*mūsīqān*', a mistake that probably originated quite early on: it is attested in a passage derived from the Ikhwān in a manuscript dated 529/1156; see Muḥammad ibn ʿAlī al-Hindī, *Jumal al-falsafa*, Publications of the Institute for the History of Arabic-Islamic Science, Series C, 19 (Frankfurt am Main: Institut für Geschichte der Arabisch–Islamischen Wissenschaften, 1985), p. 114.

28 Whereas '*āla*' is neutral (it also means 'tool'), '*malāhī*' conveys overtones of possibly frivolous pleasure, and is hence more suitable to a context of disapproval.

29 For a detailed analysis of this material, see Eckhard Neubauer, 'Der Bau der Laute' (see note 8 above).

neither, like al-Fārābī, on string and wind instruments specifically and the scales they produce nor, like al-Kindī, on the associations related to the number of strings, but, rather, they range widely in their references, with the result that the number of instruments they mention is quite considerable, including, in one or two cases, what may be the earliest attestation. In one passage, their order of presentation is suggestive of a taxonomy foreshadowing that made explicit by Ibn Sīnā (Avicenna; d. 1037), who proposes a classification that includes percussion, and, in addition, that divides string instruments into two categories, one consisting of those that have a separate, unstopped string for each pitch, such as the harp, and the other consisting of those that do not.[30] Because we cannot always be sure what type of instrument a name refers to it is difficult to make the claim with confidence, but it does appear that a similar arrangement is at work in the *risāla*. The passage in question begins by mentioning percussion, moves on to wind instruments, then strings, and ends with instruments in all or most of which each pitch has a separate sound producer, that is, instruments either with unstopped strings or unstopped pipes. Mention should also be made here of the discrimination, as specific classes of instruments, between those producing continuous or sustained (blown, bowed) as against discontinuous (plucked, struck) sounds.

Cosmology

The fundamental importance of cosmology to the *risāla* on music has already been made abundantly clear, and for full and authoritative accounts of this complex field the reader is referred elsewhere.[31] Here, it remains merely to note some of the ways musical phenomena are incorporated.

One particularly prominent hook for extra-musical associations is provided by the four strings of the lute, which are paired with the

30 Ibn Sīnā, *Kitāb al-Shifā', al-Riyāḍiyyāt*, 3: *Jawāmi' 'ilm al-mūsīqī*, ed. Zakarīyā Yūsuf (Cairo: al-Maṭba'a al-Amīriyya, 1956), p. 143.

31 In addition to Alessandro Bausani, *L'enciclopedia dei fratelli della purità*; see Seyyed Hossein Nasr, *An Introduction to Islamic Cosmological Doctrines* (Cambridge, MA: Harvard University Press, 1964), pp. 44–104; and in addition, more specifically on music, Jozef S. Pacholczyk, 'Music and Astronomy in the Muslim World', *Leonardo*, 29, 2 (1996), pp. 145–150.

elements in a straightforward correlation of (high to low) pitch and (low to high) density, and are then aligned with the humours. The sound of each string tends to stimulate one and suppress another, thereby allowing for the allopathic treatment of maladies diagnosed in terms of a disturbance of the humours. But the *risāla* also contains a much more complex fourfold set, presented as matter for reflection, and taking as its self-evident basis the division of the year into four seasons. To each season are related various astronomical, astrological and natural phenomena, and the list goes on to include human attributes and qualities, before ending with tastes, colours and scents. The musical features that appear are specific rhythmic cycles and, again, the strings of the lute: the lowest with winter; the next with autumn; the highest with summer; and the second highest with spring.

There are no further musical specifics, but there is a lengthy passage in which the proportional relationships between the heavenly spheres are stated in terms of the ratios previously used to determine interval sizes. Thus, the spheres corresponding to the four elements have diameters that are successively one third greater than the last; while the diameters of the spheres of the earth and moon, for example, are in the ratio 4:3; and the corresponding relationship of the spheres of the fixed stars and Venus is 2:1. But not everything is harmoniously arranged: the section ends with the statement that 'Mercury, Mars and Saturn have different relationships, for which reason they are said to be of ill omen'. The influence of the stars is axiomatic, but, beyond this general statement, there is no specific disquisition of astrological lore, even if its techniques (and the belief in their efficacy) are central to a brief, appended passage on talismans which refers to the manner of preparing a magic three-by-three square to alleviate labour pains.

Function

In its therapeutic function, the type of music chosen (and the optimum time for performing it) will depend on the diagnosis, and presumably any combination of rhythmic and modal resources could be called upon as required.[32] Further functions of a utilitarian nature concern the ways music can affect humans or animals in specific situations for a particular purpose: a mother's lullaby, for example, or the kinds of music that animals respond to and are encouraged by while travelling, watering, mating, being milked, or even being hunted — for we are told that they can be so charmed by the hunter's voice that they make no attempt to flee.

Other uses are dictated by needs of an emotional nature. Here, a number of broad bands are distinguished, corresponding to which relevant segments of the system will be selected. The particular genres mentioned concern recognisable types of melody, each having a specific effect or emotional range: one is saddening, and is associated with repentance; another is emboldening, inspiring martial valour; another provides consolation; while yet another is appropriate for festive occasions.[33] It is a particular mark of the consummate musician to able to adjust his style and make appropriate selections from the repertoire in order to stimulate his audience to the desired response,[34] and some idea of the developmental curve of a typical performance emerges incidentally:[35] from sober settings of verse

32 For general background on this topic, see Amnon Shiloah, 'Jewish and Muslim Traditions of Music Therapy', in *Music as Medicine: The History of Music Therapy Since Antiquity*, ed. Peregrine Horden (Ashgate: Aldershot, 2000), pp. 69–83; and, for classical antecedents, Martin West, 'Music Therapy in Antiquity', ibid., pp. 51–68. For later developments, with more specific diagnoses, see Eckhard Neubauer, 'Arabische Anleitungen zur Musiktherapie', *Zeitschrift für Geschichte der Arabisch-Islamischen Wissenschaften*, 6 (1990), pp. 227–272.

33 Similar types are distinguished by other writers: al-Kindī, for example, has a binary pleasure/sadness opposition, to which is usually added a third, intermediate, term, variously characterised as moderate, for praise, and to stimulate valour.

34 It is interesting to note that there is no mention of the *qayna*, the singing slave-girl, in this epistle. This may be because the world of the *qayna* could hardly be accommodated to the purposes advocated by the Ikhwān, as becomes abundantly clear from the no doubt partial but nevertheless fascinating analysis offered by al-Jāḥiẓ in his *Risālat al-qiyān*. See al-Jāḥiẓ, *The Epistle on Singing-Girls*, ed. and trans. Alfred F. L. Beeston (Warminster: Aris and Phillips, 1980); and Michael Stigelbauer, *Die Sängerinnen am Abbasidenhof um die Zeit des Kalifen al-Mutawakkil nach dem Kitāb al-Aghānī* (Vienna: VWGÖ, 1975).

35 For a fuller account of performance contexts and structures see Sawa, *Music Performance Practice in the Early ʿAbbāsid Era*. This is based upon al-Iṣbahānī's monumental *Kitāb al-Aghānī*, to which may be added the comparable material in Part 6 (*al-yāqūta al-thāniya*) of Ibn ʿAbd Rabbih(i), *al-ʿIqd al-farīd*, ed.

stressing manly qualities to jollier and more frivolous material and dance tunes (but with the proviso that the musician should be able to calm things down when the behaviour of the more inebriated threatens to get out of hand).

Morality

Another developmental curve implied by this account of genres may be inscribed within a standard paradigm of moral decline through the ages. The characteristic modern preference is for the festive type — that which most obviously contrasts with the austere moral purpose behind the sages' original 'invention' of music, and elsewhere we are told explicitly that this purpose has latterly been abandoned in favour of a craving for pleasure. At this point in the text, we find an echo of the perennial debate about the ethical status of music in Islam,[36] albeit one oddly muted, in that there is no reference to any of the hadith usually cited, whether in condemnation or justification; rather, the argument is couched in more general terms: people have abandoned the uses of music proposed by the sages, and instead listen to hedonistic song lyrics that undermine faith. But they also pursue 'idle entertainment', and here we do touch upon the Islamic formulation of the controversy, for the phrase translates '*lahw*', a key word that provides a Qur'anic 'weapon' in the armoury of the legists who disapprove of music. In the remark that follows, the Ikhwān effectively summarise, in a rather decorously restrained way, the nature of the charges levied in the juridical literature. Yet the seductive potential, although clearly a moral concern, is less an intrinsic feature of music than one of the wiles of Nature, conceptualised here as the baser human component that contends with the

---

Aḥmad Amīn, Ibn al-Anbārī and ʿAbd al-Salām Muḥammad Hārūn, vol. 6 (Cairo: Maṭbaʿat Lajnat al-Taʾlīf waʾl-Tarjama waʾl-Nashr, 1949).

36 The earliest surviving condemnatory text is *Dhamm al-malāhī* by Ibn Abī al-Dunyā (823–894); text and English translation in *Tracts on Listening to Music*, trans. James Robson (London: The Royal Asiatic Society, 1938). The beginning of the defence can be seen in the *Kitāb al-Malāhī* by al-Mufaḍḍal ibn Salama (d. 902); text in ʿAbbās al-ʿAzzāwī, *al-Mūsīqā al-ʿirāqiyya fī ʿahd al-mughūl waʾl-turkumān* (Baghdad, 1951).

soul. More generally, Nature entices man away from his spiritual potential through its beauty, thus raising the question of whether beauty is always viewed as contained within an ethical universe, or whether aesthetic principles have some autonomy.

Aesthetics

In fact, it might be argued that the *risāla* does not deal with aesthetics proper, at least not in any direct way: its express intention is to demonstrate the primacy of the One, God the Creator, and at the core of its concerns is what might be described as a fusion of cosmology and ethics in which there is no place for a theory of autonomous beauty.[37] Yet beauty is certainly referred to,[38] and it would, in any case, be difficult to resist the temptation to detect a consciousness of beauty — akin, perhaps, to that experienced by mathematicians — as an attribute of the 'perfect proportions' that govern the cosmos and underpin the arts. It may well be that what is important about music is its potential to affect the soul in an ultimately functional way within a predetermined scheme of causality, but the account would be incomplete without considering the vocabulary of reception, for properly constructed melodies do not just have the potential to promote equilibrium in a therapeutic sense, they do not just restore or maintain physical and psychological well-being, but have a surplus value, causing delight, pleasure and enjoyment; equally, ill-sorted ones provoke feelings of revulsion. Enjoyment, moreover, is not generated automatically and universally, but selectively, in accordance with the particular preferences of the group or the changeable inclinations of the individual.

Yet the explanation given stresses the concept of affinity, which

37 This is not to deny, as in other texts of the period, the inclusion by the Ikhwān of material with aesthetic implications, as is abundantly demonstrated in the exhaustive survey conducted in José M. Puerta Vilchez, *Historia del pensamiento estético árabe* (Madrid: Ediciones Akal, 1997).

38 Rather than the pair of neutral terms '*jamīl*' ('beautiful') and '*jamāl*' ('beauty'), which do not appear in the *risāla*, the Ikhwān prefer equivalents which also have overtones of moral approval, '*ḥusn*' and '*maḥāsin*'. Associated with Nature, the beauties of which are a temptation, we tend to find the term '*zīna*' ('adornment'), which suggests a beauty that is less intrinsic than a cosmetic addition.

may be thought to have aesthetic potential but is clearly not an overtly aesthetic criterion. In parallel to the ubiquitous proposition in literary theory that 'meaning' (*maʿnā*) can be separated from 'expression' (*lafẓ*), a melody is viewed as conveying an idea, and the relationship between the two is analogous to that of body and spirit, thereby accounting for the pleasure that the idea provokes in the kindred soul, a pleasure that is further occasioned or enhanced by the parallels to the movements in time of the heavenly bodies that can be discerned within the well-proportioned rhythmic articulation of the melody.

Proportion, together with regularity, is also the principle that links music to other forms of artistic expression, notably poetry and, above all, calligraphy, to which its application is discussed independently at some length. Poetry, on the other hand, comes into contact with music directly, and here the principle that is emphasised with regard to the meanings and emotions it conveys is that of appropriateness, specifically, a matching of semantic content not to modal parameters but to rhythm: the 'heavy' cycles are suitable for the expression of glory and generosity, 'light' cycles for valour and action, *ramal* and *hazaj* for poems arousing pleasure. A similar matching of rhythm and mood is also projected onto the social plane, the skilful musician possessing the ability not only to choose cycles appropriate to the atmosphere of the gathering and the activity in question, but also to intervene where advisable by introducing rhythms that bring about a change of mood.

If any particular aesthetic thread can be disentangled, it is thus one spun from notions of appropriateness, balance, and harmonious composition in the sense of the smooth fitting together of parts so as to conform to principles of proportion. If the centrality of this concept to music needs no further explication, what may, perhaps, be worth observing is the extent to which the Ikhwān relate these ideas to phenomena in the natural world and invoke forms of artistic expression beyond music.

Nature itself, however, is regarded, at least in one passage, as problematic, offering an alluring distraction, its 'beauty' (*zīna*) a seductive trap. On the other hand, it appears that natural objects can be neutrally and safely appreciated for their beauty, with the added positive factor that in them are to be detected traces of the spiritual world beyond, provided that they have not yet been distanced from the Creator by exposure to the ravages of time in this world. Accordingly, those things singled out as exemplary are the newborn and fresh plants: the qualities of proportion and balance, it may be noted, are offered to us as properly beautiful only in biological form, in living tissue, never in the colder symmetries of the mineral world.

Among the products of human creativity, poetry and calligraphy are those to which most attention is given, but there are also briefer references to the plastic arts and painting, particularly as providers of representations of living beings. Here the same criteria apply: such images are presumed to strive for harmony of proportion, composition and colour in the context of an implied mimetic theory of representation (although here no Platonic obloquy attaches to their status as reflections of living beings that are themselves mere reflections of celestial archetypes). It may be noted that abstract design is never mentioned in connection with these arts and the criteria deemed pertinent to them. Just as the mineral world is ignored in nature, so, too, there is no reference here to architecture or to repetitive geometric patterns. Such silence should not be found surprising, for it is a cultural constant. Yet, were such forms of creative activity to reflect, as has been insistently claimed, a theological vision of oneness perceived to lie at the heart of Islamic artistic expression, one might have hoped to find, in this of all texts, some reference to the manifestation of 'perfect proportions' in, say, mosque architecture, thereby pointing to the realisation, and recognition, of the potential to symbolise in sacred space, indeed, to made tangible on earth, the fundamental arithmetic and geometric attributes of the divine order.[39]

39 Representative of radically different understandings of this area are Keith Kritchlow, *Islamic Patterns: An Analytical and Cosmological Approach* (London: Thames and Hudson, 1976); and Oliver Leaman, *Islamic Aesthetics: An Introduction* (Edinburgh: Edinburgh University Press, 2004). A judicious analytical survey of the field is given in Gülru Necipoğlu, *The Topkapi Scroll: Geometry and Ornament in Islamic Architecture* (Santa Monica: The Getty Center for the History of Art and the Humanities, 1995).

Musical Discourses

It would be appropriate to conclude by attempting, briefly, to contextualise further this complex and fascinating work. Reference has been made, above, to the major streams of Pythagorean and Neoplatonic thought that feed into this *risāla* and feature so prominently in it, but there are also more immediate antecedents to consider, ones representing a wide range of interests and approaches. To be explored here, thus, is not the complex history of translation and the intellectual apprehension and domestication of the classical legacy,[40] but rather the position of the *risāla* in relation to the various textual genres in which music is a prominent topic.

Of the substantial preceding literature, a great deal has, unfortunately, been lost,[41] but what survives is sufficient to indicate some general characteristics. For convenience, even if the boundaries between them are not always firm, a number of genres may be distinguished: works specifically concerned with music and musicians; works of history and literature (*adab*), in which the material on music may range from the minimal to the substantial; and works attacking or defending the permissibility of music. Of these, the last is the least significant to the *risāla*: despite the constant production of tracts down to the present, this type of polemic was only just beginning to come on stream in the tenth century, and, in any case, its normal means of arguing a case, by presentation of, and extrapolation from, relevant hadiths, was one foreign to the Ikhwān. The literary material might be thought rather more relevant; yet, its musical content, which sometimes conveyed notions of origins and innovations but concerned itself in the main with narratives of events and persons of a vivid and striking nature, is again essentially alien to their concerns. If anything, one might detect in the *risāla* occasional stylistic echoes of *adab*, for although it is often quietly expository and eschews any recondite vocabulary, in its rhetorically more highly wrought passages it tends to display a

40 On which see Gutas, *Greek Thought, Arabic Culture*, (and the literature cited there).

41 Indicative is the fact that of the earliest 150 titles reported by Farmer (*The Sources of Arabian Music)*, up to and including Ibn Khurradādhbih (d. ca. 912), barely twenty are known to have survived.

conscious use of syntactic balance and morphological parallelism akin to that characteristic of epistolary prose.[42]

Predictably, then, if we set aside the more specifically philosophical elements, the most significant connections are likely to be found within what could be termed specialist musicological literature. But this, in turn, consists of a number of strands: song-text collections with or without technical information; biographical (probably largely anecdotal) accounts of prominent musicians; and theoretical works. Among the last, two further main types may be distinguished: those of a primarily indigenous cast, dealing with aspects of modal and rhythmic practice or with instruments; and those representing exposure to, and assimilation of, the Greek legacy. As many of the former category have been lost, it is impossible to know the extent to which, say, what appears to be an embryonic organological classification in the *risāla* might actually reflect previously established categories; but it is abundantly clear that the most profitable comparisons are with the earliest surviving Greek-inspired texts.

Al-Kindī

Among these works, the ones most closely comparable in thematic range and treatment are the various surviving treatises by al-Kindī (d. ca. 865), which foreshadow the *risāla* in the way they explore the analysis of intervals and scales expressed in mathematical terms, in the approach they use in the definition of the rhythmic cycles, and in their stress on cosmologically grounded sets. They are, though, strikingly different in their general brevity and eclectic nature, and, particularly, in lacking an overarching concern with the celestial world as a domain of bliss to which the soul aspires. For example, the Ikhwān stress the connection between the ratios of intervals and those of the celestial spheres in a way that al-Kindī does not. Nevertheless, it is still the case that the numerological focus displayed, say, in the lengthy inventory of fourfold sets of phenomena, is already developed

42 See, e.g., J. D. Latham, 'The Beginnings of Arabic Prose Literature: The Epistolary Genre', in *Arabic Literature to the End of the Umayyad Period*, ed. A. F. L. Beeston et al. (Cambridge: Cambridge University Press, 1983), pp. 154–179.

by al-Kindī in somewhat similar terms. This could, of course, be attributed to a reliance upon a common stock of ideas, but when one takes into account areas as different as the description of the lute, the technical analysis of the rhythmic cycles, and the final presentation of the views expressed by an assembly of philosophers, it would be difficult to deny an obvious connection with, if not in every case a direct indebtedness to, al-Kindī's theoretical texts. The ordering and expression of this material, however, is far from identical, and it is reasonable to assume, not only that its authors exercised a degree of creative elaboration in the *risāla*, but also that they may not always have been working directly from the equivalent passages in al-Kindī: unknown intermediaries may sometimes have been involved. For example, al-Kindī's version of the wise sayings of the philosophers (unfortunately truncated, so that only six items survive) has a version of items 1, 3, and 5 according to the Ikhwān, but not of 2 and 4, and contains three absent from the *risāla*.[43] The point should also be made that, although some of al-Kindī's material, however transmitted, must have been known to the Ikhwān, there is at least one text which finds no reflex in the *risāla*, even though its contents would have been highly suited for inclusion.[44]

Tenth-Century Theory

After the exploratory and eclectic mid ninth-century treatises of al-Kindī, we find in the mid tenth-century writings of al-Fārābī (d. 950) and al-Khwārizmī (d. ca. 980) a major shift of emphasis: cosmological schemes and concerns are neglected in favour of a more coherent emphasis on the descriptive, scientific and analytical.[45] The compressed

43 *Mu'allafāt al-Kindī al-mūsīqiyya*, pp. 106–108. It should be added that this material (or structure) existed in at least one other version, that of Ḥunayn ibn Isḥāq, *Ādāb al-falāsifa*.

44 This is the *Kitāb al-Muṣawwitāt al-watariyya min dhāt al-watar al-wāḥid ilā dhāt al-'asharat awtār* in *Mu'allafāt al-Kindī al-mūsīqiyya*, pp. 69–92, which deals with beliefs — including ethical and cosmological considerations — concerning the varying number of strings on different instruments, a topic that could readily have been integrated into, for example, the passage discussing the merits of the number 8 (and the demerits of dualism, etc.).

45 And by Ibn Sīnā's time (d. 1037), cosmological speculation was to be summar-

account (in the form of an encyclopaedia entry) supplied by the latter has, for example, sections on instruments (including the definition of a lute fretting), pitch (including the *systema teleion*), and rhythm, and nothing else.[46] It is also worth noting that the equivalent passages in the *risāla* treat these topics rather differently: al-Khwārizmī recognises the existence of neutral intervals, for example, while the Ikhwān do not. A much more developed treatment of theory, particularly on intervals and scales, where we find material on indigenous structures as well as an elaboration of Greek tetrachord types, is provided by the extensive and magisterial *Kitāb al-Mūsīqī al-kabīr* by al-Fārābī (d. 950), which is supplemented by additional treatises further clarifying his detailed treatment of rhythm.[47] Again, the *risāla* contains no echo of this;[48] and the reverse is also true, for the overriding Pythagorean and Neoplatonic bias of the Ikhwān fails to be reflected not only in this body of work, but also in the major literary works of the period in which music features prominently, the *ʿIqd al-farīd* and, above all, the *Kitāb al-Aghānī*. This is not to say that there are no points of congruence, no thematic similarities; rather, it is to point to the general difference of emphasis which makes the *risāla* such a unique and intriguing addition to the musicological corpus. More than any of the other surviving texts, it is a work that is consciously programmatic, constantly parading different facets of the same essential core message. For those whose primary interest lies in trying to discover the musical grammar of the day, or in learning about the modalities of performance, it is bafflingly uncommunicative, but if one wishes to

ily rejected from the very beginning. See Ibn Sīnā, *al-Shifāʾ*, *al-Riyāḍiyyāt*, 3: *Jawāmiʿ ʿilm al-mūsīqī*, pp. 3–4.

46 Al-Khwārizmī, *Mafātīḥ al-ʿulūm*, pp. 235–246.

47 See Eckhard Neubauer, 'Die Theorie vom *Īqāʿ*. Part 1: Übersetzung des *Kitāb al-Īqāʿāt* von Abū Naṣr al-Fārābī', *Oriens*, 21–22 (1968–1969), pp. 196–232; Eckhard Neubauer, 'Übersetzung des *Kitāb Iḥṣāʾ al-īqāʿāt* von Abū Naṣr al-Fārābī', *Oriens*, 34 (1994), pp. 103–173. Repr. in Eckhard Neubauer, *Arabische Musiktheorie von den Anfängen bis zum 6./12. Jahrhundert* (Frankfurt am Main: Institut für Geschichte der Arabisch-Islamischen Wissenschaften, 1998).

48 This is, accordingly, consonant with the case for an earlier dating of the *Rasāʾil*. But, quite apart from the question of whether the text of the *Kitāb al-Mūsīqī al-kabīr* would have been readily available, it could be explained quite simply by lack of sympathy: al-Fārābī's theoretical concerns are often at some considerable remove from those of the Ikhwān.

learn how, in the culture of the early ʿAbbāsid period, music could be intellectually apprehended as an integral element of cosmology, how it could be conceptualised as a reflection of, and a means of access to, a higher spiritual realm, no text is more rewarding.[49]

49 I should like to express my thanks to Anna Contadini and Stefan Sperl for their helpful remarks on a draft of this chapter.

NINE

Reading The Case of the Animals versus Man: *Fable and Philosophy in the Essays of the Ikhwān al-Ṣafāʾ*

Lenn E. Goodman

'It's hard not to write satire' (Juvenal 1.30)

Back in the Victorian era, when sensibilities were softening towards all suffering — when painters of genre pictures warmed to animal travail as well as human toil and pain, when philosophers like the recently deceased Jeremy Bentham grew in stature for treating sentience rather than reason as the locus and focus of our proper ethical regard, and when naturalists like Charles Darwin could use concern for animal suffering as an index of moral progress — those who campaigned for kindlier, gentler, and more thoughtful treatment of animals often referred to the objects of their concern as 'dumb animals'. The expression found its point in the inability of suffering creatures to speak and reason, to argue their case and voice their viewpoint to their tormentors, exploiters, and abusers. But a thousand years before the reign of Queen Victoria, just a little earlier than the writing of our one surviving manuscript of the Anglo-Saxon poem *Beowulf* — in the 960s or 970s, to be a bit more precise — a group of men who cloaked their identity under the collective name Ikhwān al-Ṣafā' wa-Khillān al-Wafā', that is, the 'Sincere [or Pure-Hearted] Brethren and True Friends', used the ancient device of putting speech into the mouths of animals to allow the birds, beasts, fish, and insects, including birds and beasts of prey, and even the swarming and crawling creatures and the seemingly vilest parasites and lowest beetles and grubs, to make the case that real animals might have made had they been given the gift of speech.

Translating the classic tale of the Ikhwān al-Ṣafā' from its original Arabic, back in the late 1970s,[1] I highlighted the ethical impact of this device under the name 'virtual subjecthood'. For, although animals are not subjects in the full sense of the term, and thus do not fall under the familiar rubrics of ethical discourse that deal with rights and wrongs in terms of reciprocity, responsibility, contracts, understandings, blood relations and personal tact, nonetheless, the inarticulacy of the animals does not deprive them of interests in an *objective* sense. Inarticulacy does not make animal pain any less real. It does not render it impossible for thoughtful and concerned human beings to imagine much that animals would say if they had the gift of articulate speech. The Ikhwān al-Ṣafā' were just the sort of thoughtful inquirers who could make that kind of moral and literary leap.

First, a few words about the reception of this work. It was highly popular in the Middle Ages, and widely read. Its manuscripts were copied for many a collection, and the text was translated into Latin, and also into Hebrew, by one Rabbi Joel and again by Rabbi Jacob ben Elazar around 1240; those two versions are not known to survive. But still extant is the 1316 Hebrew translation by the Provençal Jewish scholar and thinker Kalonymos ben Kalonymos (ca. 1286–1337), who also translated the Arabic version of the tales of *Kalila and Dimna* that loomed large in the imagination of the Ikhwān when they wrote *The*

1 See Lenn E. Goodman, *The Case of the Animals versus Man Before the King of the Jinn: A Tenth-Century Ecological Fable of the Pure Brethren of Basra* (Boston: Twayne Publishers, 1978); note that in this chapter, the author presents revised fragments of his previously published translation of Epistle 22. For the Arabic text of the epistle 'On Animals' ('*Risāla fī kayfiyyat takwīn al-ḥayawānāt*'), refer to *Rasā'il Ikhwān al-Ṣafā'*, ed. Buṭrus Bustānī, 4 vols. (Beirut: Dār Ṣādir, 1957), Epistle 22, vol. 2, pp. 178–377.

Case of the Animals.[2] In testimony to its enduring popular appeal, the Hebrew version of our fable was printed in Warsaw in 1879, and by I. Toforovski and A. M. Haberman in Jerusalem in 1948. From the Hebrew it was translated into Yiddish and German; and from the Arabic, into Urdu in Calcutta in 1810 by one Ikrām ʿAlī, at the instance of a colonial official, Abraham Lockett. English translations were made from the Urdu by John Dowson, a Sandhurst professor, and by others: James Atkinson, T. P. Manuel, John Platts, and A. C. Cavendish. An illustrated popular adaptation from the Hebrew was published by Rabbis Anson Laytner and Dan Bridge in Louisville as recently as 2005.

Friedrich Dieterici edited the text in the late nineteenth century, and the Arabic text appeared again in the twentieth, with an introduction by Buṭrus al-Bustānī (and there are other editions in print). Using the published texts of Dieterici and Bustānī, I translated the work from the Arabic in the 1970s, under the title *The Case of the Animals versus Man Before the King of the Jinn.* Alma Giese prepared a new German translation from the Arabic in 1990, and my colleague Richard McGregor and I are now completing the first critical edition of the work (based on quite a number of manuscripts). This edition, along with a fresh translation, based on the critical text, and expanded notes in commentary, will take its place within the complete collection of the epistles, forthcoming from the Institute of Ismaili Studies.

In *The Case of the Animals versus Man Before the King of the Jinn*, the Ikhwān break away from the usual format of their essays, and fly up into the realm of fable. But this fable bursts the bounds of the usual

2 Kalonymos came from Arles and returned there after studies in Barcelona and work in Naples, with visits regular to Rome. He translated for Charles Robert, Duke of Anjou, who held sway over Hungary as well as Naples and Anjou. Kalonymos' Latin version of Averroes' *Incoherence of the Incoherence* is extant, a clear mark of his philosophical outlook. Over thirty of his Hebrew translations from the Arabic survive. They include translations of works on geometry, astronomy, and philosophy from al-Kindī to Averroes; a translation of Galen's works on phlebotomy, colic, and clysters; and translations from Averroes' commentaries on Aristotle. Kalonymos himself wrote several satirical works including, as a *jeu d'esprit*, a pretended Talmudic tractate on Purim that captures the light-hearted spirit of that holiday. I am indebted to David Walker for information in this note.

brief Aesopian tale. It's longer, far broader in scope, and far more arresting to the interests of an adult reader throughout its narrative. For, once the animals are given voices, they find a great deal to say, not only about their own plight but also about the human condition. Their critical comments are goaded on by human claims to *own* the animals and, therefore, to have a perfect right to do with them as they like. In turn, these claims are founded in ideas of human superiority, and much of the fable is taken up with ripostes against such claims, arguing that most — but in the end not all — are groundless. Frequently alluding to the ancient Sanskrit fables of Bidpai, known in Arabic as *Kalila and Dimna*, and often digressing into the legendary materials linked with the Qur'an and the *Qiṣaṣ al-anbiyāʾ*, the Ikhwān highlight the limitations of human nature, human culture, and human civilisation.

The essay is the usual medium in which the Ikhwān communicate their ideas. The genre evolves from the epistolary form and was well developed in classical antiquity. An essay sets itself apart from a formal treatise by its more discursive, indeed digressive form, its love of anecdotes, side notes, and byplay, not adhering to a tight thematic structure; its chief characteristic, though, is its intimacy of tone. A treatise adopts a public, impersonal voice, at times stentorian, meant for a large audience, whereas an essay addresses the reader in a softer voice and seeks a personal rapport, built in part by presuming it, inviting the reader into the author's moral universe on terms of friendship and shared confidences, as it were. Where an orator lays out enthymemes meant for completion by his hearers, bringing them on board with appeals to common interests and commitments, the essayist, embroidering on the letter-writer's salutation and complimentary close, welcomes the reader into a smaller, closer circle of shared values and ideals. The orator presses a sense of societal unity with his public, and that standpoint is taken up in the treatise. The essayist calls on the attention of his reader in more personal, even confessional tones. He appeals not just to common fears or interests but to shared ideals and experiences. In so doing, he goes beyond presuming a sense of shared identity and actually builds a sense of fellowship. That ideal, the ideal of shared ideals, is not just a rhetorical tool for the Ikhwān. It is constitutive of their message, an enduring motif of all their writing, which could be called a subtext, were it not for the explicitness of its markers.

Seneca, the Stoic philosopher of ancient Cordoba, the tragedian and tragic tutor to Nero, was a past master of the epistolary essay. Al-Kindī (d. ca. 867), the philosopher of the Arabs, pioneered the form in the early days of Arabic prose, back in the ninth century. His essay 'On How to Banish Sorrow', for example, uses its dedicatory form to situate its problematic, building out of literary and philosophical anecdotes and allegories a vigorous case for ethical reorientation, away from the specious goods of worldly attachment that are too readily lost and too often a source of anxiety, grief, and disquiet, and towards an attachment to the abiding and readily accessible goods of the mind. Maimonides (1138–1204), another philosopher from Cordoba, though he spent his mature years in Egypt, used the essay form brilliantly in the *Guide to the Perplexed* to prepare his reader for the arduous work of wrestling with the conceptual problems at the interface between the world's determinacy and God's infinitude.

Maimonides names the two great problem areas of metaphysics (divine science) and physics (philosophical cosmology) using the Midrashic homiletic applied by the ancient rabbis of the Talmud: the Account of Genesis and the Account of the Chariot. Genesis raises the problem of creation: how could the infinite and timeless God express Himself in a determinate, temporal world whose physicality would appear to place it at an absolute remove from God's transcendence? Ezekiel's vision of a celestial chariot raises the related problem of theophany: How could the absolutely transcendent God become accessible to the finite intelligence of his creatures? Where did prophets find the daring (or the tropes) to speak of God in human language? How could the demands of the Infinite be rendered specific, human obligations, when there is no proportion between the finite and the Infinite? How is God's infinite perfection compatible with the suffering of innocents? These are the questions Maimonides seeks to tackle in *The Guide to the Perplexed*. But, sensitive to the Talmud's cautions, Maimonides did not write that work as a systematic treatise but chose to address his pages to a single individual (and others like him) in the framework of a letter. An oblique and allusive discussion, he explains, will afford just the hints that a well-primed and critical intellect will require, while avoiding needless harm to the unquestioning mind, which will only be troubled by the opening up of difficulties about creation, revelation, or theodicy, and will lack the means of closing up the wound again.

The essay form was a favourite and a natural choice for the members of the so-called secretarial class, the *kuttāb* of the early Islamic *imperium*. As officers, administrators, court ministers and officials, they quite normally communicated in letters and prose reports, lightened with elements of style but aiming at clarity and presuming familiarity with the *status quo ante*. When the members of this literate and mildly secular audience sought wisdom, learning, erudition, or edification, they did not always need the blinding obliquity of the Arabic *qaṣīda*, and did not always seek the high hilarity and *précieuse* artistry of the *maqāmāt*, the genre that Hamadhānī devised for its virtuosic delights, and for Horatian or even Juvenalian satire, and sheer literary pyrotechnics. They liked a good yarn, and did not mind if it pointed a moral. After all, *adab* ('literature') was, for the literate and the literati, along with history, a chief instrument of *taʾdīb*, culture, discipline, and self-cultivation.[3]

As Kalonymos remarks in the introduction to his translation, the satire of the Ikhwān is serious at the core and no mere work of entertainment like the popular tales of Sindbad, the *Maqāmāt* of al-Ḥarīrī, or even the fables of Bidpai that the Ikhwān themselves take as a model. It uses notions of the supernatural, but these are not its point. Rather, its focus is moral. The authors, Kalonymos writes, understandably kept their work anonymous because their views were at times heretical, and they hoped to avoid the prejudices that might impede understanding of their message, given the intensity of the intellectual divisions that were rife in their days, as in his own.

What the mode of fable adds to the *Rasāʾil* is a satiric turn. The Aesopian device of giving speaking parts to animals allows what postmodern theorists have repeatedly told us is impossible — a way of getting outside ourselves, beyond the constructs and constrictions

3 See Lenn E. Goodman, *Islamic Humanism* (Oxford and New York: Oxford University Press, 2003), pp. 83–84, 101–110, 120, 147, 199.

of our familiar culture and even the natural biases of our humanity. The Ikhwān used this device to bring to a high pitch their moral and spiritual critique of human foibles and institutions, self-deceptions and self-aggrandisements. The mildness of their satire and the distance afforded by the fanciful setting and the use of animal voices allows them to do so without losing the audience with whom they have so carefully built up a rapport. The critique *is* wide ranging and hard hitting. But it does not descend into the bitter, angry tones of Juvenalian satire. And yet it does not lose itself in Horace's frequent compromises with ambiguity, the price of speaking a bit too knowingly.

Montaigne (1533–1592), a master of the essay form, offers a nice parallel to the Aesopian critique of human foibles in the *Rasāʾil.* In his *Apology for Raymond Sebond,*[4] Montaigne undertakes 'to crush and trample underfoot human arrogance and pride' — much of it grounded in a false confidence in human reason, 'the first foundation of the tyranny of the evil spirit'.[5] His nominal aim is a defence of the Catalan theologian, whose work he had translated in 1569, at the instance of his father (who had thought the work a perfect antidote to Luther and the Reformation). But human pretensions lie closer to the bull's-eye of his real target. Animals, Montaigne argues, have keener senses and seem in the end to lack no capacity that humans have. Just as the Ikhwān find a natural piety in the birds and beasts, Montaigne shows that elephants not only perform their ablutions but pray, play music, and dance.[6] We might add that the elephants in the Nashville zoo also paint; their abstract works are on sale in the zoo gift shop.

Presumption, Montaigne argues, is mankind's endemic illness. Humans are puny, frail, and vulnerable. We live in the basement of the universe — the bowels, as some mediaevals liked to say. 'When I play with my cat,' Montaigne writes, 'who knows if I am not a pastime to her more than she is to me?' Granted animals do not speak, but 'we do not understand them any more than they do us'.[7]

---

4 Michel de Montaigne, *The Complete Essays*, trans. Donald Frame (Stanford: Stanford University Press, 1976). All subsequent citations refer to this edition.

5 Ibid., pp. 327–328.

6 Ibid., pp. 341–343.

7 Ibid., pp. 330–331.

The intimacy of the essay form allows Montaigne to exercise a degree of playfulness in the argument, as it does when the Ikhwān set down the paeans and praises of the birds, the elegies of the owl, and warning cries of the ominous raven. Hyperbole is the mask of caricature and the marker of irony. But the spirit of mockery and raillery does not obscure the seriousness of the underlying point: the hollowness of human vanity and self-assurance. Montaigne, like the Ikhwān, sings the praises of the bees:

> Is there a society regulated with more order, diversified into more charges and functions, and more consistently maintained? Can we imagine so orderly an arrangement of actions and occupations as this to be conducted without reason and foresight?

He goes on to quote Virgil, who cites those who, like the Ikhwān, have it that bees share in the mind of the divine.[8] The swallows, Montaigne argues, return each spring and, 'without judgment', they find the perfect spots to build their nests. If Nature guides animal instincts, he writes, why do we not acknowledge the superiority of the animals to ourselves? For our own works, despite all that we can bring to them 'by Nature and by art' are often surpassed by what the animals do, when Nature 'with maternal tenderness,' as he puts it, comes along to 'guide them as by the hand in all the actions and comforts of their life; while she abandons us to chance and to fortune'.[9]

Like the Ikhwān, Montaigne cites the 'shells, husks, bark, hair, wool, spikes, hide, down, feathers, scales, fleece and silk' that Nature gives all creatures, according to their needs. She 'has armed them with claws, teeth, or horns for attack and defence; and has herself instructed them in what is fit for them — to swim, to run, to fly, to sing — whereas man can neither walk nor speak, nor eat, nor do anything but cry, without apprenticeship'.[10] Montaigne quotes Lucretius here, but his argument, like that of the Ikhwān, follows the stoicising appeals of Galen, finding

8 Ibid., pp. 332–333.
9 Ibid., p. 333.
10 Ibid., pp. 333–334.

the marks of Providence in all animal adaptations. Montaigne softens the sharpness of his contrast by reaffirming that every species, ourselves included, is appropriately provided for. Even ants, he argues, quoting Dante and again paralleling the Ikhwān, communicate to one another the path they must follow to the prizes they have found. But all creatures (once again a central theme for the Ikhwān) are equal, in a sense: 'We are neither above nor below the rest: All that is under heaven, says the sage, incurs the same law and the same fortune.'[11] (This echoes Ecclesiastes 3:14–15, 19.)

Human imagination, Montaigne argues, which allows us to conceive both what is true and what is false, sets us above the beasts in this one regard only, and that at a high price. For, all our vain wishes and vicious desires are fed from that spring.[12] Even free will, Montaigne argues, playing with sophistry, sets us at a disadvantage. For it makes human virtue contingent and unsteady, where animal behaviour is fixed and reliable. Human desires, after all, far outrun the simple needs of nature.[13] Besides, as the Ikhwān agree, 'there is no animal in the world as treacherous as man'.[14]

Picturing the wisest of men freely surrendering all human wisdom and virtue rather than give up the well-beloved human form, Montaigne concludes with a triumphant condemnation of human vanity:

> Well, I accept this naive, frank confession. Indeed, they knew that those qualities about which we make so much ado are but idle fancy. Even if the beasts, then, had all the virtue, knowledge, wisdom, and capability of the Stoics, they would still be beasts; nor would they for all that be comparable to the wretched, wicked, senseless man. In short, whatever is not as we are is worth nothing. And God himself, to make himself appreciated, must resemble us, as we shall presently declare. Whereby it is apparent that it is not by a true judgment, but by foolish pride and stubbornness, that we set ourselves before

11 Ibid., p. 336.
12 Ibid.
13 Ibid., p. 346.
14 Ibid., p. 350.

> the other animals and sequester ourselves from their condition and society.[15]

I was a small boy when my father first explained to me how criticisms put into the mouths of animals are somewhat shielded against counterstrikes. It was the method of Aesop (*Aethiops*) and many another outsider — down to La Fontaine, to Joel Chandler Harris' Uncle Remus with his tales of Brer Rabbit, and Orwell's *Animal Farm*. The technique is very old, well attested in the biblical story of Balaam's ass, who turns and remonstrates with his master: 'What have I done to thee that thou hast beaten me now three times? . . . Am I not thy she-ass, whom thou hast ridden all thy life, down to this day?'[16] The ass sees what the would-be seer cannot — the angel, with brandished sword, blocking the way. In an ironic reversal, the animal becomes the speaker of the moral truth, taking over the prophetic role, as they say, of speaking truth to power.

The book of Numbers underscores the dramatic irony. Balaam's first response to his ass's remonstrance is this: 'You're playing with me! If I had a sword in my hand, I'd have killed you by now!'[17] But what follows is the sheepish admission that the ass has not been balky in the past. This sets up a higher-order dramatic irony: Balaam, in the end, comes to see and explain to his hirer that the Israelites cannot be cursed, since their just ways place them under God's protection.

Reflecting on the marvellous speech of Balaam's ass, the Midrash embroiders homiletically:

> The Holy One, blessed be He, has regard for human dignity and knows our needs. He shut the mouths of beasts. For could they speak, it would have been impossible to put them in service to man or to stand up against them. Here was this ass, the dullest of beasts, and there was the wisest of men. But no sooner had she opened her mouth than he could not hold his own against her. (Numbers Rabbah 20:15)

15 Ibid., p. 358.
16 Numbers 22:28, 22:30.
17 Numbers 22:29.

Like the biblical text, the Ikhwān use dramatic irony to expose human moral weaknesses. When the humans in the fable meet in secret to plan their brief, they contemplate bribing the jinni vizier and jurists who seem unsympathetic. The satiric mirror is brought in tight, casting human morals in a more lurid light when the litigants anticipate that their judges might demand documentary evidence of human title to the animals:

> No one in the assembly had an answer for that, except the Arab: 'We'll say, we had these documents, but they were lost in the Flood.'
>
> 'And what if he says, "Swear solemnly that they are your slaves"?'
>
> 'We'll say the burden of swearing falls on the respondent. We are the plaintiffs.'

Legalism, prevarication, and a gnawing urge to cheat justice displace honest argument in the counsels of the humans. The Ikhwān add a fillip from the *Shuʿūbiyya*, the polemics of newcomers to Islam against the Arabs, who had not so many generations previous gained hegemony in the ancient provinces of Iran. They put that self-serving and self-betraying remark into the mouth of the Arab delegate in the human caucus.

Irony is a speciality of satirical literature, and a favourite device of the urbane sophisticate, proud of catching an in-joke set against the backdrop of gravitas. Irony is also a speciality of prophetic criticism, as a prophet's audience is treated to the disparity between the aims of a wrong-doer and the outcome of his choices.[18] The animals relish the irony in the fable when the humans boast of their fine garments and furnishings, only to be reminded that their choicest fabrics are of silk, wrested from the spittle of a tiny worm; their woollens and hides are stripped from the backs of animals. Honey, the most delicious of human foods, praised as health giving by the Qurʾan, is taken forcibly from bees. As the rabbit urges in his complaint:

18 See Saadiah, *The Book of Theodicy*, trans. L. E. Goodman (New Haven, CT: Yale University Press, 1988), appendix, p. 420.

> These humans drink the milk of cattle as they drank their mother's milk and ride on beasts' shoulders as they rode on their father's shoulders when small. They use animals' wool and fleece for coats and upholstery, but in the end slaughter, flay, disembowel, and dismember them, set them to boil or roast, unfeeling and unremembering all the good, all the blessings, lavished on them.

The first pleas of the animals are against human cruelty, a litany of torture and abuse laid knowingly, even calculatedly, before the king of the jinn, aiming to arouse pity for the animals' plight at the hands of humans. The ass, the ram, camel, elephant, horse, and mule raise a chorus of complaints against the beatings, proddings, and invective suffered by beasts of burden, and the flaying, roasting and cooking of those animals that human beings use for food. Each beast in turn evokes the pity that any sensitive viewer would have felt on witnessing such mistreatment. Here is the ram's complaint:

> You would have pitied us, your Majesty, had you seen us as their prisoners, when they seized our smallest kids and lambs and tore them from their dams to steal our milk. They took our young and bound them hand and foot to be slaughtered and skinned, hungry, thirsty, bleating for mercy but unpitied, screaming for help with none to aid them. We saw them slaughtered, flayed, dismembered, disemboweled, their heads, brains, and livers on butchers' blocks, to be cut up with great knives and boiled in cauldrons or roasted in an oven, while we kept silent, not weeping or complaining. For even if we had wept they would not have pitied us. Where then is their mercy?

The beasts do not fail to mention the foul language of mule drivers, the pricks of the elephant goad, the wounds and risks of battle that horses must endure. The animals' complaint against human cruelty is not forgotten as the case moves to a broader critique of human ways. The outcome forecast at the outset seems to stand: even if domestic animals are not to be freed by the jinn, and human exploitation of living species is to continue while human beings hold sway, the demand for greater kindness and more humane consideration has survived every legal and moral challenge that the human beings make, and human supremacy has been shown to rest on no absolute and intrinsic superiority to all other creatures but solely on the grace of God, whose chief epithets are those of mercy and whose cardinal expectations of His creatures are that all share with those less fortunate than they.

The Ikhwān do not propose a radical or utopian solution to the problems raised in the animals' complaint. The fable envisages vegetarianism as a past condition of the created world. It even suggests that a time may come when that norm may be restored. But no such outcome is demanded by the story. Still, the idea of revolution is never far from the thinking of the Ikhwān. They give the idea a vivid, graphic sense, derived from the imagery of the Qur'an and referring, as the Ikhwān conceive it, first to the revolutions of the heavens about the earth.

Each heavenly body in the Neoplatonic cosmology of the Ikhwān exerts its influence on what lies below. Every being gets its share of grace from God, and acquires a corresponding obligation to pass on the excess to lesser beings, as the rabbit explains:

> The two celestial luminaries, for example, the sun and moon, received from God so bounteous a share of light, brilliance, splendour, and majesty that people often fell under the delusion that they were lords or gods, so clearly do the marks of divinity show through in them. That is why they were made liable to eclipses, to show the discerning that if they were gods they would not go dark. Likewise with the rest of the stars. They may be granted brilliant light, revolving spheres, and long lives, but they are not immune to flickering, or retrograde motion, or even falling, to reveal that they too are subordinate.

The heavenly bodies are God's delegates, vehicles of the good that emanates to what lies below. Accordingly, they are reliable signs of the fortunes of all creatures. Like the heavenly bodies, regimes and dynasties rise and fall, as the Ikhwān observe. So do the fortunes of individuals, nations, and entire species and kinds. Each has its day, and each will be displaced by others that come after.

This idea of succession is one of the reasons that I have called the

Case of the Animals versus Man an ecological fable. For succession is an ecological concept, readily observed in the natural history of a forest and other ecosystems. But, for the Ikhwān, in keeping with the scriptural tradition, the concept is morally freighted. The jinn have had their day; humans are now dominant. But no one creature retains hegemony without limit; each must decline in turn. Taking up the archaeology of the Qur'an, which treats the ruins of past civilisations as memorials to the retributions visited on societies who failed to heed the warners God sent them,[19] the Ikhwān see every decline as a judgement, visited on all natural kinds by divine justice, in consequence of the limited worth and perfection that each enjoys and exercises.

Here is the scheme, then: All perfections stem from God, but none is absolute. Each species and genus has its own distinctive strengths. The elephant has his trunk, the bee and the gnat have each his own tiny proboscis. The tiger may be powerful, but the gazelle has speed. Along with its proper, God-given strengths, each kind has its own weaknesses. The elephant fears the gnat; the lion is victim to the ant; the jackal must beware of dogs; armoured warriors are harried by wasps; and even the most brutal of human tyrants was brought down by a gnat. God always has the last laugh, as the cadence and cyclicity of the food chain and the rhythms of oppression and dependency echo on the natural and the moral plane.

Each living creature and kind, as the Ikhwān read Nature's book, survives and flourishes by its strengths, and fades and falls by its weaknesses. God's grace is fair, not arithmetically equal but proportionate, commensurate with the needs and capabilities of each kind, meted out and clocked by the equable rotations of the heavens across eras that span hundreds of thousands of years.[20] The revolutions of fate are implacable, but the response of the Ikhwān to fate's fixed determinations is not fatalistic. Dire outcomes can be averted by repentance and reform, as the parrot proposes in telling how the people of one

19 See, e.g., Qur'an 30:9; Tarif Khalidi, *Arabic Historical Thought in the Classical Period* (Cambridge: Cambridge University Press, 1996), p. 8; Goodman, *Islamic Humanism*, pp. 161–162.

20 The history of the cosmos extends far beyond six thousand or ten thousand years, one must note. The Ikhwān describe sidereal cycles of 360,000 years and more.

ancient city avoided the devastation of a flash flood. Nature still keeps its course, in accordance with God's will and judgement — as the heavens turn in instant and unquestioning obedience to His decree. The moral and spiritual postures of human beings do make a difference to their fate — all the difference in the world and beyond, as the Ikhwān would have it. But no creaturely strength endures forever. That is for God alone. For, as the Qur'an teaches, *All things are perishing but His face* (28:88).

Not only in time but also in territory do all creatures face their limitations. Part of the animals' complaint is that humans do not keep to a single habitat or breeding season but spread out over the earth — land and sea, mountain and plain. Men 'encroached on the habitats of other animal kinds that lived on the earth, wrested from them their ancestral lands,' taking the whole earth as their dominion. 'We are roused to mount but once a year,' the beasts complain, 'and not with an overwhelming passion or at pleasure's call but for the survival of our race.' Human appetites and ambitions, by contrast, seem to know no limit.

Human overreaching, as the Ikhwān present it, is both a product of and a stimulant to arrogance. Recognition that the gifts humans boast of are not of our own making but are boons of God's grace is therefore critical for the case the Ikhwān seek to make. The moral implications point not only to the candid gratitude we owe God but also to the humility that befits us as creatures *within* Nature, not apart from it. Human attainments are marks of favour, not self-made achievements. Nor are they quite as splendid as we like to think. It is here that the animals play their satirical role, comparing human powers, talents, arts, and institutions with the God-given, natural capabilities of our animal counterparts.

Humans see well, but many a bird sees better. Humans have weapons, traps, snares, and stratagems, but clearly lack the physical prowess of the beasts of prey, let alone the dragon, the sea-serpent, or the mighty Simurgh. And even the most powerful animals have a softer side that humans match all too rarely:

> How many ships tossed by the tempest on the fathomless deep have I led back on course! How many shipwrecked and

> drowning men have I bought safe to islands or shores, only to please my Lord and thank Him for the blessing of my massive frame and immense body, to show due gratitude for His bounty toward me. *For He is our destiny, our faithful Protector.* (Qur'an 3:173)

The lion, too, has his gentler aspect:

> ... largest of predators and the mightiest in frame, the strongest, fiercest, most terrible and majestic. His chest is broad, his waist narrow, his haunches shapely, his head huge, his face round, his brow ample. His jaw is square, his nostrils flared. His paws are stout; his fangs and claws, strong as iron. His eyes flash like lightning. His voice is deep, and his roar mighty. His shanks are like granite, his heart bold, his aspect terrible. He fears no one. Water buffaloes and elephants do not alarm him, nor do crocodiles, nor even men, with all their powers of injury — not even armed horsemen with weapons that can pierce a coat of mail. He is doughty and steadfast. In any affair he undertakes, he attends to it himself and asks no help from his forces or vassals. But he is generous. When he has taken a prize, he eats his share and leaves the rest liberally to his followers and dependents. He disdains worldly things and attacks neither woman nor child — nor orphan. For his nature is noble. If he sees a light far off, he approaches through the dark of night and stands at a distance, his ferocity lulled and savagery gentled. If he hears a sweet melody he comes near and settles down peacefully.

Humans are proud of their arts and industries, cities and institutions. But bees build brilliantly, ants have organised societies; all creatures praise God naturally and spontaneously — birds in their songs, and beasts in their every movement and the very construction of their bodies. The birds and insects trust God for their provender and do not hoard more than they need. Even predators hunt only what they need, and the stores of ants and bees are not a miser's treasure but prudent provisions for a season. There's method here, even when the fable becomes light-hearted. For everything in God's world has a purpose, as the Ikhwān see it. Even the venom in the fangs of vipers has its uses. For death, too, is part of God's purpose — part of Nature's cycle of revolutions.

As might be expected in the adversarial setting of a court case, some of the arguments the animals offer are self-serving. Some are sheer *jeux d'esprit* on the part of the Ikhwān. It is rather sophistical, for example, to imply that hoarding is wrong, but then to praise the bee for its ingenious and airtight hexagonal storage cells, and the ant for cutting grains in half to keep them from sprouting. The Ikhwān are just as thrilled by the carefree life of the grasshopper as they are by the industry and thrift of the ant. Here the original, celebratory, and informative intent of the essay peeks through the satirical overlay, and, for a moment, it seems clear that the critique of human foibles and institutions is only partly of the essence. To a certain extent, the critique of human culture and civilisation serves as a mere foil for the exposition of the marvels of the animal world. But the foil, for the most part, takes on greater prominence than the original complaint, and the intended contrast remains vivid, between what humans rightly prize and what they have no right to boast of. Of course, it is a bit contradictory to urge that beasts of prey do not eat much and yet marvel at the vast intake of the sea-serpent. But, even in the case of the sea-serpent, the larger theme of the divinely imparted balance and justice in Nature reasserts itself:

> All marine animals fear him and flee before his vast power and strength. When he moves, the sea itself rocks with the swiftness of his swimming. His head is huge, his eyes flashing, his teeth numerous, his mouth and gullet tremendous. He swallows countless hordes of sea animals each day, and when his belly is full and he finds it hard to digest them, he arches and bends like a bow, supporting himself on his head and tail, and raises his mid-parts out of the water into the air, gleaming like a rainbow in the sunlight, huffing and puffing about, sunning himself to aid his digestion. But sometimes, in this posture, he swoons, and the rising mists lift him up from below and bear him through the air to dry land, where he dies, and the beasts feed on his hulk for days — or he is borne to the shores of the land of Gog and Magog, who live beyond the great barrier, two nations of human form but savage spirit, who know neither

> order nor government and have no commerce or trade, industry or craft, plowing or sowing, but only hunting and fishing, plundering, raiding, and eating one another.
>
> Know, your Majesty, that all sea animals flee in terror before the sea-serpent. But he fears nothing, save only a tiny beast resembling a mosquito, that he cannot harm and against whose sting he cannot defend himself. Once it stings him, its poison creeps through his body and he dies. Then all the sea animals gather to feed and gorge on his carcass for days. For, small beasts do feed on the larger when they can. The same is true with birds: Sparrows, larks, swallows, and their ilk eat grasshoppers, ants, gnats, flies, and the like. Then sparrow hawks and falcons and their like hunt the sparrows and larks and eat them. Hawks and eagles hunt and eat these in turn. But when they die they are eaten by the smallest, the ants, flies, and worms.

Clearly, the Ikhwān are having fun when they put snatches of Arabic elegaics into the mouth of the owl and homilies into the song of the skylark. But the critique of human institutions is in earnest: Ablutions and purifications are meant to compensate for human impurity and immorality. Human trades and industries are part of an exhausting, never-ending, greedy quest. Even human piety is too often self-serving:

> But most people, you find, scurry to the doctor at the first sign of illness but turn to God if treatment is prolonged and the medicine prescribed unavailing, when they've given up hope of a medical cure. Then they pray desperately, perhaps writing on scraps of paper to stick up on the walls of mosques, churches, or synagogues. They pray privately or make public penitential vows, saying 'God have mercy on a troubled supplicant' — as He did in celebrated cases. This is the reward they expect for thieving or robbing or some such crime! Had they turned to God to start with and called on Him inwardly, not just publicly, it would have been well for them, far better than their public protestations and acts of penance.

The Aesopian framing of the tale allows the Ikhwān to make specific creatures paragons of specific virtues. The fable's kings, viziers, messengers, and subjects become models of their roles. At the court of the king of the jinn, justice and impartiality rule, ministers are incorruptible, and the monarch consults broadly and enthusiastically with wise and experienced counsellors, who are urged to speak freely and encouraged to question and critique proposed strategies and courses of action. The king of the crawling creatures, the dragon, is moved to tears by the frailty and lack of device of so many of his subjects. But the cricket teaches him that worms, grubs, and intestinal parasites are compensated for their lack of limbs and organs by the simplicity and ease of their lives, their cosseted habitats, and ready access to all that they need. For, in their case, as in every other, God's gifts are matched, species by species, by creaturely needs.

Here's what the Ikhwān have to say about messengers:

> [An emissary] must be a person of intelligence and character, well-spoken, eloquent, and articulate, able to remember what he hears and use caution in what he answers. He must be loyal, faithful, true to his word, circumspect, and discreet, adding nothing to his message but what he sees is in the sender's interest. He must not be grasping or avaricious. For a greedy person who meets with generosity from his hosts may shift his loyalties and betray the sender, adopting the new country for the good life he enjoys there, the blandishments and gratifications he finds there. Rather, he must be faithful to his sender, his brethren, countrymen, and kind, deliver his message, and return promptly to those who sent him with a full report from start to finish of what passed on his mission, omitting nothing for fear of causing displeasure. For clarity is the sole duty of a messenger.

The delegate of the swarming creatures in the dispute is Yaʿsub, the bee king.[21] He speaks freely about political theory with his fellow monarch, the jinni king. The jinni praises the loyalty and devotion of his subjects, the best of whom follow their king as the stars and planets follow the sun, each heavenly body performing its proper role, as commanded

21 Even in Aristotle, the bees have a king, not a queen; *History of Animals*, Book 9, 40, 923b.

by the angels that govern the spheres. That relationship is compared in turn to the service of the senses to the rational soul, bringing it their reports 'without lag or delay' — as expected in a Neoplatonic cosmos. The celestial ideal contrasts sharply with the human case:

> But the nature and temperament of humans are just the opposite. Their obedience to their chiefs and monarchs is mainly hypocrisy and dissembling, gulling, and grasping for stipends, payments and rewards, and vestments and prizes. If they do not get what they are after, they come out in open rebellion and defiance, shed their outward loyalty, secede from the commonwealth, and bring dissension, civil war, bloodshed, and destruction to the land.

Animal rulers care for their subjects with deep solicitude and concern, the parrot explains. When the jinni monarch senses a hint of allegory in that remark, the leading jinni philosopher explains that the reference is to the ideal: A king is not just the caretaker of his subjects, not just their shepherd. He must be their angel:

> The name 'king' [*malik*], you know, derives from 'angel' [*malak*]. And kings' names are taken from those of angels. For there is no kind of these animals, nor species, nor individual among them, great or small, over which God has not appointed a band of angels to oversee its growth, preservation, and welfare, at every stage. Every class of angels has its chief to look after it. And these chiefs are kinder, gentler, and more compassionate than mothers toward their tiny sons or infant daughters.

Here, with the help of a fanciful etymology, emanation, visualised as the work of angelic emissaries,[22] becomes both the vehicle of providence and the model of benevolent governance — a model that human monarchs and humans in general are bluntly charged with failing to follow:

22 See Lenn E. Goodman, 'Maimonidean Naturalism', in *Neoplatonism and Jewish Thought*, ed. Lenn E. Goodman (Albany, NY: SUNY Press, 1992), pp. 157–194.

> The monarchs of animal kinds, however, are better than human kings and chiefs at imitating God's ways. The king of the bees looks to the interest of his subjects, troops, and vassals, and seeks their well-being. He does not serve his own private whims or even the caprices of his people, but acts in their interest and protects them from harm, serving not even one who supports his own wishes, instead acting solely in compassion and concern, kindness and affection for his subjects, troops, and supporters. So do the king of the ants, and the king of the cranes, who oversees their flight, and the king of the sandgrouse, who leads their flight and alighting. The same holds true with all other animals who have leaders and rulers. They seek no recompense or requital from their subjects for their rule, just as they seek no reward, recompense, return, or show of filial gratitude from their offspring, as Adamites do. For we find that every animal that leaps and mounts, becomes gravid, bears, nurses, and raises its young, and every kind that mounts, lays, broods, cares for, and raises chicks or hatchlings seeks no reward, recognition, or recompense from its offspring but raises its young and cares for them kindly, tenderly, gently, and compassionately, on the model of God, who created his creatures, raised, nurtured, and cared for them, showing them goodness and generosity, asking nothing of them in return and seeking no reward or requital.

Human boasts are vain and empty, the animals argue. Even revelation is little more than the reflex of human unworthiness:

> Were it not for the ignoble nature of humans, their base characters, unjust lives, vicious mores, vile doings, foul acts, ugly, misguided and depraved customs, and rank ingratitude, God would not have commanded them *Show gratitude to Me and toward your parents, for unto Me shall ye come in the end.*[23] He gave no such command to us and our offspring. For we show no such disrespect or thanklessness. Command and prohibition, promise and threat are addressed solely to you, the human race, not to us. For you are creatures of mischief. Conflict, deceit, and disobedience are ingrained in you. You are more fit for slavery than we! We are more worthy of freedom.

23 See Qur'an 31:14.

Called upon to explain why dogs have betrayed their own kind and gone over to human habitation, the wolf, who has made no such mistake, replies:

> Dogs were drawn to the neighbourhood and habitations of men simply by their kindred nature and character. With men they found food and drink that they relish and crave — and a greedy, covetous, ignoble, stingy nature like their own. The base qualities they found in men are all but unknown among carnivores. For dogs eat putrid meat from the carcasses of slaughtered animals, dried, stewed, roasted, salted or fresh, good or bad. They eat fruit, vegetables, bread, milk sweet or sour, cheese, butter, syrup, oil, confectionary, honey, porridge, pickles, and every other sort of food that humans eat and that most carnivores would not eat and do not know. They are so gluttonous, greedy, and mean that they cannot allow a wild beast to enter a town or a village, lest it vie with them for something there. So if a fox or jackal happens to enter a village at night to steal a hen, a cock, or a cat, or even drag off some discarded carcass or scrap of meat from a dead animal, or a shrivelled piece of fruit, just see how the dogs set upon him, chase him, and drive him out! They are so wretched, lowly, abject, beggarly, and covetous that when they see a human being, man, woman, or child, holding a roll, or a scrap of bread in his hand, or a date, or any morsel, they beg for it and follow him about, wagging their tails, bobbing their heads, gazing up into his eyes, until one feels embarrassed and throws it to them. Then see how they run for it and quickly snatch it, afraid another might reach it first. All these base qualities are found in humans and dogs. So it was their kindred nature and character that led dogs to leave their own kind and shelter with men as their allies against the hunting animals who were of their own race.

One ground on which the humans claim superiority to animals, and thus the right to use and treat them as they like, is human unity, as contrasted with animal diversity: the animals vary in shape, but humans are all of a single kind. The appeal is rooted in the Neoplatonic idea that singleness rises closer to God's unity than does the multiplicity and heterogeneity of the animals. But the animals answer that although humans share a single bodily type, they are at odds in spirit, whereas animals are one in soul, although differing in outward appearance. As evidence of the underlying disparity of human minds, the nightingale cites the multiplicity of human religions, sects, and schools of thought — whereas all animals, he claims, share a single outlook, the natural monotheism of the *ḥanīf*. Here animal faith is made a kind of ideal, since it is free of wrangling, doubt, and dissension. The Persian delegate retorts that humans, too, agree in faith, at least in the core essentials, although their paths of approach to the one God clearly differ. But this leads the jinni king to ask: 'Why, then, do you slay one another, if your religions all have the same goal, of encounter with God?' The answer he receives is telling:

> 'You're right, your Majesty', said the thoughtful Persian. 'This does not come from faith, for there is no compulsion in faith.[24] It comes from the specious counterpart of faith, the state.'

As the Persian explains to the king of the jinn:

> Faith and the state are inseparable, twin brothers. Neither can survive without the other. But religion is the elder. The state is the younger brother, the follower. A realm cannot do without religion for its people to live by; and religion needs a king to command the people to uphold his institutions, freely or by force. That is why the votaries of different religions slay one another — seeking political primacy and power. Each wants everyone to follow the institutions of his own faith or sect and the rules and practices of his own religion.

He goes on to distinguish spiritual from worldly struggle:

> The slaying of selves is practised in all faiths, creeds, and confessions, and all earthly dominions. But in religion the mandate is for self-sacrifice. In politics it usually means slaying others to get power.

24 Qur'an 2:256.

The king responds that he's all too familiar with the killing engendered in struggles for power, 'But how is it that seekers in the different religions slay themselves?' The Persian delegate answers:

> 'I'll explain. You know, your Majesty, that in Islam, this is clearly and plainly one's duty. For God says, *Lo, God has bought of the faithful their substance and selves, since they shall have Paradise. Let them, then, battle for God, slay and be slain. This is His promise, preserved in the Torah, the Gospels, and the Qur'ān. And whose word is more to be trusted than God's?*[25] And, *Rejoice that in the sale of yourselves you have made, truly a great bargain!*[26] And, *God loves those who do battle for Him, in ranks like a closely knit structure.*[27] In the Torah tradition, He says, *Turn to your Creator and slay your selves. Your humbling is good in the eyes of your Creator.*[28] And Christ says, in the Gospels, *Who are my aides in the service of God?' The Disciples answered 'We are God's helpers.' He replied, 'Prepare for death and the cross if you wish to aid me. Then shall you be with me in the Kingdom of Heaven, with my Father and yours. Else you are none of mine.*[29] And they were slain but did not forsake Christ's faith.'

So the valid jihad is the battle for *self*-mastery. All the rest is politics, a sordid struggle for illusory goals that detract and distract from the spiritual battle, which is universal:

> The Brahmans of India slay themselves by burning their bodies in their spiritual quest, committed to the belief that the penitent comes closest to the Lord (exalted be He) by slaying his body and burning it, to atone for his sins, certain of resurrection. And the godliest Manichaeans and dualists deny the self all gratifications and carry heavy loads of religious obliga-

25 Qur'an 9:111, commonly taken as an exhortation to do battle in behalf of the Faith.

26 Qur'an 9:111.

27 Qur'an 61:4.

28 The reference may be to Leviticus 16:30–31, perhaps conflated with passages like Hosea 14:2. But these texts call for self-trial, not self-immolation.

29 See Matthew 10:34–38, 16:24, etc.

> tions, to slay the ego and free it from this realm of trial and degradation.
>
> The same pattern of self-sacrifice is found in the varied practices of people in all religions. All religious laws were laid down to deliver the soul, to save it from hellfire and win blessedness in the Hereafter, the realm where we abide.

It's part of the dry humour of the Ikhwān that no sooner has the Persian delegate rested his case as to human unity than the Indian speaker pleads human diversity as basis for mankind's claims to hegemony over Nature. But in diversity, of course, humans are no match for the animals. It is only when the human case has seemingly ground to a halt that an orator from the Ḥijāz mentions immortality, citing the resurrection as a distinction that animals cannot match. The nightingale gamely answers that every scriptural promise of reward is matched by the threat of chastisement. But the Ḥijāzī retorts that the asymmetry persists:

> 'How are we equal?' demanded the Ḥijāzī. 'How do we stand on a par, when either way we survive eternally, immortally? It is only among us that one finds the prophets, guardians, *imāms*, saints, sages, poets, the admirable, the virtuous, the council of the holy, the pious, the elect, the ascetic and devout, aware and insightful, people of understanding, discernment, and vision . . .

Only then do the animals and the jinni sages acknowledge that the humans have finally hit upon the truth and found something indeed to boast of — as if the idea that the speakers have stumbled onto were the answer to a riddle or solution to a puzzle that lay in plain sight all along but was overlooked, somehow too obvious to notice. All thoughts turn then to those pious and precious individuals whose very existence seems to make human life worthwhile and whose intercession is expected to atone for many of the wrongs calendared so vividly in the arguments of the animals. But no one seems able to say much about them.

These holy figures, however, are critical to the argument. Immortality gives weight to the human condition. It is the emblem of human moral and spiritual responsibility. But immortality, of course, is no blessing if all it brings is eternal wrath. So the holy elite mentioned at the fable's close, like the ideal, angel-like rulers mentioned earlier, are crucial as models, not of what human beings are, but of what they can become. It is only for that reason that their intercession would be heeded, and it is only for their special qualities, their insights and their way of life — so it is suggested — that any sufferings, animal or human, are acceptable, or that human existence is worthwhile at all. Hence the inference drawn by the final human speaker in the tale, whose birth and breeding are meant to represent human character at its cosmopolitan best:

> Finally arose a learned, accomplished, worthy, keen, pious, and insightful man. He was Persian by breeding, Arabian by faith, Ḥanafī by rite, Iraqi in culture, Hebrew in lore, Christian in manner, Damascene in devotion, Greek in science, Indian in discernment, Sufi in intimations, regal in character, masterful in thought, and divine in insight. He said, 'These saints of God are the choice flower of creation, the best and purest, persons of noble quality, fair character, pious deeds, manifold sciences, godly insights, regal ways, just and holy lives, and wondrous spirits. Fluent tongues weary to list their qualities. In lengthy sermons in public assemblies all through their lives and down through the ages preachers have sought to describe them. They have dilated on the virtues and noble character of these saintly paragons, without even reaching the pith of the matter.'

The case is resolved in favour of the humans. The question remains: why is this so? What is it in the human condition that warrants calling the inner core of human identity an immortal soul, and how is it that the human potential for such perfection is to be realised? The question is as alive today as it was a thousand years ago. The Ikhwān, speaking through the animals, rule out many a familiar answer: It is not our powers of calculation or manipulation, our tool-making or our perception, or even our powers of discrimination that place human worth above the plane of mere utility and gives human personhood an inestimable worth. Nor is it our arts and industries, nor any of the many achievements on which we pride ourselves: our business acumen and wealth, our rituals and purifications — not even our fine arts and literary refinements, I suspect — and clearly not our powers of destruction. Whatever we achieve in prolonging or enhancing human life is a gift, as the Ikhwān see it, won through the use of our God-given powers. And much of what we pursue by the use of those powers is illusory — specious goods and sham rewards, on which many a life is wasted. But consciousness gives us more than the ability to plan and build and fashion means to fit our ends. It gives us powers of choice as well. That makes us responsible, for the ends we choose and the means by which we seek them. Responsibility brings accountability in train, making human beings subject to reward and retribution, and tipping the scales of the argument: the human window on immortality lies open, the pathways it puts in view are marked by the footsteps of those paragons whose piety, insight, and generosity show that they have not just gazed towards it but have trodden its pavement.

Freedom makes life a puzzle with no trivial solution. It introduces ambiguities as to how we read the book of Nature, how we relate to one another, and where we direct our own intentions. It makes knowledge itself a realm where interpretation matters but also where interpretation can go desperately wrong or can lead beyond the obvious and the surface meanings of things to higher realms beyond the reach of the senses. Freedom is the source of all the most blameworthy faults and failings that the animals mention. But it is also the source of the one unanswerable strength that places humankind above the beasts, and even, truth be told, in one respect at least, above the angels.

Contributors

Carmela Baffioni is Professor of History of Islamic Philosophy and of History of Muslim Philosophies and Sciences at the University of Naples '*L'Orientale*'. Her publications include: several monographs on the transmission of Greek thought into Islam; translations of works by the Ikhwān al-Ṣafā', al-Farābī, and Averroes, as well as al-Shahrastānī; in addition, she has written a monograph on Aristotle's *Meteorologica*, IV (1981) and books on the history of Muslim philosophy. Besides the Ikhwān al-Ṣafā', her articles focus on al-Kindī, al-Farābī, Yaḥyā ibn 'Adī, Avicenna, Averroes, and some Latin elaborations of Arabic heritage. Her scientific interests concern logic, atomism, and embryology. She has translated books from German and Russian, and edited six collective works. Her recent publications include articles on the political views of the Ikhwān al-Ṣafā', the philosophy of nature of Ḥamīd al-Dīn al-Kirmānī, about forty items in the *Enciclopedia filosofica* (Milan: Bompiani, 2006), and the revision and edition of the catalogue of manuscripts of the Ahel Habott Foundation (Chinguetti), in Arabic and French (Milan: Nottetempo, 2006).

Farhad Daftary received his doctorate from the University of California in Berkeley, and is currently Head of the Department of Academic Research and Publications and Associate Director at The Institute of Ismaili Studies, London. He is a consulting editor for the *Encyclopaedia Iranica*, co-editor of the *Encyclopaedia Islamica*, and the General Editor of the Ismaili Heritage Series and Ismaili Texts and Translations Series. Dr Daftary is the author and editor of numerous books and articles in the field of Ismaili studies. His publications include *The Ismāʿīlīs: Their History and Doctrines* (1990; 2nd ed., 2007), *The Assassin Legends* (1994), *A Short History of the Ismailis* (1998), *Intellectual Traditions in Islam* (2000), *Ismaili Literature* (2004), and *Ismailis in Medieval Muslim Societies* (2005). Dr. Daftary's books have been translated into Arabic, Persian, Turkish, Urdu, and numerous European languages.

Godefroid de Callataÿ is Professor of Islamic Studies at the Institute of Oriental Studies of the University of Louvain. He has specialised in the history of Arabic sciences and philosophy and is the author of numerous articles and books on the Brethren of Purity, including his French translations of Epistles 7, 28, and 36, and his latest volume: *Ikhwan al-Safa': A Brotherhood of Idealists on the Fringe of Orthodox Islam*, for the 'Makers of the Muslim World' series (Oxford: Oneworld, 2005). In addition, he is also directing a project on Arabic and Latin encyclopaedias of the Middle Ages at the University of Louvain.

Nader El-Bizri is a Research Associate in Philosophy at The Institute of Ismaili Studies, London, a Chercheur Associé at the Centre National de la Recherche Scientifique in Paris, and a Visiting Professor at the University of Lincoln. He also lectures at the Department of History and Philosophy of Science at the University of Cambridge, and previously he has held a lectureship at the University of Nottingham and taught at Harvard University and the American University of Beirut. He has had numerous articles published in philosophy journals, and is the author of *The Phenomenological Quest between Avicenna and Heidegger* (Binghamton, NY: Global Publications, SUNY, 2000). He is also an elected member of the Steering Committee of the Société Internationale d'Histoire des Sciences et des Philosophies Arabes et Islamiques (CNRS, Paris), and is a co-editor of a book series on phenomenology (Kluwer Academic, Dordrecht), as well as serving on the editorial board of a series on philosophy and architecture (Lexington Books, Maryland). Besides his academic undertakings, he has nine years of professional architectural consulting experience in institutions in Geneva, London, Cambridge, New York, Boston, and Beirut.

Lenn E. Goodman is Professor of Philosophy and Andrew W. Mellon Professor in the Humanities at Vanderbilt University. His books include *Islamic Humanism* (Oxford University Press, 2003); *In Defense of Truth: A Pluralistic Approach* (Humanity Press, 2002); *Jewish and Islamic Philosophy: Crosspollinations in the Classic Age* (Rutgers and Edinburgh University Presses, 1999); *Judaism, Human Rights and Human Values* (Oxford University Press, 1998); *God of Abraham* (Oxford University Press, 1996); *Avicenna* (Cornell University Press, 2006); and *On Justice* (Littman Library, 2007). A winner of the American Philosophical Association Baumgardt Prize and the Gratz Centennial Prize, Goodman has written on most of the major Islamic philosophers and has lectured widely in the US, Israel, Australia, Britain, and France. His Gifford Lectures were published by Oxford University Press (2008) as *Love Thy Neighbor as Thyself*. His translations from the Arabic include Saadiah Gaon's commentary on the Book of Job and Ibn Ṭufayl's *Ḥayy Ibn Yaqẓān*. He is the editor of *Neoplatonism and Jewish Thought* (1992) and co-editor of *Aristotle's Politics Today*, *Jewish Themes in Spinoza's Philosophy* (2002), and *Maimonides and his Heritage*, all from SUNY Press.

Abbas Hamdani is Professor Emeritus at the University of Wisconsin-Milwaukee, having retired from a long career teaching Islamic history there and previously at the American University in Cairo and the Karachi University — a total of 50 years in academia. He has had several studies published on the *Rasāʾil Ikhwān al-Ṣafāʾ*, regarding which he maintains an early, pre-Fāṭimid, composition. The thesis for his 1950 PhD from at the University of London was on the biography of the *dāʿī* al-Muʾayyad fī al-Dīn al-Shirāzī, which led him later to publish several studies on the ʿAbbāsid–Fāṭimid conflict in different areas such as Iraq, Iran, India, Syria, North Africa, and Spain, and on the history of the Ṣulayḥid dynasty of Yemen. He is also interested in the Islamic background of the voyages of discovery, about which he has published another set of studies. Several anthologies of contemporary English poetry have his poems published in them. Professor Hamdani inherited an important collection of Ismaili manuscripts, which he has now donated to the IIS library.

Yahya J. Michot acted as the Director of the Centre for Arabic Philosophy at the University of Louvain in Belgium from 1981 until 1997, where he gave courses in Arabic, History of Arabic Philosophy, Commentary on Arabic Philosophical Texts, History of Muslim Peoples, and Institutions of Islam. His primary field of research is the history of Muslim thought, mainly on Ibn Sīnā (d. 428/1037), his predecessors, and his impact on Sunni thought. This led to his growing interest in studying the texts of the theologian Ibn Taymiyya (d. 728/1328), the era of the Mamlūks and Ilkhāns, and modern Islamic movements. Since October 1998, Professor Michot has been KFAS Fellow in Islamic Studies at the Oxford Centre for Islamic Studies, and also an Islamic Centre Lecturer in the Faculty of Theology, University of Oxford. In addition, he is a member of various international scholarly societies, and founder and director of the collection 'Sagesses musulmanes'. From April 1995 to June 1998, Professor Michot was president of the Conseil Supérieur des Musulmans de Belgique. His numerous publications (mainly in French) include several volumes on Ibn Sīnā and Ibn Taymiyya.

Ian Richard Netton is Sharjah Professor of Islamic Studies at the University of Exeter. He taught at Exeter University from 1977 to 1995 where he was latterly Reader in Arab and Islamic Civilization and Thought. In September 1995 he became the first Professor of Arabic Studies at the University of Leeds, and remained there until 2007, serving four terms there as Head of the University's Department of Arabic and Middle Eastern Studies. He was also Director of Leeds University's Centre for Mediaeval Studies from 1997 to 2002. His primary research interests are Islamic theology and philosophy, Sufism, mediaeval Arab travellers, Arabic and Islamic bibliography, comparative textuality and semiotics, comparative religion, and general Islamic studies. He is the author and/or editor of sixteen books, of which the most recent are *Islam, Christianity and Tradition: A Comparative Exploration* (Edinburgh University Press, 2006), and (ed.) *Islamic Philosophy and Theology* (4 vols., Routledge, 2007). Professor Netton is also the General Editor of *The British Journal of Middle Eastern Studies* and serves on several national committees such as the Arts and Humanities Research Council (AHRC) Research Panel 8 (Philosophy, Religious Studies and Law), and the Research Assessment Exercise of 2008.

Ismail K. Poonawala is Professor of Arabic and Islamic Studies at the University of California, Los Angeles. He has also taught at the Institute of Islamic Studies, McGill University, and at Harvard University. He is a specialist in Ismaili studies and the author of the monumental *Biobibliography of Ismāʿīlī Literature* (California, 1977), which is a comprehensive survey of Ismaili authors and their writings including manuscript holdings in public and private collections. He has edited several Ismaili texts, including the *Dīwān* of al-Ṣulṭān al-Khaṭṭāb (Cairo, 1967; revised ed. Beirut, 1999), *al-Urjuza al-mukhtāra* of al-Qāḍī al-Nuʿmān (Beirut, 1970), *Kitāb al-Iftikhār* of al-Sijistānī (Beirut, 2000), and *Adʿiyat al-ayyām al-sabʿa* of the Fāṭimid caliph-Imam al-Mu'izz (Beirut, 2006). He has also translated into English with annotations, volume IX of al-Ṭabarī's history, entitled *The Last Years of the Prophet* (SUNY, 1990), and al-Qāḍī al-Nuʿmān's *Daʿa'im al-Islām*, entitled *The Pillars of Islam: Vol. 1, Acts of Devotion and Religious Observances* (Oxford, 2002), and *The Pillars of Islam: Vol. 2, Laws Pertaining to Human Intercourse* (Oxford, 2004).

Owen Wright is Professor of the Musicology of the Middle East at SOAS, University of London. His research interests focus on the textual sources for the history of music in the Middle East. Previously, he acted as the Editor of the SOAS Musicology Book Series, which is published by Ashgate (Aldershot). He has several published studies on music, and his latest publications include the following titles: *Demetrius Cantemir, The Collection of Notations II: Commentary*, SOAS Musicology Series (Aldershot: Ashgate, 2000); 'Die melodischen Modi bei Ibn Sīnā und die Entwicklung der Modalpraxis von Ibn al-Munağğim bis zu Ṣafī al-Dīn al-Urmawī, *Zeitschrift für Geschichte der Arabisch-Islamischen Wissenschaften*, 16 (2004/2005); 'Al-Kindī's Braid', *Bulletin of the School of Oriental and African Studies*, 69, 1 (2006).

Select Bibliography

al-ʿĀmirī, Abū al-Ḥasan Muḥammad ibn Yūsuf. *Kitāb al-Iʿlām bi-manāqib al-islām*, edited by Aḥmad Ghurāb. Cairo, 1967.

Āmulī. *Nafāyis al-funūn wa-ʿarāyis al-ʿuyūn*, edited by A. Shaʿrānī. Tehran, 1377–1379/1957–1959.

Anawati, Georges Chehata. 'Les divisions des sciences intellectuelles d'Avicenne'. *Mélanges de l'Institut Dominicain d'Études Orientales*, 13 (1977), pp. 323–335.

Apollonius of Perga. *Les coniques d'Apollonius de Perge*, translated by Paul ver Eecke. Bruges: Desclée De Brouwer, 1923.

Aristotle. *History of Animals*, edited by A. L. Peck and D. M. Balme. 3 vols. The Loeb Classical Library. Cambridge, MA: Harvard University Press, 1965–1991.

—— *Metaphysics*, edited by W. David Ross. Oxford: Clarendon Press, 1997.

—— *Physics*, edited by W. David Ross. Oxford: Clarendon Press, 1998.

Arkoun, Mohammed. 'Logocentrisme et vérité religieuse dans la pensée islamique, d'après *al-Iʿlām bi-manāqib al-islām* d'al-ʿĀmirī'. *Studia Islamica*, 35 (1972), pp. 5–51.

Artmann, Benno. *Euclid, the Creation of Mathematics*. New York: Springer-Verlag, 1999.

Asʿadī, Hommān. '*Nigāhī-yi ijmālī bih sayr-i taḥawwul-i mafhūm-i "laḥn" dar mūsīqī-yi jihān-i Islām*'. *Māhūr*, 13 (2001), pp. 57–68.

Badawī, ʿAbd al-Raḥmān. *Muntakhab ṣiwān al-ḥikma et Trois Traités*. Tehran, 1974.

—— *Quelques figures et thèmes de la philosophie islamique*. Paris: G. P. Maisonneuve et Larose, 1979.

Baffioni, Carmela. *Il IV libro dei 'Meteorologica' di Aristotele*. Naples: Bibliopolis, 1981.

—— 'Euclides in the *Rasāʾil* by Ikhwān al-Ṣafāʾ'. *Études Orientales*, 5–6 (1990), pp. 58–68.

—— 'Oggetti e caratteristiche del curriculum delle scienze nell'Enciclopedia dei Fratelli della Purità', in *Studi arabo-islamici in memoria di Umberto Rizzitano*, edited by Gianni di Stefano. Trapani: Istituto di Studi Arabo-Islamici 'Michele Amari' di Mazara del Vallo, 1991, pp. 25–31.
—— 'Traces of "secret sects" in the *Rasā'il* of the Ikhwān al-Ṣafā'', in *Shī'a Islām, Sects and Sufism: Historical dimensions, religious practice and methodological considerations*, edited by Frederick De Jong. Utrecht: M. Th. Houtsma, 1992, pp. 10–25.
—— *Frammenti e testimonianze di autori antichi nelle epistole degli Ikhwān al-Ṣafā'*. Rome: Istituto Nazionale per la Storia Antica, 1994.
—— 'Valutazione, utilizzazione e sviluppi delle scienze nei primi secoli dell'Islām: Il caso degli Ikhwān al-Ṣafā'', in *La civiltà islamica e le scienze*, edited by C. Sarnelli Cerqua, O. Marra, and P.G. Pelfer. Naples: CUEN, 1995, pp. 23–35.
—— 'L'Islām e la legittimazione della filosofia. I curricula scientiarum del secolo X', in *La filosofia e l'Islām*, edited by G. Piaia. Padua: Gregoriana, 1996, pp. 13-34.
—— 'Citazioni di autori antichi nelle *Rasā'il degli Ikhwān al-Ṣafā'*: il caso di Nicomaco di Gerasa', in *The Ancient Tradition in Christian and Islamic Hellenism: Studies on the Transmission of Greek Philosophy and Sciences Dedicated to H. J. Drossaart Lulofs on his Ninetieth Birthday*, edited by Gerhard Endress and Remke Kruk. Leiden: Leiden Research School, 1997, pp. 3–27.
—— 'L'influenza degli astri sul feto nell'Enciclopedia degli Ikhwān al-Ṣafā''. *Medioevo. Rivista di storia della filosofia medievale*, 23 (1997), pp. 409–439.
—— 'Sulla ricezione di due luoghi di Platone e Aristotele negli Ikhwān al-Ṣafā''. *Documenti e studi sulla tradizione filosofica medievale*, 8 (1997), pp. 479–492.
—— 'L'embriologia araba fra astrologia e medicina. Abu Ma'šar al-Balkhī e Muḥammad ibn Zakarīyā' al-Rāzī', in *La diffusione dell'eredità classica nell'età tardo-antica e medievale: Il 'Romanzo di Alessandro' e altri scritti*, edited by R. B. Finazzi and A. Valvo. Alexandria: Edizioni dell'Orso, 1998, pp. 1–20.
—— 'From Sense Perception to the Vision of God: A Path Towards Knowledge According to the Ikhwān al-Ṣafā''. *Arabic Sciences and Philosophy*, 8 (1998), pp. 213–231.
—— 'The Concept of Science and its Legitimation in the Ikhwān al-Ṣafā'', in *Religion versus Science in Islam: A Medieval and Modern Debate*, edited by C. Baffioni. *Oriente Moderno*, 19 (2000), pp. 427–441.
—— 'Frammenti e testimonianze platoniche nelle *Rasā'il* degli Ikhwān al-Ṣafā'', in *Autori classici in lingue del Vicino e Medio Oriente*, edited by

G. Fiaccadori. Rome: Istituto Poligrafico e Zecca dello Stato, Libreria dello Stato, 2001, pp. 163–178.

—— '*Al-madīnah al-fāḍilah* in al-Fārābī and in the Ikhwān al-Ṣafā': A Comparison', in *Studies in Arabic and Islam*, edited by S. Leder, H. Kilpatrick, and B. Martel-Thoumian. Leuven: Peeters, 2002, pp. 3–12.

—— *Antecedenti greci nel concetto di 'natura' negli Iḫwān al-Ṣafā'*, in *Enôsis kai Philia (Unione e amicizia): Omaggio a Francesco Romano*, edited by M. Barbanti, G.R. Giardina, and P. Manganaro. Catania: CUECM, 2002, pp. 545–556.

—— 'The "Friends of God" in the *Rasā'il Ikhwān aṣ-Ṣafā*", in *Proceedings of the 20th Congress of the Union Européenne des Arabisants et Islamisants (UEAI), Part Two: Islam, Popular Culture in Islam, Islamic Art and Architecture*, edited by A. Fodor. *The Arabist, Budapest Studies in Arabic*, 27 (2003), pp. 17–24.

—— 'Ideological Debate and Political Encounter in the *Ikhwān al-Ṣafā*", in *Maǧāz. Culture e contatti nell'area del Mediterraneo. Il ruolo dell'Islam*, edited by A. Pellitteri. Palermo: Università di Palermo, Facoltà di Lettere e Filosofia, 2003.

—— 'The "General Policy" of the Ikhwān al-Ṣafā': Plato and Aristotle Restated', in *Words, Texts and Concepts Cruising the Mediterranean Sea: Studies on the Sources, Contents and Influences of Islamic Civilization and Arabic Philosophy and Science Dedicated to Gerhard Endress on his Sixty-fifth Birthday*, edited by A. Arnzen and J. Thielmann. Leuven, Paris, and Dudley, MA: Peeters, 2004, pp. 575–592.

—— 'The "Language of the Prophet" in the Ikhwān al-Ṣafā'', in *Al-kitāb. La sacralité du texte dans le monde de l'Islam*, edited by Daniel De Smet, Godefroid de Callataÿ, and Jan van Reeth. Brussels, Louvain-la-Neuve, and Leuven: Brepols (Acta Orientalia Belgica Subsidia III), 2004, pp. 357–370.

—— 'Temporal and Religious Connotations of the "Regal Policy" in the Ikhwān al-Ṣafā'', in *The Greek Strand in Islamic Political Thought*, edited by E. Gannagé et al. *Mélanges de l'Université Saint-Joseph*, 57 (2004), pp. 337–365.

—— *Appunti per un'epistemologia profetica. L'Epistola degli Ikhwān al-Ṣafa" Sulle cause e gli effetti*. Naples: Università degli Studi di Napoli "L'Orientale", Dipartimento di Studi e richerche su Africa e Paesi Arabi-Guida, 2006.

—— 'History, Language and Ideology in the Ikhwan al-Ṣafā' View of the Imāmate', in *Authority, Privacy and Public Order in Islam*, edited by B. Michalak-Pikulska and A. Pikulski. Leuven: Peeters, 2006.

Beaujouan, Guy. 'L'enseignement du *quadrivium*', in *La scuola nell'Occidente latino dell'alto medioevo*, Settimane 19. Spoleto: Centro italiano di studi sull'alto medioevo, 1972, vol. 2, pp. 639–723.

—— 'The Transformation of the Quadrivium', in *Renaissance and Renewal in the Twelfth Century*, edited by Robert Louis Benson and Giles Constable. Cambridge, MA: Harvard University Press, 1982, pp. 463–487.

Beitia, Angel Cortabarria. 'La classification des sciences chez al-Kindī'. *Mélanges de l'Institut Dominicain d'Études Orientales*, 11 (1972), pp. 49–76.

Bergé, Marc, ed. and trans. 'Épître sur les sciences d'Abū Ḥayyān at-Tawḥīdī'. *Bulletin d'Études Orientales*, 18 (1964), pp. 241–300.

Biesterfeldt, Hans Hinrich. 'Abū al-Ḥasan al-ʿĀmirī und die Wissenschaften'. *Zeitschriften der Deutschen Morgenländischen Gesellschaft*, Suppl. 3 (1977), pp. 335–341.

—— 'Medieval Arabic Encyclopedias of Science and Philosophy', in *The Medieval Hebrew Encyclopedias of Science and Philosophy: Proceedings of the Bar-Ilan University Conference*, edited by Steven Harvey. Dordrecht, Boston, and London: Kluwer, 2000, pp. 77–98.

—— 'Arabisch-islamische Enzyklopädien: Formen und Funktionen', in *Die Enzyklopädie im Wandel vom Hochmittelalter bis zur Frühen Neuzeit.* Akten des Kolloquiums des Projekts D im Sonderforschungsbereich 231 (1996), edited by Christel Meier. Munich: Wilhelm Fink Verlag, 2002, pp. 43–83.

Blachère, Régis. 'Quelques réflexions sur les formes de l'encyclopédisme en Égypte et en Syrie du VIIIème/XIVème à la fin du IXème/XVème siècle'. *Bulletin des Études Orientales*, 23 (1970), pp. 7–19.

Black, Deborah L. 'Estimation *(wahm)* in Avicenna: The Logical and Psychological Dimensions'. *Dialogue*, 32 (1993), pp. 219–258.

Blumenthal, David. 'A Comparative Table of the Bombay, Cairo, and Beirut Editions of the Rasā'il Iḫwān al-Ṣafā''. *Arabica*, 21, 2 (1974), pp. 186–203.

Boethius. *De Institutione Arithmetica, De Institutione Musica, Geometria*, edited by G. Friedlein. Leipzig: Teubner, 1867.

Borrmans, Maurice. 'Salvation', in *Encyclopaedia of the Qur'ān*, edited by Jane Dammen McAuliffe. Vol. 4. Leiden and Boston: E. J. Brill, 2004, p. 522.

Bosworth, Charles E. 'A Pioneer Arabic Encyclopedia of the Sciences'. *Isis*, 54, 1 (1963).

Brentjes, Sonja. 'Die erste *Risāla* der *Rasā'il Ikhwān al-Ṣafa'* über elementare Zahlentheorie'. *Janus*, 71 (1984), pp. 181–274.

Burnett, Charles S. F. 'The Planets and the Development of the Embryo', in *The Human Embryo: Aristotle and the Arabic and European Traditions*, edited by G. R. Dunstan. Exeter: Exeter University Press, 1990, pp. 95–112.

Butterworth, Charles E. 'Paris est et sagesse ouest. Du *Trivium* et *Quadrivium* dans le monde arabe médiéval' in *L'enseignement des disciplines à la Faculté des arts (Paris et Oxford, XIIIe–XVe siècles)*. Actes du colloque international,

Studia Artistarum, 4, edited by Olga Weijers and Louis Holtz. Turnhout: Brepols, 1997, pp. 477–493.

Casanova, Paul. 'Une date astronomique dans les Épîtres des Ikhwān aṣ-Ṣafā''. *Journal Asiatique*, 11, 5 (1915), pp. 5–17.

—— 'Alphabets magiques arabes'. *Journal Asiatique*, 11, 18 (1921), pp. 37–55.

—— 'Alphabets magiques arabes'. *Journal Asiatique*, 11, 19 (1922), pp. 250–262.

Clayton, Peter A. *Chronicle of the Pharaohs: The Reign-by-Reign Record of the Rulers and Dynasties of Ancient Egypt*. London: Thames and Hudson, 1994.

Conway, J. H. and Richard K. Guy. *The Book of Numbers*. New York: Springer-Verlag, 1996.

Costa, Cristina D'Ancona. *La casa della sapienza. La trasmissione della metafisica greca e la formazione della filosofia araba*. Milan: Guerini e Associati, 1996.

Crystal, David, ed. *The Penguin Encyclopedia*. 2nd ed. London: Penguin Books Ltd., 2002; repr. 2004.

Daftary, Farhad. *The Ismāʿīlīs: Their History and Doctrines*. Cambridge: Cambridge University Press, 1990; 2nd ed., Cambridge: Cambridge University Press, 2007.

Damascenus, Nicolaus. *De Plantis: Five translations*, edited and introduced by H. J. Drossaart-Lulofs and E. L. J. Poortman. Amsterdam, Oxford, and New York: North-Holland Publishing Company, 1989.

De Boer, T. J. *Geschichte der Philosophie im Islam*. Stuttgart, 1901. Translated into English by Edward Jones as *The History of Philosophy in Islam*. London: Luzac and Co., 1903; repr. 1965, pp. 81–96.

de Callataÿ, Godefroid. 'Astrology and Prophecy. The Ikhwān al-Ṣafā' and the Legend of the Seven Sleepers', in *Studies in the History of the Exact Sciences in Honour of David Pingree*, edited by Charles Burnett et al. Leiden: Brill, 2003, pp. 758–785.

—— 'Sacredness and Esotericism in the *Rasāʾil Ikhwān al-Ṣafā*'', in *Al-kitāb: La sacralité du texte dans le monde de l'Islam*. Actes du Symposium International tenu à Leuven et Louvain-la-Neuve, 2002, edited by Daniel De Smet, Godefroid de Callataÿ, and Jan van Reeth. *Acta Orientalia Belgica: Subsidia*, 3. Brussels, Louvain-la-Neuve, and Leuven: Société Belge d'Études Orientales, 2004, pp. 389–401.

—— *Ikhwan al-Safa': A Brotherhood of Idealists on the Fringe of Orthodox Islam*. Oxford: Oneworld, 2005.

De Smet, Daniel. *La quiétude de l'intellect. Néoplatonisme et gnose ismaélienne dans l'œuvre de Ḥamīd al-Dīn al-Kirmānī*. Leuven: Peeters, 1995.

De Young, G. 'The Arabic Textual Traditions of Euclid's *Elements*'. *Historia Mathematica*, 11 (1984), pp. 147–160.

Dieterici, Friedrich. *Die Philosophie der Araber im X. Jahrhundert n. Chr. aus den Schriften der Lauteren Brüder*. 8 vols. Leipzig, 1858–1872.

Diwald, Susanne, ed. and trans. *Arabische Philosophie und Wissenschaft in der Enzyklopädie Kitāb Iḫwān aṣ-ṣafā' (III): Die Lehre von Seele und Intellekt*. Wiesbaden: Otto Harrassowitz, 1975.

El-Bizri, Nader. 'La perception de la profondeur: Alhazen, Berkeley et Merleau-Ponty'. *Oriens–Occidens: Sciences, mathématiques et philosophie de l'antiquité à l'âge classique (Cahiers du Centre d'Histoire des Sciences et des Philosophies Arabes et Médiévales, CNRS)*, 5 (2004), pp. 171–184.

—— 'Variations autour de la notion d'expérience dans la pensée arabe', in *L'expérience, collection les mots du monde,* edited by N. Tazi. Paris: Editions la Découverte, 2004, pp. 39–58.

—— 'The Conceptions of Nature in Arabic Thought', in *Keywords: Nature,* edited by N. Tazi. New York: Other Press, 2005, pp. 63–92.

—— 'Ibn al-Haytham', in *Medieval Science, Technology, and Medicine: An Encyclopedia*, edited by Thomas F. Glick, Steven J. Livesey, and Faith Wallis. New York and London: Routledge, 2005, pp. 237–240.

—— 'A Philosophical Perspective on Alhazen's *Optics*'. *Arabic Sciences and Philosophy*, 15, 2 (2005), pp. 189–218.

—— 'Ikhwān al-Ṣafā' (Brethren of Sincerity)', in *Encyclopedia of Philosophy*. 2nd ed., edited by D. M. Borchert. Detroit: Macmillan Reference USA, 2006, pp. 575–577.

—— 'The Microcosm/Macrocosm Analogy: A Tentative Encounter between Graeco-Arabic Philosophy and Phenomenology', in *Islamic Philosophy and Occidental Phenomenology on the Perennial Issue of Microcosm and Macrocosm*, edited by Anna-Teresa Tymieniecka. Dordrecht: Kluwer Academic Publishers, 2006, pp. 3–23.

—— 'In Defence of the Sovereignty of Philosophy: al-Baghdādī's Critique of Ibn al-Haytham's Geometrisation of Place'. *Arabic Sciences and Philosophy*, 17 (2007), pp. 57–80.

Encyclopaedia of Islam, edited by H. A. R. Gibb et al. 2nd ed. 12 vols. Leiden and London: Brill, 1960–2004.

Englisch, Brigitte. *Die Artes liberales im frühen Mittelalter (5.–9. Jh.). Das Quadrivium und der Komputus als Indikatoren für Kontinuität und Erneuerung der exakten Wissenschaften zwischen Antike und Mittelalter*. Stuttgart: Franz Steiner Verlag, 1994.

Epstein, Isidore, ed. *Hebrew–English Edition of the Babylonian Talmud*. London: Soncino Press, 1969.

Euclid. *Stoikheia* (*Elements*), edited by J. L. Heiberg and H. Menge. Teubner Classical Library. 8 vols., with a supplement entitled: *Euclid opera omnia*. Leipzig: Teubner, 1883–1916.

—— *The Thirteen Books of the Elements*, edited by Thomas L. Heath. 3 vols. Cambridge, 1905; repr. 1925. Repr., New York: Dover, 1956.

Fakhry, Majid. 'The Liberal Arts in the Mediaeval Arabic Tradition from the Seventh to the Twelfth Centuries', in *Arts Libéraux et Philosophie au Moyen Âge*. Actes du quatrième congrès international de philosophie médiéval, Montreal, 1967. Montreal and Paris: Institut d'études médiévales, 1969, pp. 91–97.

—— *A History of Islamic Philosophy*. New York: Columbia University Press, 1983.

Al-Fārābī. *Iḥṣā' al-'ulūm*, edited and translated into Spanish by Angel González Palencia in *Alfarabi. Catálogo de las Ciencias*. Madrid: Consejo Superior de Investigaciones Cientificas, 1932.

—— *Kitāb al-Mūsīqī al-kabīr*, edited by Ghatts 'Abd al-Malik Khashaba, with revision and introduction by Maḥmūd Aḥmad al-Ḥifnī. Cairo: Dār al-Kitāb al-'Arabī li'l-Ṭibā'a wa'l-Nashr, 1967.

—— *Kitāb Ārā' ahl al-madīna al-fāḍila*, edited by Albert Nader. 2nd ed. Beirut: Dār al-Mashriq, 1968.

Farhān, Muḥammad Jalūb. 'Philosophy of Mathematics of Ikhwan al-Safa'. *Journal of Islamic Science* [Aligarh], 15 (1999), pp. 25–53.

Farmer, Henry G. *Sa'adya Gaon On the Influence of Music*. London: Probsthain, 1943.

—— *The Sources of Arabian Music*. Leiden: Brill, 1965.

Farrūkh, 'Umar. 'Ikhwān al-Ṣafā'', in *History of Muslim Philosophy*, edited by M. M. Sharif. Vol. 1. 4th repr., Delhi: Low Price Publications, 1999, pp. 289–310.

al-Faruqi, Lois I. *An Annotated Glossary of Arabic Musical Terms*. Westport, VA: Greenwood Press, 1981.

Foucault, Michel. *Les mots et les choses*. Paris: Gallimard, 1966; repr. 1992.

Gabrieli, F. 'Ibn al-Muḳaffa''. *EI2*, edited by H. A. R. Gibb et al. Leiden and London: Brill, 1960—, vol. 3, p. 883.

Gadamer, Hans-Georg. *Wahrheit und Methode, Gesammelte Werke Band 1*. Tübingen: Moh Siebeck, 1990.

Galston, M. 'Realism and Idealism in Avicenna's Political Philosophy'. *The Review of Politics*, 41 (1979), pp. 561–577.

Gardet, Louis and Georges Chehata Anawati. *Introduction à la théologie musulmane. Essai de théologie comparée*. Paris: Vrin, 1948, pp. 101–124.

Gardies, Jean-Louis. 'Sur l'axiomatique de l'arithmétique Euclidienne'. *Oriens-Occidens* 2 (1998), pp. 125–140.

Gerson, Lloyd P. *Plotinus*. The Arguments of the Philosophers. London and New York: Routledge, 1994.

Al-Ghazālī. *Al-Jawāhir al-ghawālī*. Cairo, 1353/1934.

—— *Al-Munqidh min al-ḍalāl* (*Erreur et délivrance*), edited and translated by Farīd Jabre. Beirut: Commission Internationale pour la Traduction des Chefs-d'oeuvre, 1959.

—— *Al-Munqidh min al-ḍalāl* (*Erreur et délivrance*), edited and translated into French by Farid Jabre. Beirut: Librairie Orientale, 1969.
—— *Al-Munqidh min al-ḍalāl*. Translated into English by R. J. McCarty as *Freedom and Fulfillment*. Boston: Twayne Publishers, 1980.
Gohlman, W. E. *The Life of Ibn Sina: A Critical Edition and Annotated Translation*. Studies in Islamic Philosophy and Science. Albany, NY: State University of New York Press, 1974.
Goodman, Lenn E. *The Case of the Animals versus Man Before the King of the Jinn: A Tenth-Century Ecological Fable of the Pure Brethren of Basra*. Boston: Twayne Publishers, 1978.
——, ed. *Neoplatonism and Jewish Thought*. Albany, NY: State University of New York Press, 1992.
—— *Islamic Humanism*. Oxford and New York: Oxford University Press, 2003.
Gosling, J. C. B. *Plato*. The Arguments of the Philosophers. London: Routledge and Kegan Paul, 1983.
Gutas, Dimitri. 'Paul the Persian on the Classification of the Parts of Aristotle's Philosophy: A Milestone Between Alexandria and Baghdad'. *Der Islam*, 60 (1983), pp. 231–267.
—— *Greek Thought, Arabic Culture*. London: Routledge, 1998; repr. 2002.
Guyard, Stanislas. 'Le *fetwa* d'Ibn Taymiyyah sur les Nosairis'. *Journal Asiatique*, 6, 18 (1871), pp. 158–198.
Hadot, Ilsetraut. *Arts libéraux et philosophie dans la pensée antique*. Paris: Études Augustiniennes, 1984.
al-Ḥamawī, Yāqūt ibn 'Abd Allāh. *Mu'jam al-buldān*. 10 vols. Cairo, 1906–1907.
—— *Mu'jam al-udabā'*, edited by Iḥsān 'Abbās. Beirut: Dār al-Gharb al-Islāmī, 1993.
Hamdani, Abbas. 'Abū Ḥayyān al-Tawḥīdī and the Brethren of Purity'. *International Journal of Middle Eastern Studies*, 9 (1978), pp. 345–353.
—— 'An Early Fāṭimid Source on the Time and Authorship of the *Rasā'il Ikhwān al-Ṣafā'*'. *Arabica*, 26 (1979), pp. 62–75.
—— 'Shades of Shī'ism in the Tracts of the Brethren of Purity', in *Traditions in Contact and Change*, edited by Peter Slater and Donald Wiebe. Waterloo, ON: Wilfrid Laurier University Press, 1983, pp. 447–460, with notes pp. 726–728.
—— 'The Arrangement of the *Rasā'il Ikhwān al-Ṣafā'* and the Problem of Interpolations'. *Journal of Semitic Studies*, 29, 1 (1984), pp. 97–110.
—— 'A Critique of Paul Casanova's Dating of the *Rasā'il Ikhwān al-Ṣafā'*', in *Mediaeval Isma'ili History and Thought*, edited by Farhad Daftary. Cambridge: Cambridge University Press, 1996, pp. 145–152.
—— 'Brethren of Purity, a Secret Society for the Establishment of the Fāṭimid Caliphate: New Evidence for the Early Dating of their Encyclopaedia', in

L'Égypte fatimide: Son art et son histoire, edited by Marianne Barrucand. Paris, Presses de l'Université de Paris-Sorbonne, 1999, pp. 73–82.

al-Hamdānī, Ḥusayn. '*Rasā'il Ikhwān al-Ṣafā'* in the Literature of the Ismā'īlī Ṭayyibī *Da'wat*'. *Der Islam*, 20 (1932), pp. 281–300.

—— *Baḥth tārīkhī fī Rasā'il Ikhwān al-Ṣafā' wa-'aqā'id al-Ismā'īliyya fīhā*. Bombay: The Arabic Library and Co., 1354/1935.

Hart, George. *A Dictionary of Egyptian Gods and Goddesses*. London and New York: Routledge and Kegan Paul, 1988.

Heath, Thomas L. *Diophantus of Alexandria: A Study in the History of Greek Algebra*. New York: Dover, 1964.

Hein, Christel. *Definition und Einteilung der Philosophie. Von der spätantiken Einleintungsliteratur zur arabischen Enzyklopädie*. Frankfurt am Main: P. Lang, 1985.

Heinrichs, Wolfhart. 'The Classification of the Sciences and the Consolidation of Philology in Classical Islam', in *Centres of Learning: Learning and Location in Pre-modern Europe and the Near East*, edited by Jan Willem Drijvers and Alastair A. MacDonald. Leiden: Brill, 1995, pp.119–139.

al-Hindī, Muḥammad ibn 'Alī. '*Jumal al-falsafa*'. Publications of the Institute for the History of Arabic-Islamic Science, Series C, 19. Frankfurt am Main: Institut für Geschichte der Arabisch–Islamischen Wissenschaften, 1985.

Hornung, Erik. *Der Eine und die Vielen*. Darmstadt: Wissenschaftliche Buchgesellschaft, 1971.

—— *Conceptions of God in Ancient Egypt: The One and the Many*, translated by John Baines. London, Melbourne, and Henley: Routledge and Kegan Paul, 1983.

Hugonnard-Roche, Henri. 'La classification des sciences de Gundissalinus et l'influence d'Avicenne', in *Études sur Avicenne*, edited by Jean Jolivet and Roshdi Rashed. Paris: Belles Lettres, 1984, pp. 41–75

Hume, David. *An Enquiry Concerning Human Understanding*, edited by L. A. Shelby-Bigge. Oxford: Oxford University Press, 1972.

Ḥusayn, Ṭāha and 'Abd al-Wahhāb 'Azzām, eds. *Bāb al-ḥamāma al-muṭawwaqa* in *Kalīla wa-Dimna*. Cairo: Dār al-Ma'ārif, 1941, pp. 125–146.

Ibn 'Abd al-Hādī, Abū 'Abd Allāh. *Al-'Uqūd al-durriyya min manāqib Shaykh al-Islām Aḥmad bin Taymiyya*, edited by Muḥammad Ḥamīd al-Fiqī. Cairo: Maṭba'at Ḥijāzī, 1357/1938.

Ibn 'Abd Rabbih(i). *Al-'Iqd al-farīd*, edited Aḥmad Amīn, Ibn al-Anbārī, and 'Abd al-Salām Muḥammad Hārūn. Vol. 6. Cairo: Maṭba'at Lajnat al-Ta'līf wa'l-Tarjama wa'l-Nashr, 1949.

Ibn Abī al-Dunyā. *Dhamm al-malāhī*. Translated into English by James Robson as *Tracts on Listening to Music*. London: The Royal Asiatic Society, 1938.

Ibn al-Athīr. *Al-Kāmil fī al-tārīkh*, edited by C. J. Tornberg. Leiden: E. J. Brill, 1851. Repr., Beirut: Dār Ṣādir, 1979.

Ibn Baṭṭūṭa. *Riḥla*. Beirut: Dār Ṣādir, 1964.

Ibn Jubayr. *Riḥla*. Beirut: Dār Ṣādir, 1964.

Ibn al-Nadīm. *Fihrist*. Tehran, 1971.

Ibn al-Qifṭī. *Ta'rīkh al-ḥukamā'*, edited by Julius Lippert. Leipzig: Dieterich's Verlagsbuchhandlung, 1903.

Ibn Qurra, Thābit. *Kitāb al-Madkhal ilā 'ilm al-'adad*, in *Arabische Übersetzung der Arithmêtikê Eisagôgê des Nikomachus von Gerasa*, edited by Wilhelm Kutsch. Beirut: Imprimerie Catholique, St. Joseph, 1959.

Ibn Salama, al-Mufaḍḍal. *Kitāb al-Malāhī*, in *al-Mūsīqā al-'irāqiyya fī 'ahd al-mughūl wa'l-turkumān*, by 'A. al-'Azzāwī. Baghdad, 1951.

Ibn Sīnā. *Risāla fī aqsām al-'ulūm al-'aqliyya*, in *Tis' rasā'il fī al-ḥikma wa'l-ṭabī'iyyāt*. Cairo, 1326/1908.

—— *Kitāb al-Shifā', al-Riyadhiyyāt*, 3: *Jawāmi' 'ilm al-mūsīqī*, edited by Zakarīyā Yūsuf. Cairo: al-Maṭba'a al-Amīriyya, 1956.

Ibn Taymiyya (d. 728/1328). *Nuṣayriyya*, in 'Le fetwa d'Ibn Taymiyyah sur les Nosairis', edited and translated by Stanislas Guyard. *Journal Asiatique*, 18 (1871), pp. 158–198.

—— *Majmū'at al-rasā'il al-Kubrā*. 2 vols. Cairo: al-Maṭba'a al-'Āmira al-Sharqiyya, 1323/1905.

—— *Bayān talbīs al-jahmiyya fī ta'sīs bida'i-him al-kalāmiyya aw naqd ta'sīs al-jahmiyya*, edited by Muḥammad ibn 'Abd al-Raḥmān ibn Qāsim. 2 vols. [Cairo?]: Mu'assasat Qurṭuba, 1392/1972.

—— *Dar' ta'āruḍ al-'aql wa'l-naql aw muwāfaqat ṣaḥīḥ al-manqūl li-ṣarīḥ al-ma'qūl*, edited by Muḥammad Rashad Sālim. 11 vols. Riyadh: Dār al-Kunūz al-Adabiyya, [1399/1979].

—— *Majmū' al-fatāwā*, edited by 'Abd al-Raḥmān ibn Muḥammad ibn Qāsim. 37 vols. Rabat: Maktabat al-Ma'ārif, 1401/1981.

—— *Majmū'at fatāwā*. 5 vols. Beirut: Dār al-Fikr, 1403/1983.

—— *Bughyat al-murtād fī al-radd 'alā al-mutafalsifa wa'l-Qarāmiṭa wa'l-Bāṭiniyya, ahl al-ilḥād min al-qā'ilīn bi'l-ḥulūl wa'l-ittiḥād*, edited by Musa ibn Sulayman al-Duwaysh. [Medina?]: Maktabat al-'Ulūm wa'l-Ḥikam, 1408/1988.

—— *Kitāb al-Nubuwwāt*. Beirut: Dār al-Fikr, 1409/1989.

—— *Minhāj al-sunna al-nabawiyya fī naqḍ kalām al-shī'a al-qadariyya*, edited by Muḥammad Rashad Sālim. 9 vols. Cairo: Maktabat Ibn Taymiyya, 1409/1989.

—— *Sharḥ al-'aqīda al-Iṣfahāniyya*, edited by Ḥ. M. Makhlūf. Cairo: Dār al-Kutub al-Islāmiyya, 1386/1966.

—— *Kitāb al-Istighātha fī al-radd 'alā al-Bakrī*, edited by 'Abd Allāh ibn Dujayn al-Suhaylī. 2 vols. Riyadh: Dār al-Waṭan, 1417/1997.

—— *Al-Jawāb al-ṣaḥīḥ li-man baddala dīn al-masīḥ*, edited by ʿAlī ibn Ḥasan ibn Nāṣir, ʿAbd al-ʿAzīz ibn Ibrāhīm al-ʿAskar, and Ḥamdan ibn Muḥammad al-Ḥamdān. 7 vols. Riyadh: Dār al-ʿĀṣima li'l-Nashr wa'l-Tawzīʿ, 1419/1999.
—— *Kitāb al-Ṣafadiyya*, edited by M. R. Sālim. 2 vols. Mansoura: Dār al-Hady al-Nabawī / Riyadh: Dār al-Faḍīla, 1421/2000.
—— *Kitāb al-Radd ʿalā al-mantiqiyyīn*, edited by M. Ḥ. M. Ḥ. Ismāʿīl. Beirut: Dār al-Kutub al-ʿIlmiyya, 1423/2003.
Ikhwān al-Ṣafā'. *Rasā'il*, translated into Urdu by Mawlawī Ikrām ʿAlī. Calcutta: Munshī Muḥammad Ṭāha, 1810.
—— *Tuḥfat Ichwān-oos-Suffa* [*Rasā'il*], in the original Arabic, revised and edited by Schuekh Ahmud-bin-Moohummud Schurwan-ool-Yummunee [Shaykh Aḥmad ibn Muḥammad Shurwān al-Yamanī] with a short preface in English by T. T. Thomason. Calcutta: printed by P. Pereira at the Hindoostanee Press, 1812.
—— *Kitāb Ikhwān al-Ṣafā' wa-Khullān al-Wafā'*, edited by Wilāyat Ḥusayn. 4 vols. Bombay: Maṭbaʿat Nukhbat al-Akhbār, 1305–1306/ca. 1888.
—— *Rasā'il Ikhwān al-Ṣafā' wa-Khillān al-Wafā'*, edited by Khayr al-Dīn al-Ziriklī, with two separate introductions by Ṭāhā Ḥusayn and Aḥmad Zakī Pasha. 4 vols. Cairo: al-Maktaba al-Tijāriyya al-Kubrā, 1347/1928.
—— *Rasā'il Ikhwān al-Ṣafā'*, edited by Buṭrus Bustānī. 4 vols. Beirut: Dār Ṣādir and Dār Bayrūt, 1377/1957; repr., Dār Ṣādir, 2004 and 2006.
—— *Rasā'il Ikhwān al-Ṣafā' wa-Khullān al-Wafā'*, edited by ʿĀrif Tāmir. 5 vols. Beirut and Paris: Manshūrāt ʿUwaydāt, 1415/1995.
—— *Al-Risāla al-jāmiʿa*, edited by Jamīl Ṣalībā. 2 vols. Damascus: Maṭbaʿat al-Taraqqī, 1949–1951.
—— *Al-Risāla al-jāmiʿa*, edited by Muṣṭafā Ghālib. Beirut: Dār Ṣādir, 1974.
—— *Al-Risāla al-jāmiʿa*, edited by ʿĀrif Tāmir (as the fifth volume of this edition of the *Rasā'il*, for which see above). Beirut and Paris: Manshūrāt ʿUwaydāt, 1415/1995.
—— *Risālat jāmiʿat al-jāmiʿa*, edited by ʿĀrif Tāmir. Beirut: Dār Ṣādir, 1959.
—— 'The Forty-Third Treatise of the Ikhwān al-Ṣafā'', translated into English by D. H. Yūsufjī. *Muslim World*, 33 (1943), pp. 39–49.
—— 'A Treatise on Number Theory from a Tenth-Century Arabic Source', edited and translated into English by Bernard R. Goldstein. *Centaurus*, 10 (1964), pp. 129–160.
—— 'L'Épître sur la musique des Ikhwān al-Ṣafā'', edited and translated into French by Amnon Shiloah. *Revue des Études Islamiques*, 32 (1965), pp. 125–162.
—— 'L'Épître sur la musique des Ikhwān al-Ṣafā'', edited and translated into French by Amnon Shiloah. *Revue des Études Islamiques*, 34 (1967), pp. 159–193.

—— *L'enciclopedia dei Fratelli della Purità*, edited and translated by Alessandro Bausani. Naples: Istituto Universitario Orientale (Dipartimento di Studi Asiatici), 1978.

—— *Les révolutions et les cycles (Épîtres des Frères de la Pureté, XXXVI)*, edited and translated into French by Godefroid de Callataÿ. Beirut and Louvain-la-Neuve: Al-Burāq and Academia-Bruylant, 1996.

—— 'Ikhwān al-Ṣafā': des arts scientifiques et de leurs objectifs (*Épître VII* des Frères de la Pureté)', edited and translated into French by Godefroid de Callataÿ. *Le Muséon*, 116 (2003), pp. 231–358.

—— 'Ikhwān al-Ṣafā': Sur les limites du savoir humain (*Épître XXVIII* des Frères de la Pureté)', edited and translated into French by Godefroid de Callataÿ. *Le Muséon*, 116 (2003), pp. 479–503.

'Imād al-Dīn, Idrīs. *'Uyūn al-akhbār*, edited by Muṣṭafā Ghālib. Vol. 4. Beirut, 1973.

—— *'Uyūn al-akhbār wa-funūn al-āthār*, edited by Ma'mūn al-Ṣāghirjī. Vol. 4. Damascus and London: Institut Français du Proche Orient and The Institute of Ismaili Studies, 2007.

Irani, Rida A. K. 'Arabic Numeral Forms'. *Centaurus*, 4 (1955), pp. 1–12.

Ivanow, Wladimir. 'Ismailis and Qarmatians'. *Journal of the Bombay Branch of the Royal Asiatic Society*, n.s., 16 (1940), pp. 43–85.

Ivry, Alfred. 'Al-Kindī on *First Philosophy* and Aristotle's Metaphysics', in *Essays on Islamic Philosophy and Science*, edited by George F. Hourani. Albany, NY: State University of New York Press, 1975, pp. 15–24.

al-Jābirī, Muḥammad 'Ābid. *Takwīn al-'aql al-'arabī*. 6th repr., Beirut: Markaz Dirāsāt al-Wiḥda al-'Arabiyya, 1994,.

Al-Jāḥiẓ. *'Risālat al-qiyān'*. Translated into English by Alfred F. L. Beeston as *The Epistle on Singing-Girls*. Warminster: Aris and Phillips, 1980.

Jolivet, Jean. 'Classifications des sciences', in *Histoire des sciences arabes*, edited by Roshdi Rashed. Vol. 3. Paris: Seuil, 1997.

al-Jundī, Salīm. *'Abū al-'Alā' al-Ma'arrī wa-Ikhwān al-Ṣafā'*. *Majallat al-Majma' al-'Ilmī al-'Arabī*, 16 (1941), pp. 346–351.

Kalābādhī. *Il sufismo nelle parole degli antichi*, edited and translated by P. Urizzi. Palermo: Officina di Studi Medievali, 2002.

Khalidi, Tarif. *Arabic Historical Thought in the Classical Period*. Cambridge: Cambridge University Press, 1996.

Khalīfa, Ḥājjī (a.k.a. Kâtip Çelebi). *Kashf al-ẓunūn*, edited by Gustav Flügel. 7 vols. Leipzig, 1835–1858.

Al-Khwārizmī. *Mafātīḥ al-'ulūm*, edited by Gerlof van Vloten. Leiden: Brill, 1895.

—— *Le commencement de l'algèbre*, edited and translated into French by Roshdi Rashed. Paris: Albert Blanchard, 2007.

Al-Kindī. *Risāla fī kutub Arisṭūṭālīs*, in *Rasā'il al-Kindī al-falsafiyya*, edited by Abū Rīda. Cairo: Dar al-Fikr al-'Arabi, 1369/1950.

—— *Mu'allafāt al-Kindī al-mūsīqiyya*, edited by Zakarīyā Yūsuf. Baghdad: Maṭba'at Shafīq, 1962.

Kraemer, Joel L. *Philosophy in the Renaissance of Islam: Abū Sulaymān al-Sijistānī and his Circle*. Leiden: E. J. Brill, 1986.

Kraus, Paul. *Jābir ibn Ḥayyān. Contribution à l'histoire des idées scientifiques dans l'Islam. Jābir et la science grecque*. Paris: Les Belles Lettres, 1986.

Kritchlow, Keith. *Islamic Patterns: An Analytical and Cosmological Approach*. London: Thames and Hudson, 1976.

Latham, J. D. 'The Beginnings of Arabic Prose Literature: The Epistolary Genre', in *Arabic Literature to the End of the Umayyad Period*, edited by Alfred F. L. Beeston et al. Cambridge: Cambridge University Press, 1983, pp. 154–179.

Lazard, Gilbert, ed. and trans. *Les premiers poètes persans (IX–Xe siècles): Fragments rassemblés (Ash'ār-i parakanda-yi qadīmītarīn shu'arā-yi fārsī-zabān)*. 2 vols. Tehran and Paris : Département d'Iranologie de l'Institut Franco-Iranien and Adrien-Maisonneuve, 1964.

Leaman, Oliver. *Islamic Aesthetics: An Introduction*. Edinburgh: Edinburgh University Press, 2004.

Lewis, Bernard. 'An Epistle on Manual Crafts'. *Islamic Culture*, 17 (1943), pp. 141–151.

Lloyd, A. C. 'The Later Neoplatonists', in *The Cambridge History of Later Greek and Early Medieval Philosophy*, edited by A. H. Armstrong. Cambridge: Cambridge University Press, 1970.

al-Ma'arrī, Abū al-'Alā'. *Dīwān saqṭ al-zand*, edited by Amīn Hindī. N.p., 1319/1901. Translated into English by Arthur Wormhoudt as *Saqṭ al-zand: The Spark from the Flint*. Ann Arbor, MI: University Microfilms, 1972.

Madelung, Wilferd. 'Al-Kayyāl'. *EI2*, edited by H. A. R. Gibb et al. Leiden and London: Brill, 1960—, vol. 4, p. 847.

—— 'Al-Khuramiyya'. *EI2*, edited by H. A. R. Gibb et al. Leiden and London: Brill, 1960—, vol. 5, pp. 63–65.

Mahdi, Muhsin. 'Science, Philosophy, and Religion in Alfarabi's *Enumeration of the Sciences*', in *The Cultural Context of Medieval Learning*. Proceedings of the First Colloquium on Philosophy, Science, and Technology in the Middle Ages, edited by John E. Murdoch and Edith D. Sylla. Dordrecht and Boston: D. Reidel Publishing Company, 1973, pp. 113–147.

—— 'Al-Fārābī's Imperfect State'. *Journal of the American Oriental Society*, 110 (1990).

al-Manṭiqī (al-Sijistānī), Abū Sulaymān. *Kitāb Ṣiwān al-ḥikma*, edited by 'Abd al-Raḥmān Badawī. Tehran, 1974.

Marquet, Yves. 'Ikhwān al-Ṣafā''. *EI2*, edited by H. A. R. Gibb et al. Leiden and London: Brill, 1960—, vol. 3, pp. 1071–1076.

—— 'La place du travail dans la hiérarchie ismā'īlienne d'après *L'encyclopédie des Frères de la Pureté*'. *Arabica*, 8, 3 (1961), pp. 225–237.

—— 'Imamat, resurrection et hiérarchie selon les Ikhwān aṣ-Ṣafā''. *Revue des Études Islamiques*, 30 (1962), pp. 49–142.
—— *La philosophie des Ikhwān al-Ṣafā de Dieu à l'homme*. Lille: Service de reproduction des thèses, 1973.
—— *La philosophie des Ikhwān al-Ṣafā': Thèse présentée devant l'Université de Paris IV, 1971*. Algiers: Société nationale d'édition et de diffusion, 1973. Repr., Paris and Milan: Archè, 1999.
—— *La philosophie des Iḫwân aṣ-Ṣafâ': L'imâm et la société*. Dakar: Université de Dakar, Faculté des Lettres et Sciences Humaines, Département d'arabe Travaux et documents no. 1), 1973.
—— 'Ikhwān al-Ṣafā', Ismaïliens et Qarmates'. *Arabica*, 24, 3 (1977), pp. 233–257.
—— '910 en Ifrīqiya: Une épître des Ikhwān al-Ṣafā''. *Bulletin d'Études Orientales*, 30 (1978), pp. 61–73.
—— 'Les Ikhwān al-Ṣafā' et l'ismaélisme', in *Convegno sugli Ikhwân al-Ṣafā', Roma, 1979*. Rome: Accademia Nazionale dei Lincei, 1981, pp. 69–96.
—— 'Les Ikhwān al-Ṣafā' et le Christianisme'. *Islamochristiana*, 8 (1982), pp. 129–158.
—— 'Note annexe: à propos d'un poème Ismaïlien dans les épîtres des Iḫwān aṣ-Ṣafā''. *Studia Islamica*, 55 (1982), pp. 137–142.
—— 'Les Épîtres des Ikhwān al-Ṣafā', œuvre ismaïlienne'. *Studia Islamica*, 61 (1985), pp. 57–79.
—— *La philosophie des alchimistes et l'alchimie des philosophes: Jâbir ibn Ḥayyân et les 'Frères de la Pureté'*. Paris: Maisonneuve et Larose, 1988.
—— 'Les références à Aristote dans les Épîtres des *Ikhwān aṣ-Ṣafā*'', in *Individu et société: L'influence d'Aristote dans le monde méditerranéen*, edited by T. Zarcone. Istanbul, Paris, Rome, and Trieste: Isis, 1988, pp. 159–164.
—— 'La détermination astrale de l'évolution selon les Frères de la Pureté'. *Bulletin d'Études Orientales*, 44 (1992), pp. 127–146.
—— 'Ibn al-Rūmī et les Ikhwān al-Ṣafā''. *Arabica*, 47 (2000), pp. 121–123.
—— *Les 'Frères de la Pureté', pythagoriciens de l'Islam: La marque du pythagorisme dans la rédaction des Épîtres des Ikhwān aṣ-Ṣafā'*. Paris: EDIDIT, 2006.
Marzolph, Ulrich. 'Medieval Knowledge in Modern Reading: A Fifteenth-Century Arabic Encyclopedaedia of *omni re scibili*', in *Pre-modern Encyclopaedic Texts: Proceeding of the Second COMERS Congress, Groningen, 1996*, edited by Peter Binkley. Leiden: Brill, 1997, pp. 407–419.
Massignon, Louis. 'Sur la date de la composition des *Rasā'il Ikhwān al-Ṣafā*''. *Der Islam*, 4 (1913), p. 324.
—— 'Esquisse d'une bibliographie Qarmate', in *A Volume of Oriental Studies Presented to Edward G. Browne on his 60th Birthday (1922)*, edited by T. W. Arnold and Reynold A. Nicholson. Cambridge: [Cambridge] University Press, 1922, pp. 329–338.

al-Maymanī, ʿAbd al-ʿAzīz. *Abū al-ʿAlā wa-mā ilayh*. Cairo, 1342/1923–1924.

McDannell, Colleen and Bernhard Lang. *Heaven: A History*. New Haven and London: Yale University Press, 1988.

Meredith-Owens, G. "Arūḍ'. *EI2*, edited by H. A. R. Gibb et al. Leiden and London: Brill, 1960—, vol. 1, pp. 667–677.

Michot, Yahya Jean. 'L'épître de la résurrection des Ikhwān al-Ṣafā". *Bulletin de Philosophie Médiévale*, 16–17 (1974–1975), pp. 114–148.

—— *La destinée de l'homme selon Avicenne. Le retour à Dieu* (maʿād) *et l'imagination*. Leuven: Peeters, 1986.

——, ed. and trans. *Musique et danse selon Ibn Taymiyya. Le Livre du* samāʿ *et de la danse* (Kitāb al-Samāʿ wa'l-raqṣ) *compilé par le Shaykh Muḥammad al-Manbijī*. Paris: J. Vrin, 1991.

——, ed. and trans. *Ibn Taymiyya: Lettre à Abū al-Fidā'*. Louvain-la-Neuve: Université Catholique de Louvain, 1994.

—— 'Ibn Taymiyya on Astrology: Annotated Translation of Three Fatwas'. *Journal of Islamic Studies*, 11, 2 (2000), pp. 147–208.

—— 'Vanités intellectuelles . . . L'impasse des rationalismes selon le *Rejet de la contradiction* d'Ibn Taymiyya'. *Oriente Moderno*, 19, 80 (2000), pp. 597–617.

—— 'Vizir "hérétique" mais philosophe d'entre les plus éminents: al-Ṭūsī vu par Ibn Taymiyya'. *Farhang*, 15–16 (2003), pp. 195–227.

——, ed. and trans. 'A Mamlūk Theologian's Commentary on Avicenna's *Risāla aḍḥawiyya*: Being a Translation of a Part of the *Dar' al-taʿāruḍ* of Ibn Taymiyya'. *Journal of Islamic Studies*, 14, 2–3 (2003), pp. 149–203, 309–363.

Montaigne, Michel de. *The Complete Essays*, translated by Donald Frame. Stanford: Stanford University Press, 1976.

Mueller, Ian. *Philosophy of Mathematics and Deductive Structure in Euclid's Elements*. Cambridge, MA: MIT Press, 1981.

Al-Mutanabbī. *Dīwān*, edited by Nāsīf al-Yāzijī. Vol. 2. Cairo: n.d. Repr., Beirut: Dār Ṣādir, 1964.

Nafīsī, Saʿīd. *Muḥīṭ-i zindagī-u aḥwāl u ashʿār-i Rūdakī*. 3rd ed. Tehran: Sepehr, 1958.

—— *Fihrist al-Majdūʿ*, edited by ʿAlī Naqī Munzawī. Tehran, 1966.

Nasr, Seyyed Hossein. *Islamic Science: An Illustrated Study*. Westerham: Westerham Press, 1976.

—— *An Introduction to Islamic Cosmological Doctrines: Conceptions of Nature and Methods Used for Its Study by the Ikhwān al-Ṣafā', al-Bīrūnī and Ibn Sīnā*. Revised ed. London: Thames and Hudson, 1978. Originally published by Cambridge, MA: Harvard University Press, 1964.

Al-Nawawī. *Matn al-arbaʿīn al-nawawiyya*, (Arabic-English text) translated by Ezzedin Ibrahim and Denys Johnson-Davies. 3rd ed. Damascus: The Holy Koran Publishing House, 1977.

Necipoglu, Gülru. *The Topkapi Scroll: Geometry and Ornament in Islamic Architecture.* Santa Monica, CA: The Getty Center for the History of Art and the Humanities, 1995.

Netton, Ian Richard. 'Brotherhood versus Imāmate: Ikhwān al-Ṣafā' and the Ismā'īlīs'. *Jerusalem Studies in Arabic and Islam*, 2 (1980), pp. 253–262.

—— 'Foreign Influences and Recurring Ismā'īlī Motifs in the *Rasā'il* of the Brethren of Purity', in *Convegno sugli Ikhwān al-Ṣafā', Roma, 1979.* Rome: Accademia Nazionale dei Lincei, 1981, pp. 49–67.

—— *Allah Transcendent: Studies in the Structure and Semiotics of Islamic Philosophy, Theology and Cosmology.* Richmond: Curzon Press, 1989; repr., 1994.

—— *Seek Knowledge: Thought and Travel in the House of Islam.* Richmond: Curzon, 1996.

—— *Muslim Neoplatonists. An Introduction to the Thought of the Brethren of Purity (Ikhwān al-Ṣafā').* Repr., London: Routledge Curzon, 2000. First published by George Allen and Unwin, 1982.

—— 'Private Caves and Public Islands: Islam, Plato and the Ikhwān al-Ṣafā''. *Sacred Web*, 15 (2005).

—— 'Riḥla'. *EI2*, edited by H. A. R. Gibb et al. Leiden and London: Brill, 1960—, vol. 8, p. 528.

Neubauer, Eckhard. 'Die Theorie vom *Īqā'*. Part 1: Übersetzung des *Kitāb al-Īqā'āt* von Abū Naṣr al-Fārābī'. *Oriens*, 21–22 (1968–1969), pp. 196–232.

—— 'Arabische Anleitungen zur Musiktherapie'. *Zeitschrift für Geschichte der Arabisch-Islamischen Wissenschaften*, 6 (1990), pp. 227–272.

—— 'Der Bau der Laute und ihre Besaitung nach arabischen, persischen und türkischen Quellen des 9. bis 15. Jahrhunderts'. *Zeitschrift für Geschichte der Arabisch-Islamischen Wissenschaften*, 8 (1993), pp. 279–378.

—— 'Übersetzung des *Kitāb Iḥṣā' al-īqā'āt* von Abū Naṣr al-Fārābī'. *Oriens*, 34 (1994), pp. 103–173.

—— 'Al-Khalīl ibn Aḥmad und die Frühgeschichte der arabischen Lehre von den "Tönen" und den musikalischen Metren, mit einer Übersetzung des *Kitāb al-Nagham* von Yaḥyā ibn 'Alī al-Munajjim'. *Zeitschrift für Geschichte der Arabisch-Islamischen Wissenschaften*, 10 (1995–1996), pp. 255–323.

—— *Arabische Musiktheorie von den Anfängen bis zum 6./12. Jahrhundert.* Frankfurt am Main: Institut für Geschichte der Arabisch–Islamischen Wissenschaften, 1998.

Nicomachus of Gerasa. *Theologumena arithmeticae* (The Theology of Numbers), edited by Fridericus Astius. Leipzig: Libraria Weidmannia, 1817.

—— *Nicomachi Geraseni Pythagorei Introductionis arithmeticae*, edited by Richard Hoche. Leipzig: Teubner, 1886.

—— *Introduction to Arithmetic*, translated by M. L. D'Ooge, with studies by F. E. Robbins and L. C. Karpinski. New York, 1926.

—— *Introduction arithmétique*, translated by Janine Bertier. Paris: J. Vrin, 1978.

O'Meara, Dominic J. *Pythagoras Revived: Mathematics and Philosophy in Late Antiquity*. 2nd ed. Oxford: Clarendon Press, 1990.

—— *Plotinus: An Introduction to the Enneads*. Oxford: Clarendon Press, 1993.

Pacholczyk, Jozef S. 'Music and Astronomy in the Muslim World'. *Leonardo*, 29, 2 (1996), pp. 145–150.

Palacios, Asin M. 'Un faqīh siciliano, contradictor de Al Ghazzālī (Abū 'Abd Allāh de Māzara)', in *Centenario della Nascita di Michele Amari*. Vol. 2. Palermo: 1910, pp. 216–244.

Pellat, Charles. 'Les encyclopédies dans le monde arabe'. *Cahiers d'histoire mondiale*, 9 (1966) pp. 631–658.

——, Živa Vesel, and E. van Donzel. 'Mawsū'a'. *EI2*, edited by H. A. R. Gibb et al. Leiden and London: Brill, 1960—, vol. 6, pp. 903–911.

Peters, Francis E. *Aristoteles Arabus: The Oriental translations and commentaries on the Aristotelian corpus*. Leiden: Brill, 1968.

Pinès, Shlomo. 'Some Problems of Islamic Philosophy'. *Islamic Culture*, 11, 1 (1937), pp. 66–80.

—— 'Une encyclopédie arabe du 10e siècle. Les Épîtres des Frères de la Pureté, *Rasā'il Ikhwān al-Ṣafā*". *Rivista di storia della filosofia*, 40 (1985), pp. 131–136.

Plato. *The Dialogues of Plato*, edited and translated by Benjamin Jowett. 4th ed. Oxford: Oxford University Press, 1953.

—— *Republic*, edited and translated into Hebrew by Ralph Lerner. Ithaca, NY: Cornell University Press, 1974.

—— *Plato Arabus*, edited by Paul Kraus and Richard Walzer. 2 vols. London: Warburg Institute, 1951.

Plotinus. *The Enneads*, edited and translated by Stephen MacKenna. London: Faber and Faber, 1956.

—— *Plotini Opera*, edited by Paul Henry and Hans-Rudoplh Schwyzer. Paris: Desclée de Brouwer / Brussels: Edition Universelle, 1951–1959.

—— *Enneads*, edited and translated by A. H. Armstrong. 7 vols. The Loeb Classical Library. Cambridge, MA: Harvard University Press, 1966–1988.

Poonawala, Ismail K. *Biobibliography of Ismā'īlī Literature*. Malibu, CA: Undena, 1977.

Proclus. *The Elements of Theology*, (bilingual Greek-English text) edited and translated by E. R. Dodds. Oxford: Clarendon Press, 1933. 2nd ed., 1963; repr., 1992.

Rashed, Roshdi. 'Analyse combinatoire, analyse numérique, analyse diophantienne et théorie des nombres', in *Histoire des sciences arabes*, edited by Roshdi Rashed with Régis Morélon. Vol. 2. Paris: Editions du Seuil, 1997.

—— *Les mathématiques infinitésimales du IXe au XIe siècle*. Vol. 4. London: al-Furqān Islamic Heritage Foundation, 2002.

—— and Christian Houzel. 'Thābit ibn Qurra et la théorie des parallèles'. *Arabic Sciences and Philosophy*, 15 (2005), pp. 9–55.

Ricardo-Felipe, Albert R. 'La "*Risāla fī māhiyyat al-ʿišq*" de las *Rasā'il Ikhwān al-Ṣafā*". *Anaquel de Estudios Árabes*, 6 (1995), pp. 185–207.

Rosenthal, Franz. *Das Fortleben der Antike in Islam*. Zurich and Stuttgart: Artemis Verlag, 1965.

—— *Knowledge Triumphant: The Concept of Knowledge in Medieval Islam*. Leiden: E. J. Brill, 1970.

Rudolph, U. *Islamische Philosophie. Von den Anfängen bis zur Gegenwart*. Munich: Beck, 2004.

Saadiah. *The Book of Theodicy*, translated by Lenn E. Goodman. New Haven, CT: Yale University Press, 1988.

Sabra, A. I. "Ilm al-ḥisāb'. *EI2*, edited by H. A. R. Gibb et al. Leiden and London: Brill, 1960—, vol. 3, pp. 1138–1141.

—— 'The Appropriation and Subsequent Naturalization of Greek Science in Medieval Islam: A Preliminary Statement'. *History of Science*, 25 (1987), pp. 223–243.

—— 'Situating Arabic Science: Locality versus Essence'. *Isis*, 87 (1996), pp. 654–670.

Ṣafā', Dhabīḥ Allāh. *Ganj-i-sukhan*. Vol. 1, *From Rūdakī to Anwarī*. Tehran: Ibn Sīnā Press, n.d.

Saidan, Ahmad S. 'The Earliest Extant Arabic Arithmetic: *Kitāb al-Fuṣūl fī al-Ḥisāb al-Hindī* of Abū al-Ḥasan Aḥmad ibn Ibrāhīm al-Uqlīdisī'. *Isis*, 57 (1966), pp. 475–490.

—— *The Arithmetic of al-Uqlīdisī*. Dordrecht: D. Reidel, 1978.

—— 'Numeration and Arithmetic', in *Encyclopedia of the History of Arabic Science*, edited by Roshdi Rashed with Régis Morélon. Vol. 2. London: Routledge, 1996, pp. 331–348.

Sawa, George D. *Music Performance Practice in the Early ʿAbbāsid Era*. Toronto: Pontifical Institute of Medieval Studies, 1989.

al-Shahrastānī, Abū al-Fatḥ. *Livre des religions et des sectes*, edited and translated by D. Gimaret and G. Monnot. Vol. 1. Leuven: Peeters / Paris: UNESCO, 1986.

al-Shāmī, R. Y. 'Ibn Taymiyya: *Maṣādiru-hu wa manhaju-hu fī taḥlīli-hā*'. *Journal of the Institute of Arabic Manuscripts*, 38, 1 (1994), pp. 183–269.

Shams, M. 'U. and 'A. ibn M. al-ʿImrān. *Al-Jāmiʿ li-sīrat shaykh al-Islām Ibn Taymiyya khilāl sabʿat qurūn. Āthār shaykh al-Islām Ibn Taymiyya wa-mā laḥiqa-hā min aʿmāl*, 8. Mecca: Dār ʿĀlam al-Fawā'id li'l-Nashr wa'l-Tawzīʿ, AH 1422.

Shehadi, Fadlou. *Philosophies of Music in Medieval Islam*. Leiden: Brill, 1995.

al-Shihabī, Muṣṭafā. 'Filāḥa [Middle East]'. *EI2*, edited by H. A. R. Gibb et al. Leiden and London: Brill, 1960—, vol. 2, pp. 899–901.

Shiloah, Amnon. 'Jewish and Muslim Traditions of Music Therapy', in *Music as Medicine: The History of Music Therapy Since Antiquity*, edited by Peregrine Horden. Ashgate: Aldershot, 2000, pp. 69–83.

Siorvanes, Lucas. *Proclus: Neoplatonic Philosophy and Science*. Edinburgh: Edinburgh University Press, 1996.

Sprenger, Aloys. 'Notices of Some Copies of the Arabic Work Entitled *Rasāyil Ikhwān al-Cafā*". *Journal of the Asiatic Society of Bengal*, 17 (1848), part 1, pp. 501–507; part 2, pp. 183–202. Repr. in *Islamic Philosophy, Vols. 20–21: Rasā'il Ikhwān aṣ-Ṣafā' wa-Khillān al-Wafā'*, edited by Fuat Sezgin et al. Frankfurt am Main: Institut für Geschichte der Arabisch-Islamischen Wissenshaften, 1999, vol. 20, pp. 201–228.

Stern, Samuel M. 'The Authorship of the Epistles of the Ikhwān aṣ-Ṣafā''. *Islamic Culture*, 20 (1946), pp. 367–372.

—— 'Additional Notes to the Article: "The Authorship of the Epistles of the Ikhwān aṣ-Ṣafā'"'. *Islamic Culture*, 21 (1947), pp. 403–404.

—— 'New Information about the Authors of the "Epistles of the Sincere Brethren"'. *Islamic Studies*, 3 (1964), pp. 405–428.

Stigelbauer, Michael. *Die Sängerinnen am Abbasidenhof um die Zeit des Kalifen al-Mutawakkil nach dem Kitāb al-Aghānī*. Vienna: VWGÖ, 1975.

Straface, Antonella. 'Testimonianze pitagoriche alla luce di una filosofia profetica: la numerologia pitagorica negli Ikhwān al-Ṣafā''. *Annali dell'Istituto Universitario Orientale*, 47 (1987), pp. 225–241.

Tarán, Leonardo. 'Nicomachus of Gerasa', in *Dictionary of Scientific Biography*, edited by Charles Coulston Gillispie et al. Vol. 10. New York: Charles Scribner's Sons, 1974.

al-Tawḥīdī, Abū Ḥayyān. *Kitāb al-Imtā' wa'l-mu'ānasa*, edited by Aḥmad Amīn and Aḥmad al-Zayn. 2 vols. Beirut: Manshūrāt Dār Maktabat al-Ḥayāt, 1939–1944; repr., Cairo, 1953; 2nd ed. Beirut, 1965.

—— *Risāla fī al-'ulūm*, edited and translated by Marc Bergé in 'Épître sur les sciences'. *Bulletin d'Études Orientales*, 18 (1964), pp. 286–298.

—— *Risālat al-saqīfa*, edited by Ibrāhīm al-Kaylānī in *Trois Épîtres d'Abū Ḥayyān al-Tawḥīdī*. Damascus, 1951.

Ṭībāwī, 'Abd al-Laṭīf (or, Abdul Latif Tibawi). 'Ikhwān aṣ-Ṣafā' and their *Rasā'il*: A Critical Review of a Century and a Half of Research'. *Islamic Quarterly*, 2 (1955), pp. 28–46.

—— 'Further Studies on Ikhwān aṣ-Ṣafā''. *Islamic Quarterly*, 20–22 (1978), pp. 57–67.

Vandoulakis, Ioannis M. 'Was Euclid's Approach to Arithmetic Axiomatic?' *Oriens-Occidens*, 2 (1998), pp. 141–181.

Vernet, J. 'Al-Madjrīṭī'. *EI2*, edited by H. A. R. Gibb et al. Leiden and London: Brill, 1960—, vol. 5, pp. 1109–1110.

Vesel, Živa. *Les Encyclopédies persanes: Essai de typologie et de classification des sciences*. Paris : Editions Recherche sur les Civilisations, 1986.
Vilchez, José M. Puerta. *Historia del pensamiento estético árabe*. Madrid: Ediciones Akal, 1997.
Vuillemin, Jules. *Mathématiques pythagoriciennes et platoniciennes*. Paris: Albert Blanchard, 2001.
Walker, Paul E. *Early Philosophical Shiism: The Ismaili Neoplatonism of Abū Yaʿqūb al-Sijistānī*. Cambridge: Cambridge University Press, 1993.
Wallis, R. T. *Neoplatonism*. London: Duckworth, 1972.
Weber, Edouard. 'La classification des sciences selon Avicenne à Paris vers 1250', in *Études sur Avicenne*, edited by Jean Jolivet and Roshdi Rashed. Paris: Belles Lettres, 1984, pp. 77–101.
Wensinck, A. J. et al. *Concordance et Indices de la Tradition Musulmane*. Leiden: E. J. Brill, 1967.
Werner, Karel. *A Popular Dictionary of Hinduism*. Richmond: Curzon Press, 1994.
West, Martin. 'Music Therapy in Antiquity', in *Music as Medicine: The History of Music Therapy Since Antiquity*, edited by Peregrine Horden. Ashgate: Aldershot, 2000, pp. 51–68.
White, Nicholas P. *A Companion to Plato's Republic*. Indianapolis, IN and Cambridge, MA: Hackett Publishing, 1979.
Widengren, Geo. 'Macrocosmos–Microcosmos Speculation in the *Rasāʾil Ikhwān al-Ṣafāʾ* and Some Ḥurūfī Texts'. *Archivio di Filosofia* (1980), pp. 297–312.
Wiet, Gaston. 'Les classiques du scribe égyptien'. *Studia Islamica*, 18 (1963), pp. 41–80.
Wisnovsky, Robert. *Avicenna's Metaphysics in Context*. Ithaca, NY: Cornell University Press, 2003.
Zakī (Pasha), Aḥmad. *Mawsūʿat al-ʿulūm al-ʿarabiyya wa-baḥth ʿalā Rasāʾil Ikhwān al-Ṣafāʾ* (*Étude bibliographique sur les encyclopédies arabes*). Cairo: Būlāq, 1308/1890.
al-Ziriklī, Khayr al-Dīn. *Al-Aʿlām: Qāmūs tarājim liʾashharr al-rijāl waʾl-nisāʾ min al-ʿarab waʾl-mustaʿribīn waʾl-mustashriqīn*. Sidon: al-Maṭbaʿa al-ʿAṣriyya, 1956.

Index